Cellular Antigens

Cellular Antigens

Lectures and Summaries of the
Conference on Cellular Antigens
Held in Philadelphia, June 7–9, 1971
Sponsored by Ortho Research Foundation

Edited by

Alois Nowotny

Temple University
Philadelphia, Pennsylvania

SPRINGER-VERLAG BERLIN · HEIDELBERG · NEW YORK
1972

© 1972 by Springer-Verlag New York Inc.
Softcover reprint of the hardcover 1st edition 1972

Library of Congress Catalog Card Number 72-85952.

ISBN-13: 978-1-4612-9845-8   e-ISBN-13: 978-1-4612-9843-4

DOI: 10.1007/978-1-4612-9843-4

Preface

Scientists may feel that there are too many meetings these days, and we tend to agree on this until it comes to our own field of interest. In our own areas we would like to hear about other people's achievements and learn whatever may be helpful in the search for answers to our own pet questions. Exchange of ideas, discussions, and critical evaluations are almost as essential to progress as the actual laboratory work. These were the major motives which initiated our own efforts, with the support of Dr. Earle H. Spaulding, the Chairman of the Department of Microbiology and Immunology at Temple University School of Medicine, to bring together for the first time specialists from various disciplines who are attempting to achieve the same thing, *i.e.,* the clarification of the chemical and immunological nature of different cellular antigens.

Instead of publishing the proceedings of the conference, it was decided that we would attempt to review achievements in the different subjects as well as report the latest developments from our own laboratories. Thus we hope to give scientists involved in this explosively growing field not only an up-to-date report, but also a useful source of relevant references.

Despite these efforts and aspirations, we realize that this book is not a complete survey of all known cellular antigens. We tried to present major representations of the most important cell types. By carefully selecting the most prominent specialists on the various topics, we tried to present as many aspects of the subjects as the limitations of time would allow. While immunochemistry dominated the program, many other areas were also discussed from an immunological point of view. If our present plans materialize, we will reconvene in a few years, when we will not only review the newest achievements, but also emphasize those topics of aspects that could not be discussed at our first meeting.

Finally, I would like to express our gratitude to the Ortho Research Foundation and especially to its Vice President, Dr. William Pollack, whose generous support was the sole financial aid which made this conference possible. We also wish to thank all the participants, chairmen, lecturers, and discussants for their valuable contributions, as well as all those who helped us in organizing this meeting.

Last but not least, I wish to acknowledge the superb help of Mrs. June Whitcombe, editorial assistant of our department, and of Mrs. Alice Stone for their enthusiastic and outstanding cooperation.

Philadelphia
March, 1972

ALOIS NOWOTNY

Conference Chairman,
Professor of Immunology,
Temple University School of Medicine

Introduction

The importance of cellular antigens in the homeostasis of the organism as well as in the pathogenesis and natural history of disease is well documented. Currently, there are promising leads suggesting that in addition to their usefulness in diagnosis, they may be applied to the treatment of malignant neoplasms and chronic degenerative diseases. Exploitation of cellular antigens for these purposes requires full understanding of their chemical nature and precise biological function.

The lectures and reviews of the first of a series of meetings on cellular antigens are reported in this volume. The fundamental characteristics and behavior of cellular antigens are presented in breadth and depth. Antigens of bacteria, formed blood elements, normal tissues, and malignant neoplasms, though discussed in separate sessions, are recognized as a continuum. In both the formal presentations and the panel discussions, linkages between model systems, both *in vivo* and *in vitro,* and potential clinical significance are noted. Immunology as a clinical science has, in the past, enjoyed its most spectacular success in the field of infectious diseases. More recently, progress in immunobiology and more specifically immunochemistry has been crucial in facilitating tissue and organ transplantation.

Fundamental knowledge so ably reported by these proceedings is gradually transforming immunochemistry into a rational and useful weapon in the study of the most rapidly growing causes of morbidity and mortality.

Explanations for variations in the occurrence and natural history of cancer and chronic degenerative diseases invoking immunological and host differences are moving from a "mystique" to a science. Accelerated progress from the former to the latter demands the development of models and techniques and elucidation of mechanisms which will hasten the day when the critical components of the immunological equation can be applied to the understanding and control of disease.

The participants in this conference are addressing themselves both as architects and as engineers to the creation of a solid foundation in this most exciting area on the leading edge of biomedical science.

Philadelphia
January, 1972

PAUL KOTIN, M.D.

Vice President of the Health Sciences Center,
Dean of the School of Medicine,
Temple University

Table of Contents

Part 1 Gram-Negative Bacterial Antigens

Part 2 Gram-Positive Bacterial Antigens

Part 3 Erythrocyte Antigens

Part 4 White Blood Cell Antigens

Part 5 Normal Tissue Antigens

Part 6 Malignant Tissue Antigens

Part 1

Gram-Negative Bacterial Antigens

IMMUNE RESPONSE OF MICE TO SUBCELLULAR VACCINES OF SALMONELLA TYPHIMURIUM*

L. J. BERRY and M. R. VENNEMAN

Department of Microbiology, The University of Texas at Austin, Texas

One of the major unanswered questions of immunology is why a living bacterial vaccine affords better protection than any of the currently available killed vaccines. It has been suggested that this is due either to a labile antigen (or antigens) no longer present in killed preparations or that it results from a more continuous and hence effective antigenic stimulation, probably through a limited proliferation of the pathogen. A distinction between the two possibilities would be possible only if the labile antigen could be identified.

My entry (L. J. B.) into this area of research was predicated on years of work with experimental mouse typhoid and was inspired by the important observations of YOUMANS and YOUMANS (1964, 1965, 1966) to the effect that a ribosomal vaccine derived from *Mycobacterium tuberculosis* provided impressive protection against tuberculous challenge in mice. You will hear an account of their studies, but it is appropriate for me at this time to express my indebtedness to them. Working with me at Bryn Mawr College at the time of our initial experiments were Ruth Levy, Toby Eisenstein, and Stuart Winston.

At Ohio State University, Dr. Nancy Bigley and Martin Venneman were similarly motivated by the Youmans' work to initiate studies with *Salmonella typhimurium*. Their findings (VENNEMAN and BIGLEY, 1969; VENNEMAN et al., 1970) and those of ours (EISENSTEIN et al., 1968; WINSTON and BERRY, 1970a, b; EISENSTEIN and BERRY, unpublished observations) were in essential agreement. Dr. Venneman first joined me at Bryn Mawr College as a postdoctoral fellow and then made the move with

* The work described in this report was supported in part by grants from the National Institute of Allergy and Infectious Diseases, the National Science Foundation, and Smith, Kline, and French Laboratories.

me to Texas. In my remarks, I will try to highlight our results to date (VENNEMAN and BERRY, 1971a, b, c).

Table 1 summarizes our most important initial results. CF-1 mice or a similar Swiss Webster strain were used throughout the studies. Vaccines were given subcutaneously and challenge was via the intraperitoneal route. The use of an adjuvant resulted in a slight increase in level of immunity achieved with the preparations used here but not with the highly purified RNA, as will be demonstrated below.

Table 1.* Protective Effect in Mice of a Single Immunizing Dose of Different Vaccines Followed 15 Days Later by Challenge with 1000 LD_{50}'s of *Salmonella Typhimurium*

Vaccine	Dose	Living/Total after 30 days	Percent survival
RIA	$0.1\ LD_{50}$	37/37	100
Heat-killed *S. typhimurium*	50 μg	13/62	21
Ribosomal fraction	100 μg	36/36	100
	10 μg	39/45	82
RNA† fraction	100 μg	83/90	92
	10 μg	139/150	92
Control		55/280	20

* Modified from VENNEMAN and BIGLEY (1969) and VENNEMAN *et al.* (1970).

† Derived from the ribosomal fraction.

RIA is the designation of a strain of *S. typhimurium* of low virulence. It was obtained originally—as was our highly virulent challenge strain, SR-11—from Dr. Howard Schneider, then at the Rockefeller University. The heat-killed vaccine was prepared by heating the SR-11 strain to boiling point for 30 minutes, or to 63°C for 30 minutes. Both gave similar results.

The ribosomal fraction was obtained by a procedure similar to that used by YOUMANS and YOUMANS (1965). The flow diagram presented in Figure 1 indicates the steps employed.

Cells of *S. typhimurium*, SR-11 (Figure 1), grown overnight, were ruptured with a French pressure cell. Whole cells and debris were sedimented at 22,000 × g for 10 minutes and 43,000 × g for 15 minutes, and the supernatant was sedimented at 105,000 × g for 3 hours. Following treatment of the pellet with 0.5% sodium dodecylsulfate (SDS) for 30 minutes at room temperature, the material obtained after 3 hours centrifugation at 105,000 × g was designated the ribosomal fraction.

The initial RNA fraction was obtained, as indicated by the flow diagram in Figure 2, by extracting the first high-speed pellet (the ribosomal

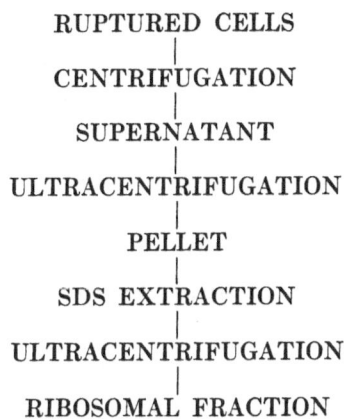

RUPTURED CELLS
|
CENTRIFUGATION
|
SUPERNATANT
|
ULTRACENTRIFUGATION
|
PELLET
|
SDS EXTRACTION
|
ULTRACENTRIFUGATION
|
RIBOSOMAL FRACTION

Fig. 1. Flow diagram for preparation of ribosomal fraction.

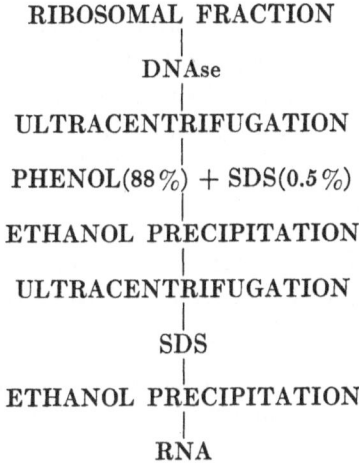

RIBOSOMAL FRACTION
|
DNAse
|
ULTRACENTRIFUGATION
|
PHENOL(88%) + SDS(0.5%)
|
ETHANOL PRECIPITATION
|
ULTRACENTRIFUGATION
|
SDS
|
ETHANOL PRECIPITATION
|
RNA

Fig. 2. Flow diagram for extraction of ribosomal RNA.

fraction) for 10 minutes at 65°C with 88% phenol and 0.5% SDS. Extraction with phenol was repeated four times at 25°C. The RNA was precipitated with ethanol, collected by high-speed centrifugation, treated again with SDS to decrease further the protein and endotoxin content of the material, and again precipitated with ethanol.

The challenge dose for the data presented in Table 1 was 1000 LD_{50} of SR-11. High levels of protection were afforded by the living vaccine and by both the ribosomal and RNA preparations each at two dose levels. Even with 1 μg of the ribosomal or the RNA vaccine, highly significant levels of immunity, not shown in Table 1, were evident.

The lack of protection seen with the killed vaccine is explainable in part by the high challenge dose. Even though the survival time of mice

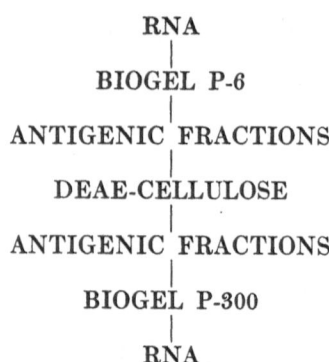

RNA
|
BIOGEL P-6
|
ANTIGENIC FRACTIONS
|
DEAE-CELLULOSE
|
ANTIGENIC FRACTIONS
|
BIOGEL P-300
|
RNA

Fig. 3. Flow diagram for purification of ribosomal RNA.

immunized with this vaccine was greater than that of controls, the percentage alive at 30 days was not.

We shall omit description of earlier studies and analyses aimed at identifying the active component in the RNA vaccine and go directly to the latest work, which makes it look as if the active immunogen is, in fact, RNA. Figure 3 is a flow diagram of the procedures used in obtaining a highly purified RNA. The last Biogel excludes substances of 300,000 or more molecular weight.

Table 2 summarizes the results of chemical analyses of the purest RNA preparation so far obtained. It is evident that any material other than RNA (with the possible exception of carbohydrate) would have to

Table 2.* Chemical Analyses of the Purified Ribosomal RNA from *Salmonella Typhimurium*

Material sought	Assay method	Amount of RNA assayed (dry wt)	Amount present
Protein	Analytical Polyacrylamide Disc Electrophoresis	2 mg	$<0.5\ \mu g$ (none detected)
DNA	Lowry Diphenylamine	50 mg 50 mg	$<0.5\ \mu g$ (none detected)
Lipid	Thin-layer chromatography	300 μg	$<0.5\ \mu g$ (none detected)
RNA	Orcinol	50 mg	"100%"
Carbohydrate	Pentose equivalents	50 mg	20–25%

* Modified from Venneman (submitted for publication).

be a superantigen, were it responsible for protection. This becomes apparent from the data in Table 3. The "purified" RNA protects extremely well two weeks after an immunizing dose of 50 ng. Were protein responsible, the most that could be present would be less than 0.5 ng. It is apparent that incorporating the purified RNA in Freund's incomplete adjuvant failed to increase its protective effect. To this extent, our results are different from those of YOUMANS and YOUMANS (1966), who find adjuvants essential for protection.

Table 3.* Comparison of the Protective Effect of a Living Vaccine with Ribosomal RNA Vaccines in Mice

Vaccine	Diluent	Living/Total after 30 days	Percent survival
RIA (living)	Saline	35/40	90
"Crude" RNA⎱	Saline	35/40	90
(50 μg) ⎰	Adjuvant†	32/40	80
"Purified" RNA⎱	Saline	38/40	95
(50 μg) ⎰	Adjuvant	34/40	85
Controls		4/40	10

    * Modified from Venneman (submitted for publication).
    † Freund's incomplete adjuvant.

With the above evidence implicating RNA as an active antigen capable of conferring a level of protection equal to that obtained with a living vaccine, attention should now be directed to another aspect of the work.

One of the long-sought goals of immunology has been to distinguish between the relative contribution of cellular and humoral immunity to an animal's resistance to disease. Can there, in fact, be cellular immunity without antibody involvement? One of the major obstacles to obtaining an answer is methodology. How can one clearly separate one types of immunity from the other? Some outstanding investigators have addressed themselves to this problem. ROWLEY *et al.* (1964) and JENKIN and ROWLEY (1963) in Australia, ELBERG (1960) in Berkeley, California, MACKANESS (1964) and his group at the Trudeau Institute, in Saranac Lake, N.Y., and SAITO *et al.* (1962) in Japan serve as an incomplete but illustrative list.

The ideal way to separate cellular immunity from humoral immunity requires a technique for the initiation of one without the other. Assuming that this were possible, what test of immunity should be used to evaluate the relative resistance of the host to challenge? There are, in general, two approaches. One relies on survival and the other on the ability of immunized-versus-control animals to suppress the *in vivo* growth of the pathogen. We have done both, but the latter procedure saves time—at least three weeks per experiment.

Our design was to infect mice two weeks or 15 days postvaccination with multiple LD_{50}'s of the SR-11 (virulent) strain of *S. typhimurium*. Viable pathogen counts from whole mouse homogenates were then followed with time postinfection, a technique described some years ago by BERRY *et al.* (1956). Since controls began to die after 5 days, this time was chosen as the standard interval when all counts were made. The data presented in Table 4 are typical when mice were challenged 15 days post-

Table 4.* Viable Counts of *Salmonella Typhimurium* in Whole Mouse Homogenates 5 Days Postchallenge in Control and Vaccinated Mice

Vaccine	Log of mean number of bacteria/g mouse
None	5.7
Heat-killed S. typhimurium	3.0
Ribosomal RNA	3.1

* Modified from VENNEMAN and BERRY (1971a,b).

immunization. They had been injected with 20 μg of either a killed vaccine or 20 μg of the "purified" ribosomal RNA. The challenge dose was 40 LD_{50}'s. Each value is the log of the mean of 10 to 15 separate determinations. The suppression of growth was equal in vaccinated mice, regardless of the immunogen employed, and in each group there were about 1/500 as many salmonellae as in controls. The ability of mice given the killed vaccine to control the pathogen count is probably related to the use of the small challenge dose of 40 LD_{50}'s. In Table 1, mice that were immunized with the killed vaccine failed to survive better than controls, but the challenge dose in that series of experiments was 1000 LD_{50}'s. The level of protection may vary, therefore, depending on the immunogen administered, but only with large infectious doses may the differences become evident.

The next step in our studies was to evaluate the ability of serum or of peritoneal cells from mice previously immunized with one of the vaccines to passively protect normal recipient animals against challenge as determined by pathogen count. In Figure 4, a diagram of the experimental design is presented. Peritoneal macrophages were removed without any attempt to separate them into cell types. The mice, it should be recalled,

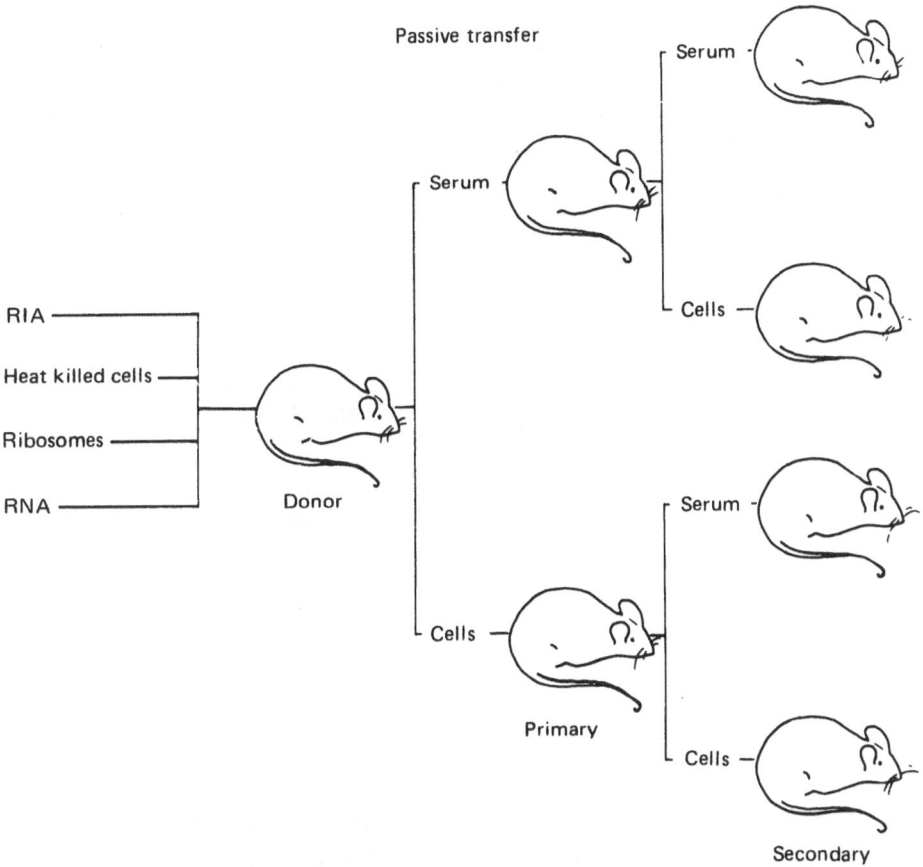

Fig. 4. Diagram of procedure for primary and secondary passive transfer of immunity with peritoneal cells or serum.

are outbred. Cells were injected intraperitoneally, and the recipients were challenged 15 days later via the same route. Pathogen counts were made 5 days after that.

An 0.1 ml amount of serum was injected intraperitoneally, and the mice were challenged 7 days later via the same route. Pathogen counts were made 5 days after challenge.

For secondary transfer, cells were removed from primary recipients 15 days posttransfer, and the new recipients were challenged 15 days later. For serum, transfer from primary recipients to secondary recipient was at 7 days and challenge at 4 days. Table 5 shows the log of average number of pathogens in mice challenged with 45 LD_{50}'s after the mice in all groups had received 10^5 peritoneal cells 15 days before infection. The controls received peritoneal cells from normal mice. The donor mice were vaccinated 15 days earlier with 20 μg of one or the other antigenic materials. The pathogen counts were made 5 days postinfection. Each value is the average of 10 mice. Passive transfer with cells from mice vaccinated with RNA conferred a level of immunity just about as good as that found for the mice that served as cell donors (see Table 4). Passive transfer of immunity with cells from donors vaccinated with heat-killed cells, resulted in a significant reduction in pathogen number, but less than that seen in the group immunized with the ribosomal RNA (Table 5).

Table 5.* Viable Counts of *Salmonella Typhimurium* in Whole Mouse Homogenates Five Days Postchallenge in Mice Passively Immunized 15 Days Previously with an Intraperitoneal Injection of 10^5 Peritoneal Cells from Controls and Immunized Mice

Vaccine	Log of mean number of bacteria/g mouse
None	6.4
Heat-killed S. typhimurium	4.8
Ribosomal RNA	3.7

* Modified from VENNEMAN and BERRY (1971a,c).

Table 6 contains the results of experiments where 0.1 ml of serum from either normal mice or from mice immunized with heat-killed cells or RNA were infected 7 days posttransfer with 52 LD_{50}'s of *S. typhi-*

Table 6.* Viable Counts of *Salmonella Typhimurium* in Whole Mouse Homogenates 5 Days Postchallenge in Mice Passively Immunized 7 days Previously with an Intraperitoneal Injection of 0.1 ml of Serum from Control and Immunized Mice

Vaccine	Log of mean number of bacteria/g mouse
None	6.3
Heat-killed S. typhimurium	5.0
Ribosomal RNA	6.1

* Modified from VENNEMAN and BERRY (1971a,b).

Table 7.* Viable Counts of *Salmonella Typhimurium* in Whole Mouse Homogenates 5 Days Postchallenge in Mice Passively Immunized 15 Days Previously through Secondary Transfer of 10^5 Peritoneal Cells

Vaccine used in original donors	Log of mean number of bacteria/g mouse
None	5.9
RIA	4.7
Heat-killed S. typhimurium	5.3
Ribosomal RNA	4.2

* Modified from VENNEMAN and BERRY (1971c).

murium. Bacterial counts were made 5 days after infection. The important observation in this experiment was the failure of serum from RNA-vaccinated mice to protect recipient animals from infection, whereas serum from mice immunized with heat-killed cells did.

Table 7 shows that the RIA and RNA vaccines permitted secondary transfer of immunity via peritoneal cells, whereas a heat-killed vaccine provided only marginal protection at best.

Table 8 shows that none of the vaccines permitted secondary transfer of immunity via serum.

Table 8.* Viable Counts of *Salmonella Typhimurium* in Whole Mouse Homogenates 5 Days Postchallenge in Mice Passively Immunized 4 Days Previously through Secondary Transfer of 0.1 ml of Serum. Different Vaccines were Administered the Original Donors

Vaccine used in original donors	Log of mean number of bacteria/g mouse
None	5.8
RIA	5.7
Heat-killed S. typhimurium	5.7
Ribosomal RNA	5.7

* Modified from VENNEMAN and BERRY (1971c).

The results presented in this report permit the following conclusions:

(1) *S. typhimurium* RNA derived from a crude ribosomal pellet is a highly effective immunogen. As an alternative interpretation of our results, a superantigen associated with the RNA is active in exceedingly small amounts, *i.e.,* in amounts smaller than have been observed in mice up to the present time.

(2) The level of immunity established by this material is fully as effective as a living vaccine.

(3) If a viable count of challenge organisms 5 days postinfection is accepted as a valid assessment of the immune status of an animal, then the ability of mice vaccinated with the purified RNA preparation to suppress pathogen growth depends on cells and not on serum. This conclusion is based on the results obtained in passive transfer experiments.

(4) The secondary transfer of immunity with cells and not with serum is consistent with the concept that some type of cellular transformation is responsible for the immunity stimulated by the RNA vaccine.

Additional work is needed before these conclusions are fully substantiated. On the other hand, we are convinced that the highly purified RNA—about as pure, in fact, as an RNA preparation can be—emerges as a highly promising subcellular vaccine. It affords levels of protection, when assessed by all methods available, as good as any immunogen yet described.

References

Berry, L. J., M. K. deRopp, M. H. Fair, and E. H. Schur (1956). Dynamics of bacterial infections in mice under conditions known to alter survival time. *J. Infec. Dis., 98*:198–207.

Eisenstein, T., S. H. Winston, and L. J. Berry (1968). Ribosomal extracts as protective antigens against *Salmonella typhimurium* and *Staphylococcus aureus* in sections. *Bacteriol. Proc.,* p. 87.

Elberg, S. S. (1960). Cellular immunity. *Bacteriol. Rev., 24*:67–95.

Jenkin, C. R., and D. Rowley (1963). Basis for immunity to typhoid in mice and the question of cellular immunity. *Bacteriol. Rev., 27*:391–404.

Mackaness, G. B. (1964). The immunologic basis of acquired cellular resistance. *J. Exp. Med., 120*:105–109.

Rowley, D., K. J. Turner, and C. R. Jenkin (1964). The basis for immunity to mouse typhoid. II. Cell bound antibody. *Aust. J. Exp. Biol. Med. Sci., 42*:237–248.

Saito, K., M. Nakamo, T. Akeyma, and D. Ushiba (1962). Passive transfer of immunity to typhoid by macrophage. *J. Bacteriol., 84*:500–511.

Venneman, M. R., and L. J. Berry (1971a). Experimental salmonellosis: Differential passive transfer of immunity with serum and cells obtained from mice immunized with ribosomal and ribonucleic acid preparations. *J. Reticuloendoth. Soc., 9*:491–502.

—— (1971b). Serum mediated resistance induced with immunogenic preparations of *Salmonella typhimurium. Infect. and Immunity,* in press.

—— (1971c). Cellular mediated resistance induced with immunogenic preparations of *Salmonella typhimurium. Infect. and Immunity,* in press.

Venneman, M. R., and N. J. Bigley (1969). Isolation and partial characterization of an immunogenic moiety obtained from *Salmonella typhimurium. J. Bacteriol., 100*:140–148.

Venneman, M. R., N. J. Bigley, and L. J. Berry (1970). Immunogenicity of ribonucleic acid preparations obtained from *Salmonella typhimurium. Infect. and Immunity, 1*:574–682.

Winston, S. H., and L. J. Berry (1970a). Antibacterial immunity induced by ribosomal vaccines. *J. Reticuloendoth. Soc., 8*:13–24.

—— (1970b). Immunity induced by ribosomal extracts from *Staphylococcus aureus. J. Reticuloendoth. Soc., 8*:66–73.

Youmans, A. S., and G. P. Youmans (1964). Nature of labile immunogenic substance in the particulate fraction isolated from *Mycobacterium tuberculosis. J. Bacteriol., 88*:1030–1037.

—— (1965). Immunogenic activity of a ribosomal fraction obtained from *Mycobacterium tuberculosis. J. Bacteriol., 89*:1291–1298.

—— (1966). Preparation of highly immunogenic ribosomal fractions of *Mycobacterium tuberculosis* by the use of sodium dodecyl sulfate. *J. Bacteriol., 91*:2139–2145.

THE COMMON ANTIGEN OF
GRAM-NEGATIVE ENTERIC BACTERIA*

E. NETER and H. Y. WHANG

*Departments of Microbiology and Pediatrics, State
University of New York at Buffalo, Medical School,
and Laboratory of Bacteriology, Children's
Hospital, Buffalo, New York*

During the past eight decades the extensive research dealing with bacterial antigens has been concerned largely with those antigens that differentiate genera, species, serogroups, and serotypes from each other, rather than with antigens that are shared by otherwise unrelated microorganisms. The direction of this research has been stimulated to a considerable extent by problems of classification, etiology of disease, and epidemiology. Without doubt, extraordinary progress has been made in this field. Nonetheless, it is surprising to note how little is known about common antigens of various microorganisms, in spite of the fact that heterogenetic antigens have been known for half a century—since the classical discovery of the Forssman antigen. The early literature was reviewed by LANDSTEINER (1945) in his fundamental volume on *The Specificity of Serological Reactions*. Among more recent studies of heterogenetic antigens are those of Springer dealing with blood group specific antigens produced by enteric bacteria (SPRINGER and HORTON, 1964; SPRINGER *et al.,* 1966). It is not surprising that two common bacterial antigens, one shared by gram-positive and the other by gram-negative bacteria, were discovered by serendipity rather than design. RANTZ and associates (1956), while exploring the usefulness of a passive hemagglutination test for the detection of streptococcal antibodies, discovered an antigen that is shared by varied gram-positive microorganisms, including cocci and bacilli, aerobes and anaerobes. KUNIN *et al.* (1962) discovered a corresponding antigen characteristic of certain enteric gram-negative bacteria, while establishing the basic features of a hemagglutination test for the titration of antibodies to the O antigens of *Escherichia coli*. This antigen (CA) is produced by *Escherichia, Klebsiella, Enterobacter, Salmonella, Shigella, Serratia,* and *Proteus,* but not by *Pseu-*

* The investigations of the authors were supported by U.S. Public Health Service Grant AI00658 from the National Institute of Allergy and Infectious Diseases.

14

domonas and *Flavobacterium*. Of further interest is the observation of KUNIN (1962) that, although this antigen is widely distributed, only a few of these serotypes of enterobacteriaceae engender antibodies in rabbits. Our own studies, summarized here, have been concerned with (1) the source and nature of this antigen, (2) its serologic detection, (3) its antigenicity and immunogenicity, and (4) its biologic significance, particularly as an immunogen against infection by varied microorganisms sharing the serologically identical antigenic determinant.

I. Source of the Common Enterobacterial Antigen

In their original report KUNIN *et al.* (1962) documented production of CA by various genera and species of enterobacteriaceae. As yet, the production of CA by other microorganisms has not been fully elucidated. It has been established that enteric bacteria produce CA in a completely synthetic culture medium (PERLMANN *et al.,* 1967; GORZYNSKI and NETER, 1970). In our own study, we examined whether or not CA production by pigment-producing *Serratia* and *Flavobacterium* is related to the taxonomic position of these organisms. The former is currently classified among the *Enterobacteriaceae,* and the latter among the *Achromobacteriaceae.* Indeed, it was found that all strains of *Serratia* studied, including both pigmented and nonpigmented isolates, produced CA, and that none of the strains of *Flavobacterium* did so (DIAZ and NETER, 1969). This observation suggests the possibility that production of certain heterogenetic or common antigens, such as CA, may be used as one of many characteristics for taxonomic purposes.

Dr. Hubert Mayer of the Max-Planck-Institut, Freiburg, Germany, and our group are presently engaged in a study of genetically and biochemically well-characterized rough mutants regarding the production of CA. It has become evident that certain mutants of *Salmonella typhimurium* fail to produce this antigenic determinant, thus opening the way for relating genetic characteristics and biochemical composition to the production of this antigen.

Extensive studies carried out by PERLMANN *et al.* (1967) indicate that an antigen related to CA is present in the colon of rats. These authors also demonstrated antibodies to this colon antigen in the serum of patients with ulcerative colitis. Additional studies on the production of this antigenic determinant by various species of animals are clearly needed.

II. Serologic Detection of the Common Enterobacterial Antigen

The CA-antiCA system presents a challenge to immunologists in view of the unusual findings regarding the detection of the antigen by serologic means. Antibodies against CA fail to cause agglutination of CA containing bacteria, although CA is present on the surface of the bacterial cells, as demonstrated by the FA technique (AOKI *et al.*, 1966). Nor do these antibodies cause agglutination of antigen-coated latex particles (WHANG and NETER, 1962). Even when highly purified latex particles (provided by Dr. Jack Singer) were utilized, agglutination by the corresponding antibodies could not be observed. Nonetheless, latex particles do adsorb the antigen to their surface, and these antigenically modified particles do remove CA antibodies from CA antiserum. Even more surprising is the observation that antibodies against CA fail to precipitate the antigen in solution, either in test tubes or by means of the gel diffusion technique; nor are CA antibodies bactericidal (DOMINGUE and NETER, 1966). Whereas these conventional serologic reactions have yielded negative results, antibodies against CA can be detected by the FA technique (AOKI *et al.*, 1966), by hemagglutination of antigenically modified erythrocytes, as well as by hemolysis in the presence of complement and suitable erythrocyte carriers of the antigen (KUNIN, 1963; KUNIN *et al.*, 1962; WHANG and NETER, 1962). Further, the Jerne plaque technique is useful for the detection of CA antibody-producing cells (DOMINGUE and NETER, 1967). As is expected from the foregoing, complement fixation can be documented. Of particular interest is the observation that these antibodies also opsonize CA containing bacteria and CA modified latex particles (DOMINGUE and NETER, 1966). The basis for these somewhat unusual findings has not been elucidated. However, the antibodies present in both rabbit immune sera and in sera of human subjects certainly are not univalent, since they readily induce agglutination of antigenically modified erythrocytes. Thus far, on the basis of Sephadex filtration and susceptibility to 2-mercaptoethanol, the antibodies have been identified as 19S IgM immunoglobulins (WHANG *et. al.*, 1967), but it remains for future studies to determine whether IgG and IgA antibodies of identical specificity can be engendered in suitable hosts.

III. Antigenicity versus Immunogenicity

As originally reported by KUNIN *et al.* (1962), only a few strains of enterobacteriaceae, notable *E. coli* 014, engender antibodies *in vivo,* although many others produce CA of identical specificity. The reasons for this difference in immunogenicity and antigenicity of this antigen has attracted the attention of our group. For the sake of clarity, it may be pointed out that "immunogenicity," as used in this report, reflects the capacity of CA to elicit the production of antibodies in a suitable animal. The term "antigenicity" connotes the capacity of the antigen to interact *in vitro* with the corresponding antibody. It has been shown in our laboratory that CA produced by enterobacteriaceae exists in two forms: immunogenic CA produced by *E. coli* 014 is ethanol-insoluble, whereas nonimmunogenic CA produced by most other enterobactericeae is ethanol-soluble (NETER, 1969). In addition, the former is resistant and the latter is susceptible to an enzyme produced by *Pseudomonas* species (WHANG and NETER, 1964). Subsequently, it was shown that nonimmunogenic CA present in the supernates of cultures can be rendered highly immunogenic to the rabbit by removal of the simultaneously present O antigen (lipopolysaccharide) (NETER, 1969). The latter acts as an antigen-associated immunosuppressant. The mechanism by which homologous and heterologous lipopolysaccharides and other substances as well—such as lipid A, cardiolipin, chlorphenesin, and certain serum fractions—render immunogenic CA nonimmunogenic, has not been fully elucidated. However, it is clear that this immunosuppressive effect is not due to toxicity, antigen competition, general immunosuppressive action, or alteration of the antigenic determinant. Immunosuppression is observed only when the inhibitor is mixed *in vitro* with the antigen prior to immunization, and not when both are injected separately. It is important to emphasize that these substances do not alter the antibody-combining capacity of the antigen. Recent observations indicate that immunogenicity may depend upon the state of aggregation of the antigen. Heating of immunogenic CA just prior to injection reduces immunogenicity, although it does not affect antigenicity. In turn, immunogenicity is largely restored by repeated freezing and thawing (WHANG *et al.*, 1971). Thus it may be postulated that *E. coli* 014 and other immunogenic microorganisms produce CA in an aggregated form. Recently, MAYER and SCHMIDT (1971) have undertaken a detailed study of rough strains of *E. coli* and observed that even heated cultures of R1 mutants—in contrast to R2 mutants—engender CA antibodies in high

Table 1. CA Antibody Response of Rabbits Following Injection of Rough *Escherichia Coli* Strain Isolated from Patient with Peritonitis Associated with Dialysis

Materials for immunization* *E. coli* (rough)	Mean CA† hemagglutinin titers (reciprocal) Days			
	0	7	10	14
Viable bacterial suspension	<10	1493	2133	4693
Killed bacterial suspensions				
100°C 1ʰ	<10	320	2133	1920
60°C 1ʰ	<10	320	1920	1920
56°C 3ʰ	<10	267	1067	1280
0.5% formalin	<10	320	960	1920

* Injections on days 0, 3, and 7. Suspensions 5%, 5%, and 10% (1 ml each).

† CA = common antigen of enterobacteriaceae.

titers in the rabbit. Based on this information it was postulated that a rough strain of *E. coli,* repeatedly isolated from the peritoneal fluid of a patient undergoing dialysis, might be of the R1 type, provided that it induced CA antibodies. Indeed, it was shown that the CA antibody titers of this patient, as determined by hemagglutination, rose from <1:10 to 1:640. The results of immunization studies with this isolate in rabbits are recorded in Table 1 and indicate that it readily induced CA anitbodies in high titers when either living or killed organisms were injected. Determination of the R type of this strain is being carried out by Drs. Mayer and Schmidt, and the preliminary results indicate that, as expected, this organism belongs to the R1 category. Whether CA antibodies, which are frequently present in serum of healthy subjects (albeit in low titers), are engendered by such R strains, remains to be determined.

IV. Immunologic Priming and Antibody Response

The observation that CA produced by most enteric bacteria (supernates of heated suspensions) fails to engender antibody production in the rabbit could be due to the induction of immunologic unresponsiveness (tolerance). Surprisingly, challenge with a minimally effective dose of immunogenic (aggregated) CA results in a typical secondary immune response, characterized by an accelerated and intensified production of circulating CA antibodies (NETER *et al.,* 1966; WHANG *et al.,* 1971). Essentially identical findings were obtained in animals injected with non-aggregated (heated or millipore-filtered) CA. The results of a representative experiment are recorded in Table 2. Two groups of rabbits were immunized with the supernate of an agar-grown culture of *Salmonella minnesota* on days 0, 3, and 7. A third group served as the nonprimed control group. The latter two groups were given a challenge injection of minimally effective amounts of aggregated CA from *S. typhimurium* (ethanol-soluble frac-

Table 2. Immunologic Priming by Nonimmunogenic CA of
Salmonella Minnesota

Materials for immunization (in days)		Mean CA† hemagglutinin titers (reciprocal) Days				
Primary	Secondary	0	7	14	17	21
S.M.* (0, 3, 7)	—	<10	<10	<10	<10	<10
S.M. (0, 3, 7)	CA† (14)	<10	<10	<10	140	360
—	CA (14)	<10	—	<10	13	20

* S.M. = supernate of *S. minnesota* culture.
† CA = ethanol-soluble CA of *S. typhimurium.*

tion) on day 14. It is evident from Table 2 that antibodies in significant titers were engendered 3 days after the secondary (booster) injection only in animals which had been primed with nonaggregated CA which, by itself failed to stimulate the production of CA antibodies. This system, then, offers the opportunity of dissecting experimentally the immune response into its two major components, namely the emergence of memory cells responsible for immunologic priming (presumably T cells) from the other component of the immune response, the proliferation of antibody-producing B cells.

V. Biologic Significance of the Common Enterobacterial Antigen

The biologic significance of this particular bacterial antigen and its corresponding antibody has not yet been fully elucidated. It has been demonstrated, however, that injecting rabbits intravenously with their own erythrocytes, which had been modified by CA *in vitro,* results in immunologic hemolysis, provided that the recipients have antibodies specific for the antigen; otherwise, erythrocyte survival is normal (Suzuki *et al.,* 1964). It remains to be determined whether CA (or other bacterial antigens), during natural infection, become attached to the surface of cells; if this were to occur, cell injury could follow the appearance of the corresponding bacterial antibodies. In the absence of information on the source of the antigen, the pathologic changes might be ascribed erroneously to autoimmunity.

Immunologic injury may also take place if a bacterial antigenic determinant is shared with host cells. Reference may be made to the possible role of antibodies directed against a CA-related antigen present in the colon in the pathogenesis of ulcerative colitis (Lagercrantz *et al.,* 1968).

Since numerous enteric bacteria share the antigenic determinant characteristic of CA, the possibility presents itself that immunization with a single antigen might protect against infection by these microorganisms. This speculation is in accord with the following observations. Domingue and Neter (1966) found that antibodies against CA opsonize enteric bacteria for phagocytosis. More recently, Gorzynski *et al.* (1971) observed that CA antibodies provide transient protection to mice challenged with a highly virulent strain of *S. typhimurium.* Similar transient protection was observed in Swiss albino and C57 BL/6Ha black mice actively immunized with CA and challenged with the same virulent microorganism (Gorzynski, Priore, and Neter, unpublished observations). The most incisive results on the protective effect of CA as immunogen and of CA antibodies have been obtained by Domingue *et al.* (1970), who showed that rabbits thus immunized were protected to a significant degree from experimental pyelonephritis due to CA-positive but not CA-negative challenge strain. With the possibility in mind that human subjects, too, might be effectively immunized with CA, exploratory studies were undertaken by Neter *et al.* (1970). It was shown that the ethanol-soluble fraction prepared from *E. coli* 0111, upon both intradermal and subcutaneous injection, is well tolerated, even when undiluted antigen is administered. Under these circumstances, only a modest immune response could be documented in 4 out of 10 subjects. More recent observations have shown that this antigen, when given intravenously in dilutions of 1:100 or 1:200, readily engendered CA antibodies in significant titers (up to 1:1280) in 10 out of 12 subjects without producing significant side-effects in any of the volunteers. Van Oss, Ambrus, Gorzynski, and Neter (unpublished observations) have shown recently that these human CA antibodies opsonize enteric bacteria for phagocytosis. In view of these findings, further studies on the possible protective effect of the common enterobacterial antigen for active immunization and of CA antibodies for passive protection in enterobacterial infections of man are warranted.

References

Aoki, S., M. Merkel, and W. R. McCabe (1966). Immunofluorescent demonstrations of the common enterobacterial antigen. *Proc. Soc. Exp. Biol. Med., 121*:230–234.

Diaz, F., and E. Neter (1969). Production of common enterobacterial antigen by pigment-producing, Gram-negative bacilli. *J. Bacteriol., 97*:453–454.

Domingue, G. J., and E. Neter (1966). Opsonizing and bactericidal activity of antibodies against common antigen of *Enterobacteriaceae. J. Bacteriol., 91*:129–133.

—— (1967). The plaque test for the demonstration of antibodies against the enterobacterial common antigen produced by spleen and lymph node cells. *Immunology, 13*:539–545.

Domingue, G., A. Salhi, C. Rountree, and W. Little (1970). Prevention of experimental hematogenous and retrograde pyelonephritis by antibodies against enterobacterial common antigen. *Infect. and Immunity, 2*:175–182.

Gorzynski, E. A., J. L. Ambrus, and E. Neter (1971). Effect of common enterobacterial antiserum on experimental *Salmonella typhimurium* infection of mice. *Proc. Soc. Exp. Biol. Med., 137*:1209–1212.

Gorzynski, E. A., and E. Neter (1970). Production of common antigen by enteric bacteria grown in a synthetic culture medium. *Infect. and Immunity, 2*:767–771.

Kunin, C. M. (1963). Separation, characterization and biological significance of a common antigen in enterobacteriaceae. *J. Exp. Med., 118*:565–586.

Kunin, C. M., M. V. Beard, and N. E. Halmagyi (1962). Evidence for a common hapten associated with endotoxin fractions of *E. coli* and other enterobacteriaceae. *Proc. Soc. Exp. Biol. Med., 111*:160–166.

Lagercrantz, R., S. Hammarström, P. Perlmann, and B. E. Gustafsson (1968). Immunological studies in ulcerative colitis. IV. Origin of autoantibodies. *J. Exp. Med. 128*:1339–1352.

Landsteiner, K. (1945). The Specificity of Serological Reactions. Cambridge, Mass.: Harvard University Press.

Mayer, H., and G. Schmidt (1971). Hämagglutinine gegen ein gemeinsames enterobacteriaceen-Antigen in *E. coli* R1-Antiseren. *Zentbl. Bakt.,* *216*:299–313.

Neter, E. (1969). Endotoxins and the immune response. *Curr. Topics Microbiol. Immunol.* *47*:82–124.

Neter, E., E. A. Gorzynski, and J. L. Ambrus (1970). Common enterobacterial antigen: Studies in man. *10th Int. Congr. Microbiol.,* August 1970, Mexico.

Neter, E., H. Y. Whang, O. Lüderitz, and O. Westphal (1966). Immunological priming without production of circulating bacterial antibodies conditioned by endotoxin and its lipoid A component. *Nature,* London, *212*:420–421.

Perlmann, P., S. Hammarström, R. Lagercrantz, and D. Campbell (1967). Autoantibodies to colon in rats and human ulcerative colitis: Cross reactivity with *Escherichia coli* 0:14 antigen. *Proc. Soc. Exp. Biol. Med.,* *125*:975–980.

Rantz, L. A., E. Randall, and A. Zuckerman (1956). Hemolysis and hemagglutination by normal and immune serums of erythrocytes treated with a nonspecies specific bacterial substance. *J. Infect. Dis.,* *98*:211–222.

Springer, G. F., and R. E. Horton (1964). Erythrocyte sensitization by blood group-specific bacterial antigens. *J. Gen. Physiol.,* *47*:1229–1250.

Springer, G. F., E. T. Wang, J. H. Nichols, and J. M. Shear (1966). Relations between bacterial lipopolysaccharide structures and those of human cells. *Ann. N.Y. Acad. Sci.,* *133*:566–579.

Suzuki, T., E. A. Gorzynski, H. Y. Whang, and E. Neter (1964). Hemolysis in immune rabbits of autologous erythrocytes modified with common enterobacterial antigen. *Experientia,* *20*:75–76.

Whang, H. Y., H. Mayer, and E. Neter (1971). Differential effects on immunogenicity and antigenicity of heat, freezing and alkali treatment of bacterial antigens. *J. Immunol.,* *106*:1552–1558.

Whang, H. Y., and E. Neter (1962). Immunological studies of a heterogenetic enterobacterial antigen (Kunin). *J. Bacteriol.,* *84*:1245–1250.

—— (1964). Selective destruction by *Pseudomonas aeruginosa* of common antigen of enterobacteriaceae, *J. Bacteriol.,* *88*:1244–1248.

Whang, H. Y., Y. Yagi, and E. Neter (1967). Characterization of rabbit antibodies against common bacterial antigens and their presence in the fetus. *Int. Arch. Allergy Appl. Immunol.,* *32*:353–365.

AMINO SUGARS OF *PSEUDOMONAS AERUGINOSA* ATCC 7700 LIPOPOLYSACCHARIDE

ROBERT BARROW and ROBERT W. WHEAT*

*Department of Microbiology and Immunology
and Department of Biochemistry,
Duke University Medical Center,
Durham, North Carolina*

I. Abstract

The amino sugar content of *Pseudomonas aeruginosa* lipopolysaccharide was investigated. Glucosamine, galactosamine, quinovosamine, and fucosamine were isolated and characterized from the lipopolysaccharide of *P. aeruginosa* ATCC 7700. Two unidentified reducing cationic compounds were observed.

II. Introduction

The amino sugars, glucosamine, galactosamine, and either DL-fucosamine (FENSOM and GRAY, 1969; SUZUKI, 1969) or D-quinovosamine (SUZUKI *et al.*, 1970) have been reported in the lipopolysaccharides of different strains of *P. aeruginosa*. Various 2-amino-2,6-dideoxyhexosamines, including D-quinovosamine, both D- or L-fucosamine, L-rhamnosamine and L-6-deoxy-talosamine, have been reported in both gram-positive and gram-negative bacterial polysaccharides, but again, these compounds occur singly (BARKER *et al.*, 1961; CRUMPTON and DAVIES, 1958; ERLER, 1969; JANN and JANN, 1968; KELETI *et al.*, 1971; LÜDERITZ *et al.*, 1968a, b; RAFF and WHEAT, 1970; SHARON *et al.*, 1964; SHARON, 1965; SMITH, 1964; WHEAT *et al.*, 1964). The present work reports the occurrence of glucosamine, galactosamine, quinovosamine, and fucosamine in hydrolysates of the lipopolysaccharide (LPS) isolated from *P. aeruginosa* ATCC 7700. The occurrence of both quinovosamine and fucosamine

* This work was supported in part by an Afgrad grant (R. B.) and in part by NIAID research grant 01659, National Institutes of Health.

in the same organism is unusual and has not previously been reported. Two additional unidentified reducing cationic compounds were also observed. These and similar unidentified reducing cationic compounds observed in lipopolysaccharides of *Citrobacter freundii* ATCC 10053 (KORCZYNSKI and WHEAT, 1970) and other bacteria (Bowser and Wheat, unpublished observations) are currently under investigation. A different reducing cationic compound, 4-keto-norleucine, which may or may not be a strong mineral acid degradation product of an unknown amino sugar, was previously reported from several enterobacterial polysaccharides by BARRY and ROARK (1964).

A variety of amino sugars, as well as pentoses and hexoses, have been found in the O-specific repeating units of different lipopolysaccharides. These include 6-deoxyhexoses with an amino group at the C-3 or C-4 position of the carbon chain as well as at the above-mentioned 2-amino-6-deoxyhexoses (ASHWELL and VOLK, 1965; HICKMAN and ASHWELL, 1966; JANN *et al.*, 1967; KELETI *et al.*, 1971; LÜDERITZ *et al.*, 1966, 1967 1968a, b; LÜDERITZ, 1970; RAFF and WHEAT, 1967, 1970; SMITH *et al.*, 1962; SMITH, 1964; STEVENS *et al.*, 1963; VOLK *et al.*, 1970; WHEAT *et al.*, 1962). Of these various amino sugars, only glucosamine has been found in the core polysaccharides (HAMMERLING *et al.*, 1970; HEATH, 1971; KELETI *et al.*, 1971; LÜDERITZ, 1970; NIKAIDO, 1970; OSBORN, 1969; SCHMIDT *et al.*, 1969; SIMMONS, 1971), and until recently, only glucosamine was found in the lipid-A region of lipopolysaccharides (NOWOTNY, 1961; ADAMS and SINGH, 1969, 1970; BURTON and CARTER, 1964; DROGE *et al.*, 1970; GMEINER *et al.*, 1969; HEWETT *et al.*, 1971; KASAI and NOWOTNY, 1967; OSBORN, 1969; RIETSCHEL, 1970). The glucosamine appears as phosphate ester in partial hydrolysate of lipid-A preparations (NOWOTNY, 1961). Phosphodiester-linked glucosamine disaccharide units (similar to the teichoic acid-like polymers of gram-positive bacteria cf. BADDILEY, 1970) esterified with both saturated fatty acids and β–OH fatty acids at the hydroxyl and amino groups of the glucosamines, and phosphoryl-ethanolamine appear to be the major constituents of the lipid-A moiety of enteric bacteria, as described first by NOWOTNY (1961, 1971). Other investigators assumed similar structures, but disagree on the existence of phosphodiester bridges between glucosamine units (ADAMS and SINGH, 1970; BURTON and CARTER, 1964; LÜDERITZ, 1970; RIETSCHEL, 1971; SIMMONS, 1971). Covalent linkage of the lipid-A to amino acids has also been suggested (NOWOTNY, 1961; NOWOTNY *et al.*, 1968; WOBER and ALAUPOVIC, 1971a, b). Various C-10 to C-16 β–OH fatty acids are found in the lipid-A moieties of endotoxins as well as other toxins (ADAMS and SINGH, 1969, 1970; BOYLAN and SCHEUER, 1967; FENSOM and GRAY, 1969; GMEINER *et al.*, 1969; HEWETT *et al.*, 1971; LÜDERITZ, 1970; RIETSCHEL, 1971). That the lipid-A represents the active principle of LPS was first claimed by WESTPHAL and LÜDERITZ (1954). In the light of recent investigations, it appears that the presence of β–OH, as well as other fatty acids, in conjuntion with the phosphorylated glucosamine disaccharide, is important for the toxicity of the LPS (KASAI and NOWOTNY, 1967; KIM and WATSON, 1967; LÜDERITZ, 1970). For reviews on this see NOWOTNY, 1969, 1971.

The involvement of unusual amino compounds in the endotoxic components of bacteria remains to be demonstrated, but the occurrence of alternatives to glucosamine or ethanolamine seems possible. The unusual 4,5-diamino-eicosane derivative, necrosamine, has been reported to occur in at least one *Escherichia coli* LPS (IKAWA, 1966). And more recently, VOLK *et al.* (1970) isolated, after mild hydrolysis, 4-amino-L-arabinose from both the complete wild type *Salmonella* LPS and heptoseless LPS (glycolipid) obtained from *Salmonella mutants.* 4-Aminoquinovose has been isolated from *Chromobacterium violaceum* 7917 (SMITH *et al.*, 1962; STEVENS *et al.*, 1962; WHEAT *et al.*, 1962), and both 4-aminoquinovose and 4-aminofucose have been isolated from *E. coli* (JANN and JANN, 1967). In addition, these acid-labile amino sugars presumably occur in polysaccharides of various salmonella and pasteurella where their biosynthesis is known to take place through the intermediate, TDP-4-keto-6-deoxyglucose (GILBERT *et al.*, 1965; MATSUHASHII and STROMINGER, 1964; OHASHI *et al.*, 1971; OKAZAKI *et al.*, 1962). This raises the question of whether these 4-amino sugars, the 2,4-diamino-trideoxyhexoses so far described only in the gram-positive diplococcus C-polysaccharide (DISTLER *et al.*, 1966), and *Bacillus licheniformis* (SHARON and JEANLOZ, 1960), the various reducing cationic compounds we describe here and those described by BARRY and ROARK (1964), are widely distributed but remain undetected in materials which exhibit endotoxin activity.

III. Experimental

P. aeruginosa ATCC 7700 was grown on 0.8% nutrient broth (Difco) at 37°C for 18 hours, harvested by centrifugation at 4000 $\times$ g 30 minutes at 5°C, and cells sedimented were acetone-powdered. A 5% aqueous cell suspension was extracted with 45% aqueous phenol according to WESTPHAL *et al.* (1952). Aqueous phase material was dialyzed against hot running tap water for 24 hours, then against four changes of deionized water, and was lyophilized (yield 359 mg, 7.5%). The LPS was sedimented by ultracentrifugation at 100,000 $\times$ g for 8 hours to remove soluble RNA. The pellet was resuspended, centrifuged at 15,000 $\times$ g to remove debris, and lyophilized (yield 210 mg, 4.3%). Purified LPS, 136 mg, was hydrolyzed in 2 N HCl for 6 hours at 100°C in sealed evacuated tubes, the acid was removed by evaporation, and the hydrolysate residue was dissolved in water and sorbed onto a 2.5 $\times$ 7.0 cm column of Dowex 50.H⁺. After washing with 40 vol. of water, amino sugars were eluted batchwise with 200 ml 0.3 N HCl. Acid was removed by distillation and the residue dissolved in 10 ml H_2O. This fraction, 0.2 ml of 10 ml, yielded

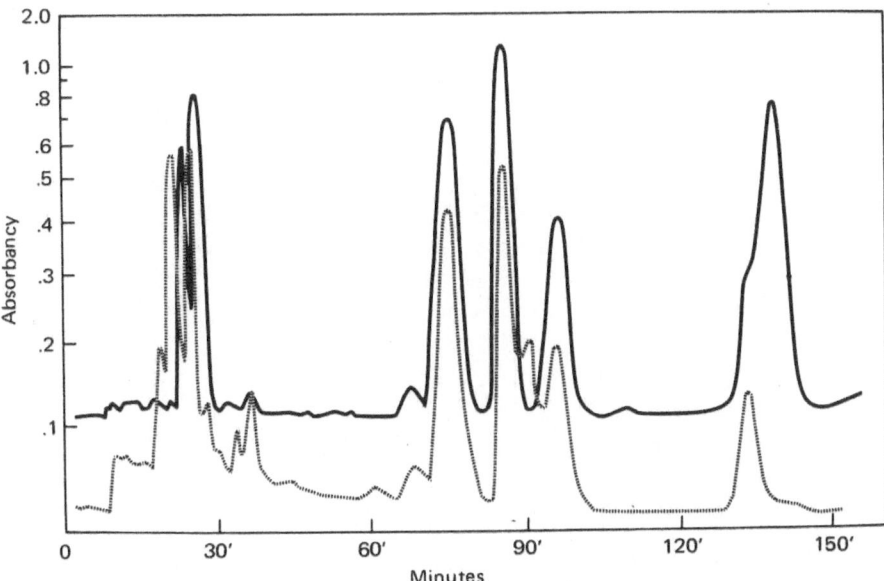

Fig. 1. Amino sugar analyzer profile of 1.58 mg hydrolysate chromatographed according to Steele *et al.* (1970). Upper trace, ninhydrin. Lower trace, reducing group activity.

the amino sugar elution profile at pH 4.3 on the amino acid analyzer shown in Figure 1. Amino sugars yielding both ninhydrin and reducing group activities were identified as glucosamine (67 minutes), galactosamine (75 minutes), quinovosamine (86 minutes), and fucosamine (96 minutes) by retention times of corresponding ninhydrin and reducing group reactivities previously determined (STEELE *et al.,* 1970). Two unidentifiable reducing compounds were observed, one at 90 minutes between quinovosamine and fucosamine, and another at 133 minutes in the region where ethanolamine is eluted just before ammonia (137 minutes). The remainder of the 0.3 N HCl eluate, 9.8 ml, was sorbed onto a 2 × 40 cm column of Dowex 50.H⁺ and eluted with 0.3 N HCl, collecting 5 ml fractions according to the procedures of GARDELL (1953) and CRUMPTON (1959). Amino sugars were determined in fractions both as 2-amino sugars (BROWNLEE and WHEAT, 1966) and as reducing sugars (PARK and JOHNSON, 1959). The four amino sugars identified above were again observed (Table 1). The fractions containing them were separately pooled, evaporated to remove acid, and assayed for 2-amino sugar content, using a glucosamine standard. Yields were 0.18 μmoles, peak 1; 2.8 μmoles, peak 2; 4.05 μmoles, peak 3; and 0.64 μmoles, peak 4. Each 2-amino sugar fraction was then checked by amino sugar analyzer chromatography alone and by co-chromatography with either glucosamine (peak 1), galactosamine (peak 2), quinovosamine (peak 3), or fucosamine (peak 4),

Table 1.

	Cation exchange column chromatography (0.3 N HCl)		Amino sugar analyzer*		Thin layer cellulose chromatography		Ninhydrin degradation products	
	Elution volume	R_{GlcN}	Retention time	R_{GlcN}	Distance moved	R_{GlcN}	Distance	R_{Arab}
1	130	1.0	72	1.0	not detected	—	5.75	1.0
2	153	1.18	80	1.11	5.6	0.92	6.5	1.13
3	183	1.41	93	1.30	9.0	1.47	9.5	1.65
4	229	1.76	103	1.43	8.1	1.33	8.9	1.55
GlcN	130	1.0	72	1.0	6.1	1.0	5.75	1.0
GalN	154	1.19	80	1.11	5.6	0.92	6.5	1.13
QuiN	186	1.43	94	1.30	9.1	1.50	9.5	1.65
FucN	230	1.77	103	1.43	8.1	1.33	9.0	1.57
Ara							5.75	
Lyx							6.65	

* Retention times shown are for individually isolated peak materials and standards run separately and co-chromatographically on a different column than used for data presented in text for whole hydrolysate.

respectively. Each compound corresponded chromatographically with the respective known (Table 1, column 2).

Thin-layer cellulose chromatograms (Table 1), run in ethylacetate-pyridine acetic acid-water (5:5:1:3) and in butanol-acetic acid-water (5:1:2), confirmed the identities established by cation exchange column chromatography with 0.3 N HCl and by the amino sugar analyzer at pH 4.3, 0.3 *M* Na citrate (STEELE *et al.*, 1970). Ninhydrin degradation (STOFFYN and JEANLOZ, 1954) of peak 1 yielded arabinose, confirming glucosamine, while peak 2 yielded lyxose, confirming galactosamine, and the corresponding deoxypentose derivatives were obtained from peak 3, confirming quinovosamine, and peak 4, confirming fucosamine. The unusual reducing cationic materials are being investigated.

References

Adams, G. A., and P. P. Singh (1969). The linkage between the D-glucosamine units of lipid A isolated from the lipopolysaccharide of *Serratia marcescens*. *Biochim. Biophys. Acta, 187*:457–459.

—— (1970). Structural features of lipid A preparations isolated from *Escherichia coli* and *Shigella flexneri*. *Biochim. Biophys. Acta, 202*:553–555.

Ashwell, G., and W. A. Volk (1965). Isolation and identification of N-acetyl-3-amino-3,6-dideoxy-D-galactose, a cell wall constituent of *Xanthomonas campestris*. *J. Biol. Chem., 240*:4549–4555.

Baddiley, J. (1970). Structure, biosynthesis and function of teichoic acids. *Accounts Chem. Res., 3(3)*:98–105.

Barker, S. A., J. S. Brimacombe, M. J. How, M. Stacey, and J. M. Williams (1961). Two new amino-sugars from an antigenic polysaccharide of pneumococcus. *Nature, Lond., 189*:303–304.

Barry, G. T., and E. Roark (1964). L-Fucosamine and 4-oxo-Norleucine as constituents in mucopolysaccharides of certain enteric bacteria. *Nature, Lond., 202*:493–494.

Boylan, D. B., and P. J. Scheuer (1967). Pahutoxin: A fish poison. *Science, 155*:52–56.

Brownlee, S. T., and R. W. Wheat (1966). On the determination of galactosamine-uronic acid. *Anal. Biochem., 14*:414–420.

Burton, A. J., and H. E. Carter (1964). Purification and characterization of the lipid A component of the lipopolysaccharides from *Escherichia coli*. *Biochemistry, 3*:411–418.

Crumpton, M. J. (1959). Identification of amino sugars. *Biochem. J., 72*:479–486.

Crumpton, M. J., and D. A. L. Davies (1958). The isolation of D-fucosamine from the specific polysaccharide of *Chromobacterium violaceum* (NCTC 7917). *Biochem. J., 70*:729–736.

Distler, J., B. Kaufman, and S. Roseman (1966). Enzymatic synthesis of a diaminosugar nucleotide by extracts of type XIV *Diplococcus pneumoniae*. *Arch. Biochem. Biophys., 116*:466–478.

Dröge, W., V. Lehmann, O. Lüderitz, and O. Westphal (1970). Structural investigations on the 2-keto-3-deoxyoctonate region of lipopolysaccharides. *Eur. J. Biochem., 14*:175–184.

Erler, W. (1968). Serologische, chemische und immunchemische Untersuchungen an Ratlaufbacterien. IV. Das Vorkammen von Fukosamin (2,6-dideoxy-2-amino-galaktose) in den Rotlaufbakterien. *Arch. Exp. Veterinärmed., 22*:1155–1163.

Fensom, A. H., and G. W. Gray (1969). The chemical composition of the lipopolysaccharide of *Pseudomonas aeruginosa. Biochem. J., 114*:185–196.

Gardell, S. (1953). Separation on Dowex 50 ion exchange resin of glucosamine and galactosamine and their quantitative determination. *Acta Chem. Scand., 7*:207–215.

Gilbert, J. M., M. Matsuhashi, and J. L. Strominger (1965). Thymidine diphosphate-4-acetamido-4,6-dideoxyhexoses. II. Purification and properties of a thymidine diphosphate D-glucose oxidoreductase. *J. Biol. Chem., 240*:1305–1308.

Gmeiner, J., O. Lüderitz, and O. Westphal (1969). Biochemical studies on lipopolysaccharides of *Salmonella* R mutants. 6. Investigations on the structure of the lipid A component. *Eur. J. Biochem., 7*:370–379.

Hämmerling, G. (1970). Vergleichende Strukturuntersuchungen an den Kernpolysaccariden der Lipopolysaccharide von *Salmonella typhimurium* und *E. coli* 0100. Ph.D. Dissertation, Universität zu Freiburg im Breisgau.

Hämmerling, G., O. Lüderitz, and O. Westphal (1970). Structural investigations on the core polysaccharide of *Salmonella typhimurium* and the mode of attachment of the O-specific chains. *Eur. J. Biochem., 15*:48–56.

Heath, E. C. (1971). Complex polysaccharides. *Ann. Rev. Biochem., 40*:29–56.

Hewett, M. J., K. W. Knox, and D. G. Bishop (1971). Biochemical studies on lipopolysaccharides of *Veillonella. Eur. J. Biochem., 19*:169–175.

Hickman, J., and G. Ashwell (1966). Isolation of a bacterial lipopolysaccharide from *Xanthomonas campestris* containing 3-acetamido-3,6-dideoxy-D-galactose and D-rhamnose. *J. Biol. Chem., 241*:1424–1428.

Ikawa, M. (1966). Studies on a lipopolysaccharide from *Escherichia coli. Ann. N.Y. Acad. Sci., 133*:476–485.

Jann, B., and K. Jann (1967). 4-amino-4,6-dideoxyhexoses isolated from lipopolysaccharides of *Escherichia coli. Eur. J. Biochem., 2*:26–31.

—— (1968). 2-amino-2,6-dideoxy-L-mannose (L-rhamnosamine) isolated from the lipopolysaccharide of *Escherichia coli* 03:K2ab(L):H2. *Eur. J. Biochem., 5*:173–177.

Jann, B., K. Jann, and E. Müller-Seitz (1967). A 3-amino-3,6-dideoxyhexose from the lipopolysaccharide of *Escherichia coli* 071. *Nature, Lond., 215*:170–171.

Kasai, N., and A. Nowotny (1967). Endotoxic glycolipid from a heptoseless mutant of *Salmonella minnesota. J. Bacteriol., 94*:1824–1836.

Keleti, J., O. Lüderitz, D. Mlynarčík, and J. Sedlák (1971). Immunochemical studies on *Citrobacter* O antigens (lipopolysaccharides). *Eur. J. Biochem., 20*:237–244.

Kim, Y. B., and D. W. Watson (1967). Biologically active endotoxins from *Salmonella* mutants deficient in O- and R-polysaccharides and heptose. *J. Bacteriol., 94*:1320–1326.

Korczynski, M. S., and R. W. Wheat (1970). Environmental influence on lipopolysaccharide composition. *Bacteriol. Proc.*, p. 56.

Lüderitz, O. (1970). Recent results on the biochemistry of the cell wall lipopolysaccharides of *Salmonella* bacteria. *Angew. Chem.*, *9*:649–663.

Lüderitz, O., J. Gmeiner, B. Kickhöfen, H. Mayer, O. Westphal, and R. W. Wheat (1968a). Identification of D-mannosamine and quinovosamine in *Salmonella* and related bacteria. *J. Bacteriol.*, *95*:490–494.

Lüderitz, O., K. Jann, and R. W. Wheat (1968b). Somatic and capsular antigens of Gram-negative bacteria. *Comp. Biochem.*, *26a*:105–228.

Lüderitz, O., E. Ruschmann, O. Westphal, R. Raff, and R. Wheat (1967). Occurrence of 3-amino-3,6-dideoxy-hexoses in *Salmonella* and related bacteria. *J. Bacteriol.*, *93*:1681–1687.

Lüderitz, O., A. M. Staub, and O. Westphal (1966). Immunochemistry of O and R antigens of *Salmonella* and related *Enterobacteriaceae*. *Bacteriol. Rev.*, *30*:192–255.

Matsuhashi, M., and J. L. Strominger (1964). Thymidine diphosphate 4-acetamido-4,6-dideoxyhexoses. *J. Biol. Chem.*, *239*:2454–2463.

Nikaido, H. (1970). Structure of the cell wall lipopolysaccharide from *Salmonella typhimurium*. Further studies on the linkage between O side chains and R core. *Eur. J. Biochem.*, *15*:57–62.

Nowotny, A. (1961). Chemical structure of a phosphomucolipid and its occurrence in some strains of *Salmonella*. *J. Am. Chem. Soc.*, *83*:501–503.

—— (1969). Molecular aspects of endotoxic reactions. *Bacteriol. Rev.*, *33*:72–98.

—— (1971). Structure and function of bacterial endotoxins. *Naturwissenschaften*, *58*:397–409.

Nowotny, A., A. M. Nowotny, and M. P. Brigham (1968). Amino acids of endotoxins and their role in toxicity. *Bact. Proc.*, M.178.

Ohashi, H., M. Matsuhashi, and S. Matsuhashi (1971). Thymidine diphosphate 4-acetamido-4,6-dideoxyhexoses. IV. Purification and properties of thymidine diphosphate 4-keto-6-deoxy-D-glucose transaminase from *Pasteurella pseudotuberculosis*. *J. Biol. Chem.*, *246*:2325–2330.

Okazaki, T., R. Okazaki, J. L. Strominger, and S. Suzuki (1962). Thymidine diphosphate N-acetylamino sugar compounds from *Escherichia coli* strains. *Biochem. Biophys. Res. Commun.*, *7*:300–305.

Osborn, M. J. (1969). Structure and biosynthesis of the bacterial cell wall. *Ann. Rev. Biochem.*, *38*:501–538.

Park, J. T., and M. J. Johnson (1949). A submicrodetermination of glucose. *J. Biol. Chem.*, *181*:149–151.

Raff, R. A., and R. W. Wheat (1966). Occurrence of a 3-amino sugar in the cell wall of *Citrobacter freundii* 8090. *Biochim. Biophys. Acta*, *127*:271–273.

—— (1967). Characterization of 3-amino-3,6-dideoxy-D-glucose from a bacterial lipopolysaccharide. *J. Biol. Chem.*, *242*:4610–4613.

—— (1970). The naturally occurring amino sugars, pp. D-69-81. In *Handbook of Biochemistry*, 2nd ed., H. Sober (ed.), Cleveland: Chemical Rubber Co.

Rietschel, E. T. (1971). Lipoid als aktives Prinzip der Endotoxinwirkungen. Ph.D. Dissertation, Albert-Ludwig Universität zu Freiburg im Breisgau.

Schmidt, G., B. Jann, and K. Jann (1969). Immunochemistry of R-lipopolysaccharides of *Escherichia coli*. Different core regions in the lipopolysaccharides of O group 8. *Eur. J. Biochem., 10*:501–510.

Sharon, N. (1965). Distribution of amino sugars in micro-organisms, plants, and invertebrates, pp. 1–45. In *The Amino Sugars*, vol. IIA, R. Jeanloz and A. Balazs (eds.), New York: Academic Press.

Sharon, N., and R. W. Jeanloz (1960). The diaminohexose component of a polysaccharide isolated from *Bacillus subtilis. J. Biol. Chem., 235*:1–5.

Sharon, N., I. Shif, and U. Zehavi (1964). The isolation of D-fucosamine (2-amino-2,6-dideoxy-D-galactose) from polysaccharides of bacillus. *Biochem. J., 93*:210–214.

Simmons, D. A. R. (1971). Immunochemistry of *Shigella flexneri* O-antigens: a study of structural and genetic aspects of the biosynthesis of cell-surface antigens. *Bacteriol. Rev., 35*:117–148.

Smith, E. J. (1964). The isolation and characterization of 2-amino-2,6-dideoxy-D-glucose (D-quinovosamine) from a bacterial polysaccharide. *Biochem. Biophys. Res. Commun., 15*:593–597.

Smith, E. J., J. M. Leatherwood, and R. W. Wheat (1962). Isolation of a new amino sugar from *Chromobacterium violaceum. J. Bacteriol., 84*:1007–1010.

Steele, R. S., K. Brendel, E. Scheer, and R. W. Wheat (1970). Ion-exchange separation and automated assay of complex mixtures of amino acids and hexosamines. *Anal. Biochem., 34*:206–225.

Stevens, C. L., P. Blumbergs, F. A. Daniher, R. W. Wheat, A. Kiyomoto, and E. L. Rollins (1963). The identification and synthesis of the 4-amino sugar from *Chromobacterium violaceum. J. Am. Chem. Soc., 85*:3061.

Stoffyn, P. J., and R. W. Jeanloz (1954). Identification of amino sugars by paper chromatography. *Arch. Biochem. Biophys., 52*:373–379.

Suzuki, N. (1969). Isolation of DL-fucosamine from *Pseudomonas aeruginosa* N10. *Biochim. Biophys. Acta, 177*:371–373.

Suzuki, N., A. Suzuki, and K. Fukasawa (1970). Distribution of 2-amino sugars in *Pseudomonas aeruginosa:* Isolation of D-quinovosamine from *P. aeruginosa* P14. J. *Jap. Biochem. Soc., 42*:130–134.

Volk, W. A., C. Galanos, and O. Lüderitz (1970). Isolation of 4-amino-4-deoxy-L-arabinose from S and R form bacterial lipopolysaccharides. *FEBS Lett., 8*(3):161–163.

Westphal, O., and O. Lüderitz (1954). Chemische Enforschung von Lipopolysacchariden gramnegativer Bacterien. *Angew. Chem., 66*:407–417.

Westphal, O., O. Lüderitz, and F. Bister (1952). Über die Extraktion von Bakterien mit Phenol/Wasser. *Z. Naturforsch., 7b*:148–155.

Wheat, R. W., E. L. Rollins, and J. M. Leatherwood (1962). Characterization of a 4-amino-4,6-dideoxy-aldohexose from *Chromobacterium violaceum. Biochem. Biophys. Res. Commun., 9*:120–125.

Wober, W., and P. Alaupović (1971a). Studies on the protein moiety of endotoxin from Gram-negative bacteria: Characterization of the protein moiety isolated by acetic acid hydrolysis of endotoxin from *Serratia marcescens* 08. *Eur. J. Biochem., 19*:357–367.

Wober, W., and P. Alaupović (1971b). Studies on the protein moiety of endotoxin from Gram-negative bacteria: Characterization of the protein moiety isolated by phenol treatment of endotoxin from *Serratia marcescens* 08 and *Escherichia coli* 0141:K85(B). *Eur. J. Biochem., 19*:340–356.

THE NATURE OF FLAGELLAR ANTIGENS*†

Henry Koffler and R. W. Smith

*Department of Biological Sciences, Purdue University,
Lafayette, Indiana*

I. The Nature of the Bacterial Flagellum and its Constituent Proteins

The bacterial flagellum, the organelle responsible for the locomotion of flagellated bacteria, is differentiated into at least three morphologically distinct regions. The basal structure of the flagellum appears to consist of four discs connected by a central rod, intimately associated with the cell wall and the cytoplasmic membrane (Van Iterson et al., 1966; Hoeniger et al., 1966; Ritchie et al., 1966; Cohen-Bazire and London, 1967; Abram, 1968; Remsen et al., 1968; Ritchie and Bryner, 1969; Vaituzis and Doetsch, 1969; Tauschel and Drews, 1969; DePamphilis and Adler, 1971a, b). The question as to whether an internal membranous vesicle is an integral part of this structure (Abram, 1968) or an artifact produced by specimen preparation (DePamphilis and Adler, 1971a, b) still remains to be settled. The bulk of the flagellum lies external to the cell body and consists of the hook and the filament.

The hook, which constitutes about 1% of the flagellum, differs from the filament in morphology and fine structure (Glauert et al., 1963; Lowy, 1965; Abram et al., 1965, 1967, 1970; Smith and Koffler, 1971; DePamphilis and Adler, 1971b), relatively greater stability (Abram et al., 1966, 1967, 1970; Tauschel, 1970; Mitchen, 1969, 1971; Mitchen and Koffler, 1969; Mitchen et al., 1970a, b; DePamphilis and Adler, 1971b; Dimmitt and Simon, 1971), and antigenicity (Lawn, 1967; Mitchen et al., 1970; Dimmitt and Simon, 1970; McGroarty, 1971). As will be shown in this paper, digestion with trypsin results in the release of 85% of the hook mass in the form of

* Portions of the research described in this report were carried out by Drs. Estelle McGroarty, J. R. Mitchen, Frank Kocka, L. R. Yarbrough, and Mr. L. W. Oiler. We acknowledge the capable technical assistance of Judith Lawton, Cathlene Shanke, Florence Shen, and Willa Mae Curry.
† This work was supported in part by NIH grant AI00685 and NSF grant GB-6329.

partial peptides that chromatographically are similar or identical to those found in tryptic digests of filaments; this suggests that the hook protein or proteins are similar to those found in the filament (MITCHEN *et al.,* 1970a, b; MITCHEN, 1971).

The flagellar filament is about 120 to 200 A in diameter, several microns in length, and constitutes approximately 95% or more of the flagellar mass. It is a helical microtubule constructed of the protein flagellin (ASTBURY, 1951; KERRIDGE *et al.,* 1962; GLAUERT *et al.,* 1963; ABRAM *et al.,* 1964a; LOWY and HANSON, 1964, 1965; HOENIGER, 1965; RITCHIE *et al.,* 1966; CHAMPNESS and LOWY, 1967; LOWY and SPENCER, 1968; CHAMPNESS, 1971), polarly oriented ovoid subunits which are held together by noncovalent linkages (KOFFLER, 1957; LACEY, 1961; SMITH and KOFFLER, 1971). This polymer of flagellin can be disintegrated to the monomeric form by a variety of agents (for example, pH < 4, heat, urea, sodium laurylsulfate, etc.). The filament grows unidirectionally, probably at the distal end (ASAKURA *et al.,* 1968; IINO, 1969; EMERSON *et al.,* 1970; O. Kerridge, personal communication). Some cultures produce more than one molecular variety of flagellin. For example, a given cell of *Salmonella typhimurium* may be genetically capable of producing two different flagellar antigens, but it actually synthesizes only one of these at a given time (LEDERBERG and IINO, 1956; IINO and LEDERBERG, 1964). Cells of *Bacillus pumilus,* on the other hand, produce two flagellins at the same time, in the ratio of 7:3; these flagellins are located in the same flagellum (SULLIVAN, 1968; SULLIVAN *et al.,* 1969; OILER *et al.,* 1971).

Flagellin is a unique protein in that it contains few, if any, residues of cys/2, his, pro, try, and tyr (SMITH and KOFFLER, 1971) and, under appropriate conditions, is capable of assembling *in vitro* into helical filaments that are morphologically similar to normal filaments, although longer (ABRAM and KOFFLER, 1963, 1964; ASAKURA *et al.,* 1964). Flagellin is capable of assuming more than one conformation under different environmental conditions (ERLANDER *et al.,* 1960; YAGUCHI *et al.,* 1964; KLEIN *et al.,* 1967a, b; YARBROUGH *et al.,* 1969, 1971a, b; YARBROUGH, 1971). Since this structural versatility has bearing on the immunogenic and antibody-binding properties of this protein, this property will later be discussed in greater detail.

II. Demonstration and Differentiation of Flagellar and Somatic Antigens

The existence of two distinguishable classes of bacterial surface antigens has long been recognized (JOOS, 1903; SMITH and REAGH, 1903). These were considered to be somatic (O antigens) and flagellar (H antigens) in nature, since conditions that either remove or destroy flagella (*e.g.,* heat above 65°C, acid below about pH 3.5, or mechanical agitation) also destroy the H antigen. The H-type agglutination is characterized by the formation of large flocculent particles that are easily dispersed by shaking; the O type is identified by the formation of a small, compact agglutinate (BALTEANU, 1926). The fact that mutations that bring about changes in the H antigen can be mapped within the structural genes for flagellar proteins (LEDERBERG and EDWARDS, 1953; FURNESS, 1958; JOYS and STOCKER, 1963, 1966; YAMAGUCHI and IINO, 1966) speaks convincingly for the location of the H antigens within the flagellum.

Flagellar antigens are considered to be distinct from other cellular antigens, although there have been reports that antisera prepared specifically against purified flagellar filaments also react with the cell membrane (TOMCSIK and BAUMANN-GRACE, 1956; VENNES and GERHARDT, 1959; BHARIER and RITTENBERG, 1971). Most probably this reaction is due to the presence of surface materials in the antigen preparation, since it is difficult to isolate filaments without at least traces of O antigens (WEIBULL, 1948, 1949, 1950a, b, 1953; WEINSTEIN, 1959).

(More highly purified antigens can be obtained by solubilization of filaments by acid or heat treatment (pH 2, 25°C, for 30 minutes, or 60°C for 30 minutes) and reconstitution of filaments either with or without purification of the constituent flagellin(s) by diethylaminoethyl (DEAE) cellulose chromatography; these processes can be repeated as often as necessary to achieve the desired purity.)

Of course a reaction of cell membranes with antibodies against flagellar filaments might also indicate that the basal structure, at least in part, contains proteins of the flagellin type or that flagellin exists in the membrane before it becomes polymerized. Because of the uncertainties regarding the immunogenic homogeneity of the antigens used, it is not possible to make this assessment.

III. Purification of Flagellar Antigens

Since the basal structure of the flagellum appears to be built into the ectoplasmic layers of the cell, it can be regarded as part of the bacterial surface. While nothing is known, as yet, of the antigenic properties of the basal structure, such information can now reasonably be expected, since it should be possible to purify the components of the basal structure, albeit in minute amounts. When cells are shaken vigorously, the external portions of the flagellum break away from the cell bodies and can be isolated by fractional centrifugation (KOFFIER, 1957; STOCKER, 1957; KOBAYASHI et al., 1959). In the case of cells of *B. pumilus,* about 40% of the hooks remain with the cells, and 60% accompany the filaments. Since hooks are somewhat more stable than filaments, as was first demonstrated by ABRAM et al., (1966), they can be purified after selective solubilization of the filaments, as discussed below. Instead of removing the filaments and hooks by shaking, one can isolate "intact" flagella by cesium chloride density gradient centrifugation after preparation of spheroplasts, lysis of the spheroplasts with Triton X-100, precipitation of the organelles with 25% ammonium sulfate, removal of the salt by dialysis and of the debris by centrifugation at 4000 × g (DePamphilis and Adler, 1971c). If one takes advantage of the differential stability of the various flagellar components, one can isolate basal structure-hook complexes by dissolving the filaments of intact flagella at pH 3.5, with 5 M urea, or by treatment at 60°C for 10 minutes; the hooks can be disintegrated at pH 2.5, and the basal structures are left behind (DePamphilis and Adler, personal communication).

IV. Methods for Determining Antigen-Antibody Reaction

When motile cells are reacted with an antiserum prepared against flagellar filaments, such filaments are cross-linked by the antibody molecules and the cells are immobilized (Gard, 1937; Gard and Erikson, 1939). No irreversible damage is done to the cells, since motility resumes if the antibody molecules are subsequently digested by trypsin, pepsin, or bromelin (DiPierro and Doetsch, 1968). This cross-linking of flagellar filaments makes possible specialized techniques that have proven valuable in studies on bacterial flagella. For example, one is able to measure the antigen-antibody reaction in terms of immobilizing titers or doses (Mäkelä and Nossal, 1961; Ada (et al., 1964; Greenbury and Moore, 1966). Also, one can incorporate an immobilizing antiserum into a solid growth medium and select motile mutant cells that have flagella altered to the extent that the filaments are not cross-linked by antibody molecules (Iion and Mitani, 1964; Joys and Stocker, 1966).

Other techniques for determining antigen-antibody reactions with either intact flagella or the solubilized components are similar to those used in other antigenic systems and involve precipitin, agglutinin, double-diffusion, hemagglutination, and complement fixation reactions. The nature of the various reactions between flagellar components and specific antibodies has been examined by use of either ^{131}I-labeled antigen (Ada et al., 1963) or ^{131}I-antibody (Greenbury and Moore, 1966). The results of these studies will be discussed later.

One exceptionally useful technique for measuring antigen-antibody reaction was developed by Grant and Simon (1968). Antibody molecules against either solubilized flagellar protein or intact filaments are labeled with [125]I. After reaction with antigen, the mixture is filtered through DEAE-cellulose filter paper under pH and ionic conditions that cause both the free antigen and the antigen-antibody complex, but not the unreacted antibody molecules, to bind to the filter. The amount of radioactivity retained by the filter, then, is a measure of the degree of antigen-antibody reaction. Using this and similar techniques, one can determine the relative reactivities of different flagellar proteins, as well as of given flagellins in different conformational states.

V. The Occurrence of Two Similar Flagellins in Flagellar Filaments from Cells of *Bacillus pumilis*

Dependent upon the objectives of given scientific inquiries, the homogeneity of the protein may be significant. The following is offered as an interesting case history that may enable others to deal with this question more expeditiously. Flagellin of the cells of *B. pumilus* 101 purified by self-assembly (pH 5.4, 0.02 M KH_2PO_4–K_2HPO_4, 23°C, 18 hours; ABRAM and KOFFLER, 1964) appears to be homogeneous when examined by ultracentrifugation, and has only one N-terminal amino acid, met. These observations led us to believe that these flagellar filaments contained a single protein, until we noticed the presence of two protein bands, an A band and a more rapidly moving B band, when such a preparation was exposed to disc gel electrophoresis, according to the procedure of ORNSTEIN (1964) and DAVIS (1964), except that the gels contained 4 M urea. At approximately the same time we observed a fine flocculent precipitation in solutions of flagellin when the pH was adjusted from 2 to 5.4 during the reassembly procedure (ABRAM and KOFFLER, 1964). This precipitate formed most rapidly at pH 4.7 and represented 20 to 30% of the total flagellin. Eventually this material turned out to be essentially homogeneous flagellin B, and 90 to 95% of the protein in the supernatant liquid was flagellin A. Flagellins A and B can be conveniently separated by DEAE-cellulose chromatography (Figure 1) in a constant weight ratio of 7:3; the material contained in the pooled fractions within each peak, when rechromatographed, behaves homogeneously. Interestingly, when flagellins are synthesized enzymically *in vitro* by a system involving enzymes and ribosomes from cells of *Escherichia coli* and m-RNA from cells of *B. pumilus,* they are also produced in this ratio (SUZUKI and KOFFLER, 1969, 1970); most likely this reflects an *in vivo* regulatory mechanism that results in the formation of m-RNA for A and B in these relative amounts.

Apparently strain 101 is not the only culture producing 2 flagellins. Among 18 strains of *B. pumilus* studied, 9 produce 2 flagellins that appear to be similar to A and B on the basis of disc gel electrophoresis. Four of these 9 strains produce proteins immunologically identical to flagellins A and B when examined by cross-agglutination tests or Ouchterlony double-diffusion techniques (KOFFLER and SMITH, 1971). Each of the other 9 strains synthesizes only 1 flagellin, which moreover does not react with antisera against either flagellin A or B.

Each of these 2 flagellins is capable of assembling into normal-appear-

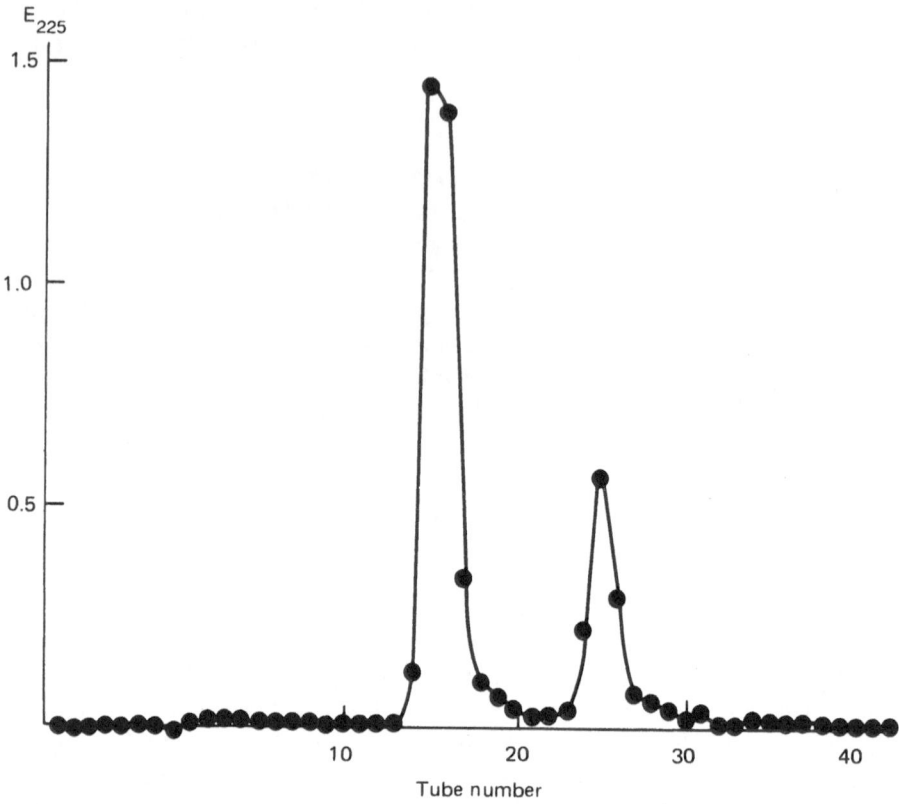

Fig. 1. Elution of flagellins A and B of *Bacillus pumilus* from DEAE-cellulose column. Flagellar filaments were solubilized with 0.01 N HCl and the insoluble material removed by centrifugation at 104,000 × g for 30 minutes. The pH was adjusted to 8.0 with KOH. The proteins were eluted from the column with 0.01 M K₂HPO₄ (pH 8) using a NaCl gradient of 0.01 to 0.10 M.

ing flagellar filaments. Furthermore, they are similar in amino acid composition and primary structure. Table 1 shows the results of amino acid analyses performed on flagellins A and B, separated by precipitation of B at pH 4.7, and purified further by DEAE cellulase column chromatography. These analyses show the 2 flagellins to be similar in composition, with the greatest differences occurring in arg and thr.

Differences in the amino acid sequences in flagellins A and B were examined by a comparison of peptide maps. Solutions of flagellin A and B at pH 9 were heated at 100°C for 10 minutes, cooled at 37°C, and then reacted with trypsin for 12 hours at a pH kept constant at 9 with the aid of a pH stat. After digestion, the partial peptides formed were separated by paper chromatography (descending, top phase of n-butanol:acetic acid:water, 4:1:5) followed by high voltage electrophoresis (pyridine:acetic acid:water, 1:10:289, pH 3.7, 2000 volts, 90 minutes) according to the procedure of Katz *et al.* (1959). Superimposition of maps

Table 1. Amino Acid Composition of Flagellins A and B from Cells of *Bacillus Pumilus*†

NUMBER OF RESIDUES PER 100,000 G

Amino acid	A	B	Amino acid	A	B
trp	0	0	gly	77	86
ile	65	62	ser	52	59
tyr	0	0	his	9	8
phe	20	20	lys	50	45
pro	7	10	arg	39	28
leu	79	71	asp + asn (A)	140	137
val	44	51	glu + gln (G)	130	120
met	14	14	a + g	270	257
cys/2	0	0	amide	152	147
ala	124	107	asp + glu	118	110
thr	52	73	tot residues	902	889

† Amino acid composition of flagellins A and B from cells of *Bacillus pumilus*. Approximately one mg protein was dissolved in three ml 6 N HCl, sealed *in vacuo*, and suspended above boiling toluene for the desired length of time. The sample was then taken to dryness and the residue dissolved in 0.5 ml of distilled water. The analyses were performed with a Beckman 120 C amino acid analyzer. Each figure represents the mean of 12 analyses, 4 at each of 3 times of hydrolysis (18, 24, and 32 hours). Values for met, thr, and ser are extrapolated to zero hours of hydrolysis. The absence of tryptophan was determined after alkaline hydrolysis and by spectrophotometric examination of the intact protein. The absence of cysteine was verified by analysis of performic acid-oxidized protein and of protein carboxy-methylated in the presence of dithiothreitol.

for A and B indicates that of 30 peptides in A, 3 were distinct for A, and of 31 peptides in B, 7 appeared to occur only in B. Peptides were also separated by passage through the short column of the amino acid analyzer and the eluate monitored automatically with ninhydrin reagent (Beckman PA 35 resin; pH gradient, 2.9 to 5.0, established with pyridine:acetic acid buffer; MOORE and STEIN, 1954). Of 24 peptides in flagellin A, 10 were characteristic for A; of 28 in B, 15 did not occur in the digest from A. This is the greatest number of differences that was observed by any of the methods used, and probably is largely due to the difficulty in aligning specific peptides when overlaying elution profiles from separate experiments. To minimize this problem, ^{3}H–A and ^{14}C–B, and vice versa, were mixed before these proteins were digested with trypsin; the peptides released were separated by ion-exchange chromatography on the short column of the analyzer, as described above. These experiments were similar in principle to the one for which results are depicted in Figures 9 and 10. The results indicated that among 25 peptides for A and 29 peptides for B, all but 3 in A and 3 in B behaved chromatographically

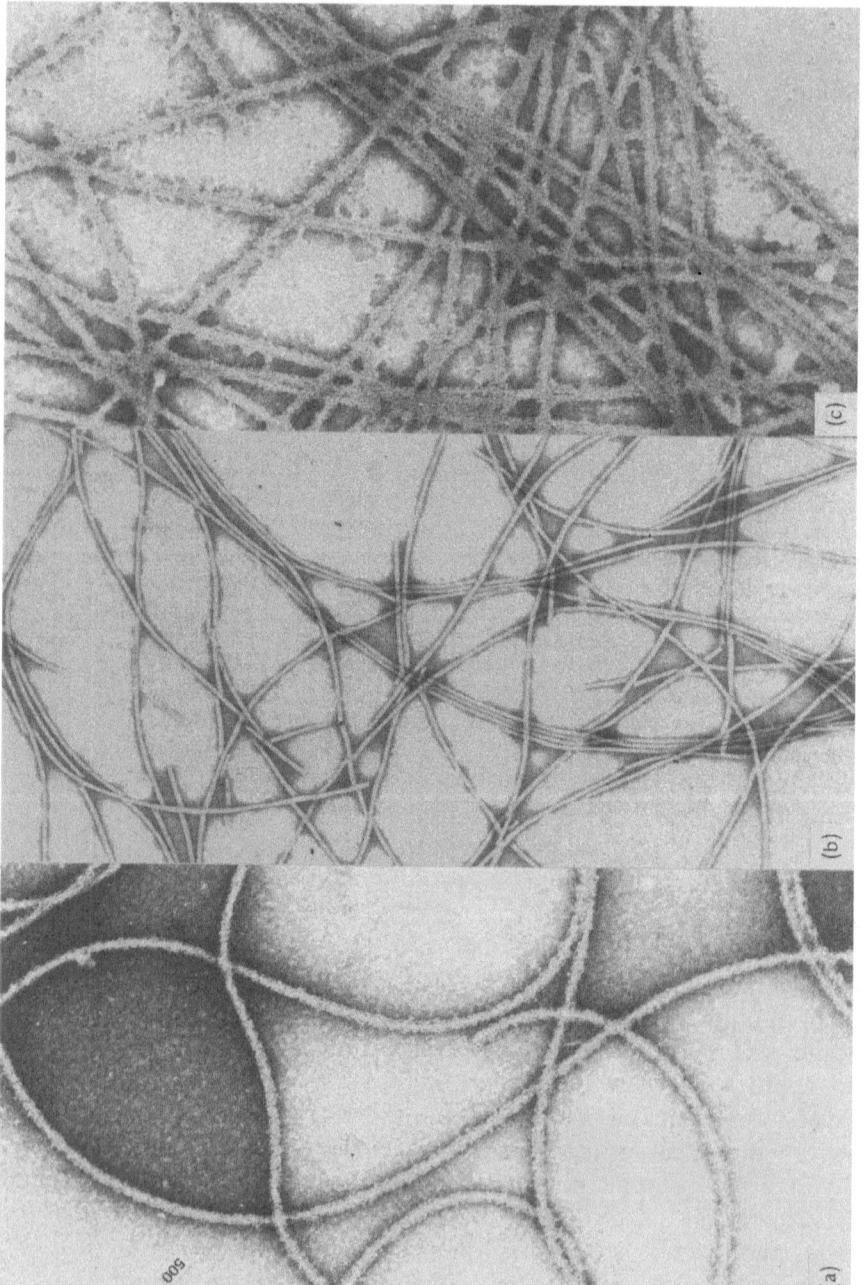

Fig. 2. Reaction of flagellar filaments with antiserum prepared against purified A flagellin. Section A shows that filaments formed by the polymerization of purified A are completely coated, whereas filaments formed by the polymerization of purified B (section B) do not react. Native filaments (section C) are coated over their entire length. Magnification of all sections is 75,000 X. Negatively stained with potassium phosphotungstate. (L. Oiler, H. Koffler, and R. W. Smith, unpublished observations).

identically. Since in these double-label experiments each peptide is exposed to identical conditions, we regard the information as being more reliable than that from the other experiments in which comparisons of separately obtained results were attempted. Therefore, our tentative conclusion is that flagellins A and B are similar. However, tryptic digestion solubilizes only 90% of the molecule. Because of its extreme insolubility the remaining material is difficult to study, and more pronounced differences may exist in those portions of the flagellin structure reflected by the trypsin-resistant materials. This possibility is now under investigation.

That flagellins A and B are produced by the same type of cells was indicated when cultures grown from cells selected from 15 separate colonies always yielded flagellins A and B in the constant weight ratio of 7:3. This was confirmed when it could be shown that both proteins are indeed copolymerized in the same flagellar filament. Each flagellin induces the formation of specific antibodies. Antisera were prepared from rabbits immunized against either flagellin A or B and reacted with filaments formed by *in vitro* self-assembly of either A or B, and with native filaments. As can be seen in Figure 2, filaments formed by polymerization of purified A are completely coated after reaction with anti-A serum (left). Filaments containing B when reacted with anti-A serum remain uncoated (middle). Native filaments are coated over the entire length (right). As expected, when anti-B serum is reacted with reconstituted B filaments, A filaments, and native filaments, this antiserum reacts only with filaments containing the homologous flagellin and with native filaments.

VI. Conformational Versatility of Flagellin

The conformation of flagellin strongly influences its immunogenic and antibody-binding characteristics. Architecturally, flagellin is a versatile molecule and undergoes repeated cycles of spatial changes in response to environmental conditions, especially pH and temperature. By measuring optical rotatory dispersion (ORD) characteristics or circular dichroism (CD), one can obtain indications regarding the amount of α-helix, β-structures and/or random coil in the molecule. Table 2 indicates the percent of α-helix in flagellins from various strains of *Bacillus* (KLEIN *et al.*, 1967a, b, 1968, 1969). In the case of mesophile flagellins, the helical portion of the molecule unfolds as the pH is lowered from 4 to 2; this transition is reversible. The α-helix of thermophile flagellins does not appear to be affected by lowering of the pH. Short of a complete crystallographic analysis, no single method describes the complete structure of the molecule. Thus, while conformational changes between pH 4 and 2 in thermophile flagellins are not detectable by spectropolarimetric methods, such transitions do occur. This is indicated by the generation of difference spectra under different pH and solvent conditions and the different degree

Table 2. Relative pH Stability of the Helical Region(s) in Various Flagellins†

Strain	Maximum growth temperature	Percent helix	
		pH 2	pH 4–11
Mesophilic (4)	<45–57	5–16	20–34
"Intermediate" (1)	66	12	20
Thermophilic (3)	71–78	18–39	18–35

† Relative pH stability of the helical regions in various flagellins. The number in parentheses indicates the number of strains examined. The mesophilic bacilli used were *P. vulgaris, Bacillus* species Xl, *B. pumilus,* and *B. licheniformis.* The intermediate strain was 194 and the thermophilic strains were 10, FJW, and 2184. The percent helix was determined by measuring optical rotatory dispersion parameters with either a Bendix or a Cary 60 recording spectropolarimeter (D. Klein, J. F. Foster and H. Koffler, unpublished observations).

to which certain amino acid residues are accessible to specific reagents under various pH conditions. Flagellin from cells of the thermophilic strain 2184 of *Bacillus stearothermophilus* contains 6 tyr residues. Acid difference spectra and solvent perturbation studies show that 3 of these residues are exposed at pH 4.8 and 3 are buried. As the pH is lowered to 2, one of the buried residues becomes exposed; this suggests structural changes in certain regions of the molecule. By the same token, when this flagellin is treated with tetranitromethane, a reagent capable of nitrating exposed tyr residues at pH values above 8.5 (SMITH *et al.,* 1968; YARBROUGH *et al.,* 1969, 1971a, b; YARBROUGH, 1971), 3 residues are accessible to nitration at pH 8.5, 4 to 5 at pH 10, and 5 at pH 11.5. Again, this indicates that certain parts of the molecule experience spatial alterations as the pH is raised above 8.5.

Of course, increases in temperature also cause unfolding of flagellin; upon cooling, the molecule appears to assume its "normal" conformation. ORD and/or CD measurements give some clues as to what happens to α-helical regions as flagellin is heated. By making ORD (at 233 mμ) or CD (at 228 mμ) measurements at various increasing temperatures, transition curves are obtained, the midpoints (T_t values) of which can be considered to represent the relative stability of these regions in given flagellin species. Mesophile flagellins have lower T_t values (44 to 45°C) than thermophile flagellins (55 to 55.7°C).

Figure 3 shows ORD measurements at pH 7 for flagellin from cells of strain 2184. The Cotton effects at room temperature (solid triangles) indicate the presence of α-helix; after having been heated at 56°C (solid circles), the molecule loses most of its helical conformation, but α-helix

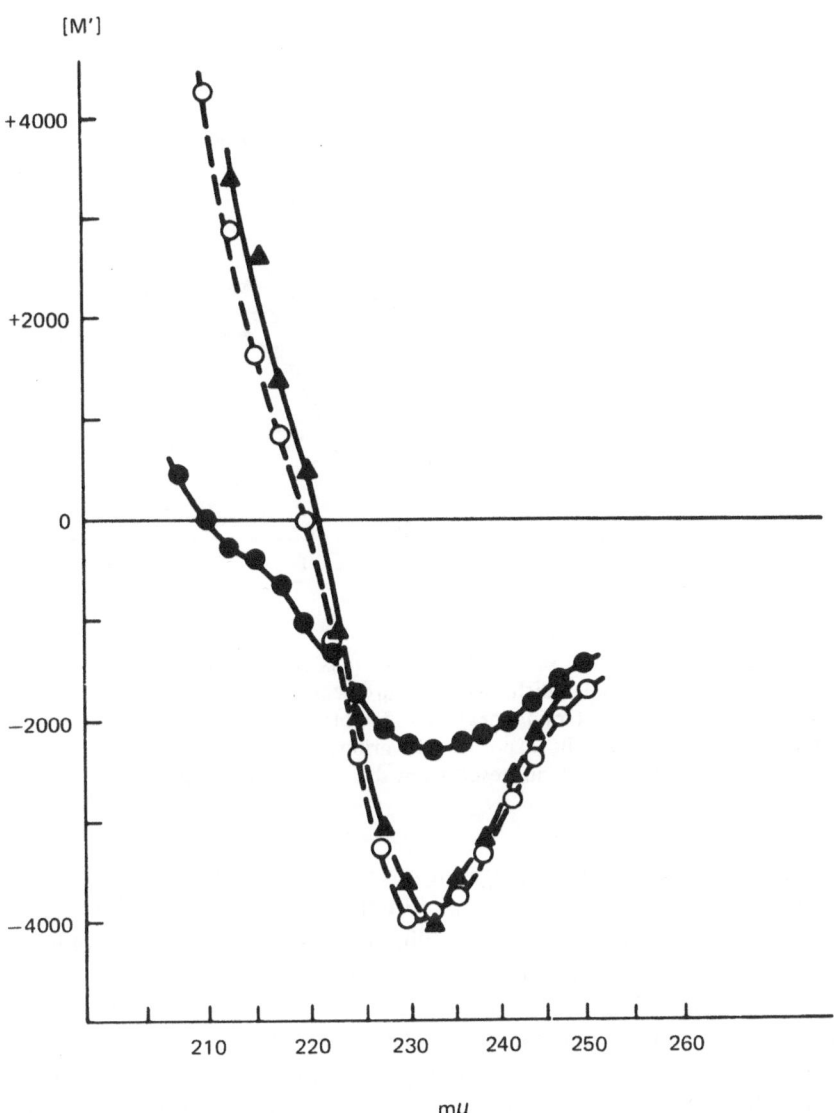

Fig. 3. Optical rotatory dispersion measurements on flagellin from *Bacillus stearo-thermophilus* 2184 at pH 7. ▲——▲, control flagellin at room temperature; ●——●, flagellin at 57°; ○——○, flagellin heated to 57° then cooled to room temperature (D. Klein, H. Koffler, and J. F. Foster, unpublished observations).

is restored aagin on cooling (open circles). As flagellin unfolds, different portions apparently have different relative stabilities. From temperature transition curves, using (1) CD measurements to monitor the behavior of α-helical regions and (2) temperature difference spectra to follow the stability of tyr- and trp-containing regions, we concluded that (1) the

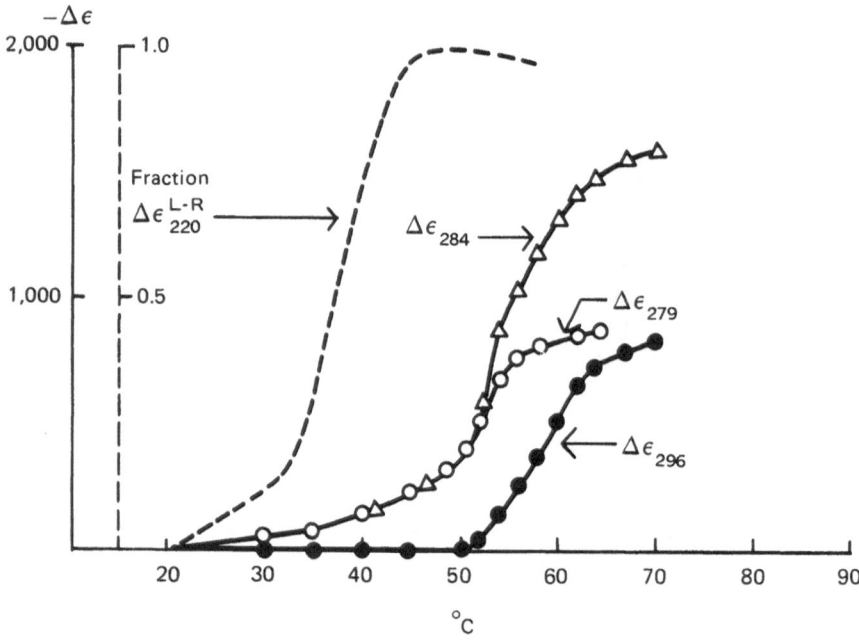

Fig. 4. Effect of temperature on flagellin of *Bacillus stearothermophilus* 2184 at pH 2.4 as determined by circular dichroism measurements at 220 nm (broken line) and by difference spectroscopy (solid lines) at different wavelengths. The measurements at 284 and 279 nm show the exposure of various tyr residues as the temperature is increased. The values at 296 nm result from the exposure of a trp residue.

α-helical region or regions in flagellin are the least stable; (2) tyr-containing regions are next in stability; and (3) the trp-containing region is most stable (Figure 4). Apparently, flagellin unfolds (and probably folds) in a stepwise rather than in an all-or-none manner.

As mentioned, flagellin molecules in solution are capable of reassembling to form flagella-like filaments. The concentration of flagellin, pH, temperature, and the ionic environment are critical factors in determining not only the rate of assembly, but also the yield and nature of the polymer. Some of these are illustrated in Table 3. In general reconstitution of helical filaments without "seed" (*i.e.,* small flagellar pieces obtained by sonication of long filaments) occurs at fairly specific pH conditions and at low ionic concentration. When a primer is used, the optimum range of conditions is considerably widened; this suggests that nucleation is more sensitive to environmental conditions than are the binding and incorporation of additional monomers. Presumably, the polymerized monomers of the seed fragments induce newly bound monomers to assume the appropriate conformation necessary for incorporation within the filament. Polymerization by "salting out" is much less specific and tends to yield relatively short, straight filaments, probably not as similar to native filaments as those syn-

Table 3. The effect of environmental conditions on the nature of flagellin polymers.

Organism	Monomer	Seed	Conditions	Product	References
B. pumilus	From filaments at pH 2-3,	None	Flagellin 1-5 mg/ml; 0.0275 M K-PO$_4$, pH 5.4-5.6; "instantaneous" assembly at 5 mg/ml	Helical filaments; normal but longer	Abram and Koffler, 1963, 1964
			pH 4.1-5.1, 26° pH 5.4-5.6, 8°	"Ribbon"-like structures; when incubated under optimum conditions can be fully converted to helical filaments	
S. adelaide, B. pumilus, etc.	From filaments at pH 2-3	None	0.35 saturated ammonium sulfate, pH 6.7-7.8, 23°	Short, relatively straight filaments	Ada et al., 1964
P. vulgaris	From filaments	None	0.4-0.5 M K-PO$_4$, pH 7-8.5	Helical filaments	McGroarty, 1971
S. typhimurium etc.	From ammonium sulfate precipitated polymers at 65°	None	0.15 M NaCl, 0.01 M PO$_4$; pH 6.5 plus one of the following at 0.38-1.8 M: Sodium or ammonium sulfate, fluoride, carbonate, phosphate, citrate, but not calcium or magnesium salt	Helical filaments	Wakabayashi et al., 1969
S. typhimurium	From filaments at 60°	0.2-0.3 μ fragment from sonicated filaments	0.15 M NaCl, 0.01 M K-PO$_4$, pH 7, room temperature	Helical filaments	Asakura et al., 1964, 1966; Asakura, 1968
B. stearothermophilus	From filaments at pH 2-3, then purified by Bio-Gel P100 column chromatography	None	Wide range of conditions 2-50° / >55°	Straight filaments / Helical filaments	Abram and Koffler, 1964; Yarbrough, 1971

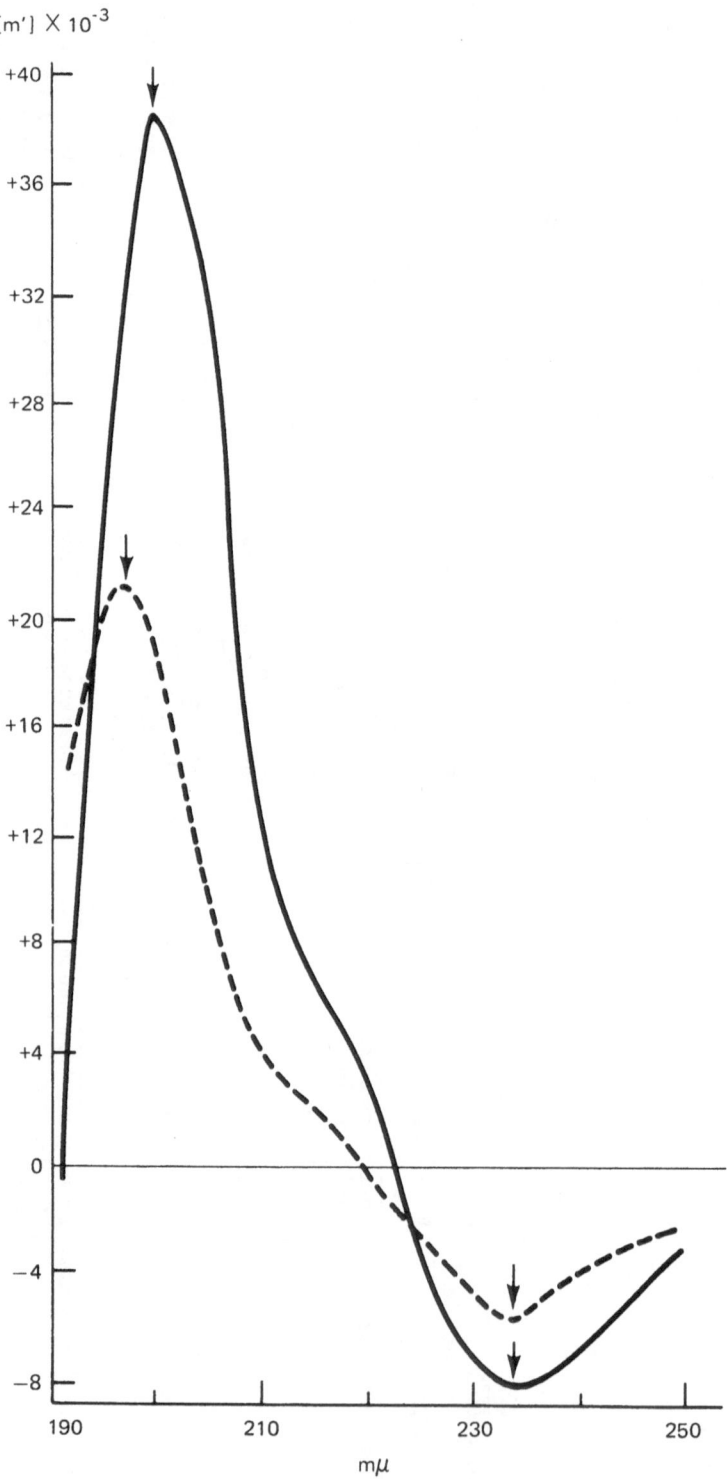

Fig. 5. See legend on opposite page.

thesized under more stringent reconstitution conditions. This should be kept in mind when one compares the immunological properties of material polymerized *in vitro* to those of native material. The nature of the product is affected by pH and temperature. For instance, thermophile flagellin assembles to form straight filaments at $<50°C$, forms a mixture of straight and helical filaments at 50 to 55°C, and forms helical filaments at $>55°C$.

All indications are that assembly of monomers to form the flagelllar filaments is preceded, accompanied, or followed by conformational changes. From electron microscopic observations (KERRIDGE *et al.*, 1962; ELEK *et al.*, 1964; ABRAM *et al.*, 1964b; LOWY and HANSON, 1964, 1965) the flagellin subunits within the filament appear to be ovoid, almost spherical. However, flagellins in solution appear to be more elongated. The axial ratio for flagellin from cells of *Proteus vulgaris* was estimated to be 15 at pH 6.1, and for several flagellins from *Bacillus* to be about 20 (ERLANDER *et al.*, 1960; STENESH and KOFFLER, 1971). Since the conformation of flagellin is responsive to the environment, one would expect such conformation to be different in solution, especially in low concentration, where each flagellin molecule is surrounded by water and more accessible to various ions and molecules than in polymeric form, where its characteristics are strongly influenced by flagellin-flagellin interactions. As will be discussed more fully later, soluble flagellin binds markedly fewer antibody molecules against filaments than homologous antibodies. This observation can be interpreted to mean that more antigenic sites come into being due to the adjacency of monomers in the polymer. However, one can also argue that this speaks for the occurrence of conformational changes which result in the exposure of groups that are inaccessible in the soluble form of the monomer. The most compelling data come from a comparison of ORD and CD characteristics of the soluble and polymeric forms of flagellin (KLEIN *et al.*, 1967b, 1969). Cotton effects and CD curves for flagellin and reconstituted flagellar filaments from cells of *B. pumilus* are obvious in Figures 5 and 6, respectively. The effect of polymerization on the spectral properties of flagellin results in a marked increase in rotational strengths accompanied by shifts of the peaks and troughs toward the red end of the spectrum. Analysis of dispersion curves and 233 mμ trough values suggests that the α-helix content of flagellins from *P. vulgaris* and from strains of *Bacillus* doubles as the subunits are polymerized; comparable results were obtained for flagellin from *Salmonella* species (OOSAWA and HIGASHI, 1967; HOTANI, 1971; URATANI *et al.*, 1971). Since the spectral changes may be due to aggregation rather than to an increase

---

Fig. 5. Optical rotatory dispersion curves for flagellin and filaments of *Bacillus pumilus*. Flagellin at pH 4.3 (----); flagellar filaments at pH 6.2 (——). Data were obtained with the Cary 60 spectropolarimeter continuously flushed with nitrogen. All measurements were made at 24°. The concentrations of flagellin in soluble and polymerized form were 60 to 90 μg per ml and 40 to 60 μg per ml, respectively. Reprinted with permission from KLEIN *et al.* (1969).

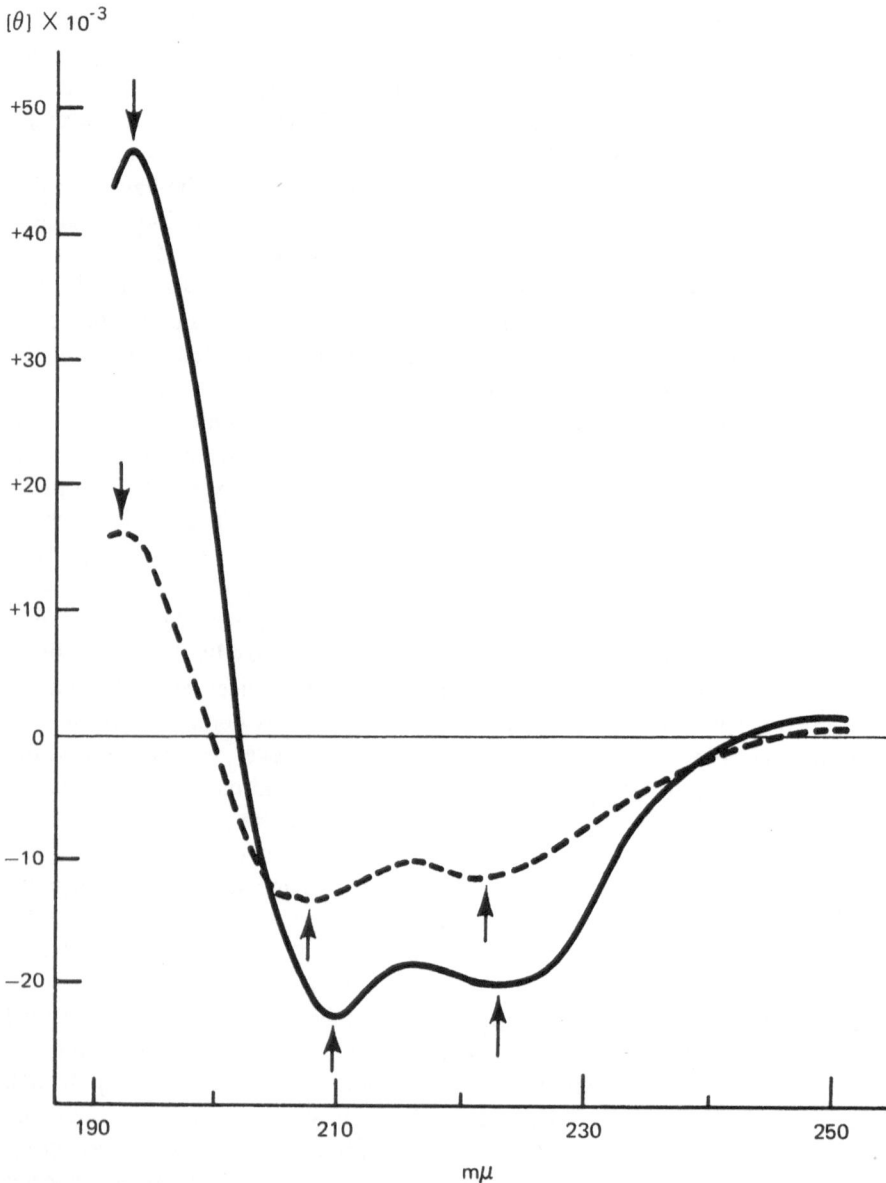

Fig. 6. Circular dichroism masurements for flagellin and flagellar filaments of *Bacillus pumilus.* Flagellin at pH 4.3 (----); flagellar filaments at pH 6.2 (——). Conditions were described for Figure 4-5. Reprinted with permission from KLEIN *et al.* (1969).

in helical content, it is yet too early to take for granted that conformational changes occur during assembly. This has been discussed more fully by KLEIN *et al.* (1969).

In theory, the ability of flagellin to assume several conformations suggests that this versatility may also be significant in influencing its immunologic properties. In practice, the conformational options probably are decreased by the standardization used in immunizing animals or in performing serologic reactions.

Since our knowledge regarding the conformation of flagellin comes from *in vitro* situations, we know very little about the structure of flagellin after injection into experimental animals. To the extent that soluble and polymeric flagellin appear to induce different immunologic responses, as is discussed below, one can assume that even *in vivo* the soluble antigen is either maintained in this form or polymerizes differently from the native or *in vitro* reconstituted filamentous form.

VII. Characteristics of the Immunologic Response

The choice of animal to be used for antibody production appears to be important. Reports of precipitin or agglutinin titers of several thousand are not unusual after injection of either intact flagellar filaments or solubilized flagellin. Invariably, however, these studies involved the use of rats. We have never observed a precipitin titer of greater than 40 after injection of soluble flagellin into either Purdue Dutch or New Zealand white rabbits when a variety of injection routes and schedules were used. PARISH *et al.* (1969) also observed that flagellin was only slightly immunogenic in rabbits as compared to rats. The reason for this difference in response is not known; perhaps the antigen is destroyed or excreted more rapidly in the rabbit or perhaps a state of tolerance is more readily established. It is clear, however, that we cannot assume that all species are equivalent producers of precipitating antibodies against flagellar antigens.

One of the major questions concerning the immunologic response deals with the fate of the antigen after injection into the test animal. Is the antigenic protein concentrated in a specific organ or type of cell, or is it perhaps digested or altered by enzymes? Apparently the protein does not circulate in the blood indefinitely since NOSSAL *et al.* (1967) observed that within 24 hours following injection, most soluble flagellin is removed from the bloodstream of the rat. It has not been possible to demonstrate clearly the presence of radioactive flagellar protein in the antibody-forming cells (NOSSAL *et al.*, 1964b), although ADA and WILLIAMS (1966) after injection of ^{131}I-flagellar filaments found radioactivity associated with large granules in a lysosome-like residue from fractionated lymph nodes.

The kinetics of antibody production as well as the effect of amount of antigen, injection schedule, and presence of adjuvant on the rate of antibody production and final titers achieved have been studied in several laboratories. NOSSAL *et al.* (1963) examined the rate of increase of the serum antibody titers after injection of filaments of *Salmonella adelaide* and found the increase to be exponential for the first 2 days, after which period the titer slowly decreases. The smaller the antigen dose administered, the more rapidly antibodies appeared in the serum. The dose range was from 0.01 to 10 μg protein per rat. LIND (1968) found that the minimum dose of flagellin of *S. adelaide* required to induce the formation of antibodies in rats was 10 ng, regardless of whether an adjuvant was present. However, the final titers were 10^3-fold higher when an adjuvant was used.

WINEBRIGHT and FITCH (1962) found a maximum agglutinin response (*i.e.,* agglutinin titer of about 8000) to a 10 μg dose of flagellar filaments of *Salmonella typhosa* in rats. Neither larger doses nor subsequent injections increased the size of the maximum titer. In contrast, a maximum precipitin titer of 2560 was achieved 4 days following 2 injections of 10 μg soluble flagellin (FITCH and WINEBRIGHT, 1962). The soluble flagellin appears to be less immunogenic than the filament, since a larger amount of protein and a longer period of time are required to reach a maximum titer. Moreover, this titer is significantly lower than the maximum titer achieved upon injection of filaments. Other investigators have also concluded that filaments are more immunogenic than flagellin (ADA *et al.,* 1963, 1964; NOSSAL *et al.,* 1964a; NOSSAL and AUSTIN, 1966; LIND, 1968; McDOWELL and LASCELLES, 1969).

Flagellar filaments also differ from soluble flagellin in the optimal route of injection (WINEBRIGHT and FITCH, 1962; FITCH and WINEBRIGHT, 1962). Highest titers for filaments are observed when the antigen is administered intravenously; for flagellin, intraperitoneal injection gives the highest titer. While this is speculative, it seems plausible that flagellin is more readily absorbed from the peritoneum than the larger filament. However, this does not explain why, for flagellin, an intraperitoneal rather than an intravenous route is more effective. These and other differences in the immunologic responses between flagellin in the soluble and polymerized forms are summarized in Table 4.

Injection of flagellar filaments of *S. adelaide* into rats induces the formation of two types of antibody molecules, a mercaptoethanol-sensitive macroglobulin (19S) and a mercaptoethanol-resistant γ-globulin (7S) (ADA *et al.,* 1963; NOSSAL *et al.,* 1964b). The 19S globulin is synthesized predominantly during the first week, whereas the synthesis of the 7S globulin dominates after 11 days. Although the significance of this finding cannot be evaluated at present, the filaments were found to be at least 10^5-fold more effective than flagellin at inducing the synthesis of the 19S globulin and 10^2- to 10^3-fold more effective at inducing the 7S primary response.

Antifilament antibodies have been isolated and purified (GRANT and SIMON, 1968). Intact filaments of *Bacillus subtilis* flagella were allowed to react with homologous antiserum; the agglutinate was washed by centrifugation and dissolved in 8 M urea. The flagellar protein was then separated from the antibody protein by passing the mixture through a DEAE-cellulose column equilibrated with 8 M urea—0.01 M phosphate buffer at pH 7.4. Fortunately, the flagellar protein but not the antibody protein is retained on the column, and purified antibodies are readily obtained. When the above procedure was performed with soluble flagellin, another distinction between soluble and polymerized flagellin became evident. It was not possible to purify the antibodies against flagellin, since the flagellin-antiflagellin complex apparently is not dissociated, and passes through the DEAE-cellulose column intact (Simon, personal communication). We have observed similar results while attempting to purify antibodies against *P. vulgaris* flagellin (McGroarty *et al.*, 1971). In soluble form each flagellin monomer may react with several antibody molecules, which may mask the sites in flagellin capable of binding to DEAE-cellulose. In polymerized form, only a portion of the flagellin molecule is exposed to the antibody, and after disintegration of the filament by urea, other portions of the molecule are exposed that may bind to the DEAE-cellulose. Furthermore, from steric consideration probably only one antibody molecule can react with each flagellin monomer in the filament, an insufficient number to cover all the DEAE-cellulose-binding sites.

VIII. Induction of Tolerance

Flagellins readily induce a state of tolerance in rats (Nossal *et al.*, 1965). In fact, flagellin is 10^6 to 10^7 times more effective than bovine serum albumin at inducing tolerance (Nossal *et al.*, 1967; Shellam and Nossal, 1968). Austin and Nossal (1966) found that different flagellins do not have to be antigenically similar to affect each other's tolerogenic capacity. Injection of *Salmonella* flagellin with the f, g antigens altered the response to subsequent injection of flagellins with either the i or the d antigen. Rats rendered tolerant to the f, g antigens were also found to be tolerant to flagellins with either the i or d antigen, although no cross-reaction occurs between antibodies prepared against the f, g antigens and flagellins with the i or d antigen. Apparently the different flagellins have some common characteristic, distinct from the antigenic determinants, that is responsible for inducing the state of tolerance.

Whether an animal becomes tolerant depends upon the amount of flagellin injected relative to the body weight of the animal. Two types of tolerance, termed "low zone" and "high zone," have been defined (Nossal *et al.*, 1967; Ada and Parish, 1968; Shellam and Nossal, 1968). The low-zone tolerance occurs upon injection of 10^{-7} μg flagellin per gram body weight, and the high-zone occurs upon injection of 10^{-3} μg flagellin per gram body weight. Dosages intermediate to these induced a typical immune

Table 4. Antigenic properties of intact flagellar filaments, filaments formed by the salt-induced polymerization of soluble flagellin, and solub flagellin.

Monomeric f	State of aggregation of flagellin — Polymeric		References
	By "salting out" $(f)_n$	Native flagellar filaments (F)	
Immunogenicity			
Less immunogenic than $(f)_n$ or F Peak titer of: 64 640 160 2560 —	More immunogenic than f; less immunogenic than F Peak titer of: — 2560 1280 — —	More immunogenic than f or $(f)_n$ Peak titer of: 512 5120 5120 — ca. 8000	McDowell and Lascelles, 1969 Nossal et al., 1964a Ada et al., 1963 Fitch and Winebright, 1962 Winebright and Fitch, 1962
IP > IV > SC in titer production; maximum titer obtained 4 days after *second* injection of 10 μg; rise to peak titer slower than for F		Rise to peak titer faster than F or f	Fitch and Winebright, 1962
		IV > IP > SC maximum titer after single injection of 10 μg	Winebright and Fitch, 1962
		F is 10^5-fold better than f and 10^2-fold than $(f)_n$ in inducing 19 S primary response and 10^2-10^3-fold better than f or $(f)_n$ at inducing 7 S primary response	Ada et al., 1963

Does not induce 19 S IgM globulin	Induces 19 S IgM	Induces 19 S IgM	Nossal et al., 1964
"Less" priming (i.e. memory as determined by subsequent challenge)		Primes	Nossal and Austin, 1966
Induction of Tolerance			
f 10-fold more effective inducer of tolerance than $(f)_n$	"Nearly" complete tolerance with multiple injections	Only "slight" tolerance	Nossal et al., 1965
f does not induce IgM in tolerant animal	$(f)_n$ does induce IgM in tolerant animal		Nossal and Austin, 1966
Antigen-Antibody Reaction			
f vs. anti-f AB N: AgN = 12.7 / f reacts with only 20% of anti-F (*P. vulgaris*)		F vs. anti-F AbN: AgN = 1.1 / F reacts with 100% of antibodies against anti-f	Read, 1957
f reacts with only 30% of anti-F (*B. subtilis*)			Grant and Simon, 1968
f does not compete with F for binding anti-F		F competes with f to bind anti-f	Ichiki and Martinez, 1968
Only 25–50% as effective as $(f)_n$ or F at neutralizing immobilizing activity of anti-F	$(f)_n$ equally as good as f at neutralizing anti-f	F equally as good as f at neutralizing anti-f	Ada et al., 1963, 1964
Inhibits complement fixation in F-anti F system		Does not inhibit complement fixation in f-anti f system	Emerson and Simon, 1971

response. After 6 weeks of daily injections, the level required to elicit the low-zone tolerance increased to 10^{-5} μg flagellin per gram body weight.

While these observations do not permit conclusions regarding the nature of the tolerant state *per se,* this phenomenon does permit further distinction between flagellin as it occurs in the soluble and polymerized form. Using three forms of flagellar protein, namely soluble flagellin from acid-disintegrated filaments, polymerized flagellin formed by the ammonium sulfate induced polymerization of soluble flagellin, and purified intact flagellar filaments, Nossal *et al.* (1965) examined the ability of each to induce tolerance in rats. Soluble flagellin was found to be 10-fold more effective at inducing tolerance than polymers formed by the salting-out procedure. Intact filaments induced only a "slight degree of tolerance" when injected in doses that would cause ammonium sulfate-induced polymers to bring about a state of complete tolerance. Again, these observations may reflect differences in conformation, as flagellin exists in these three states of aggregation. The polymerized flagellin particles have a diameter of 90 Å as compared to the 112 Å of the purified intact filaments (ADA *et al.,* 1964); this may indicate either different packing arrangements and/or different conformational states of the monomers. The possibility also remains that the intact filaments contain some unidentified material not present in the polymerized flagellin that alters the chemical nature of the surface of the filaments. However, such material has never been detected.

IX. Reaction of Soluble and Polymerized Flagellin with Antibodies

When flagellar filaments are mixed with an antiserum prepared against filaments, the viscosity of the solution increases rapidly (READ *et al.,* 1956; READ, 1957; SMITH and KOFFLER, 1971). This apparently is due to cross-linking of the filaments (GIESBRECHT *et al.,* 1964; DiPIERRO and DOETSCH, 1967, 1968). The physical nature of the agglutinate prevents the reaction of all antigenic sites on the surface of the filaments with antibody molecules, and some antibodies always remain in the supernatant liquid. At maximum agglutination the ratio of antibody nitrogen to antigen nitrogen in the precipitate is about 1.1 (READ, 1957), which corresponds to 3 to 4 flagellin molecules per IgG molecule, or 22 to 23 per IgM. This may be interpreted to indicate that each of the two antigen-binding sites of the IgG molecule reacts with at least 2 or perhaps 4 flagellin subunits in the filament. As is characteristic of particulate systems, the amount of agglutinate does not decrease in antibody excess.

If filaments are to be cross-linked, the antibody molecules with which they are reacting must be at least divalent. The 50% immobilizing dose of divalent antibodies for *S. typhimurium* generally represents fewer than 200 antibody molecules per cell (GREENBURY and MOORE, 1966). Monovalent antibody, even at 1.37×10^5 molecules per cell, has no effect on motility. Specific antibodies against flagellar filaments immobilize motile cells of either *E. coli* or *Pseudomonas fluorescens* (DIPIERRO and DOETSCH, 1968). The cells become motile when treated with proteolytic enzymes, which apparently fragment the globulin molecule into monovalent pieces. Monovalent fragments do not affect motility when added to motile cells; however, they must be bound to filaments since filaments previously treated with monovalent fragments cannot be immobilized by subsequently added divalent antibodies. Thus, the reaction of antibody with antigenic sites on the filaments does not in itself cause either agglutination or immobilization. These reactions apparently result from the binding together of individual filaments.

The nature of antibody binding to the surface of the filament has been examined electron-microscopically by several investigators. ELEK *et al.* (1964) speculated that the binding sites lie at either end of an elongated globulin molecule since they observed that the minimum interfilament distance in preparations of filaments coated with antibodies and collapsed by drying on an electron microscope grid is about 180 Å. At that time the length of the 7S IgG molecule was thought to be 190 to 200 Å. However, more recent findings indicate the maximum span of the IgG and IgM molecules to be 140 Å and 300 Å, respectively, (DORRINGTON and TANFORD, 1970).

GOTO *et al.* (1967) observed that the surfaces of flagellar filaments were coated with a layer of antibodies 95 Å thick, which at that time they considered to be about one-half the length of the IgG molecule. They proposed that the antibody molecule was looped on the surface of the filament so that both ends of the molecule were bound to the same filament. Such an explanation may hold true for dilute suspensions of filaments (<0.01 mg protein per ml), because a low concentration of filaments reduces the opportunity for cross-linking. This was probably the case, since formalin-treated cells in growth medium were used. Although the conclusions are probably correct, the reasoning should be reconsidered in view of more recent data concerning the size and shape of the immunoglobulins. DORRINGTON and TANFORD (1970) concluded that the flexible model proposed by NOELKEN *et al.* (1965) for the IgG molecule more closely fits the multitude of data available. Based on this model, both the maximum thickness of a layer of IgG molecules and the maximum interfilament distance would be 140 Å if the filaments were cross-linked. This would also be true if one filament were coated by IgG molecules through both antigen-binding sites, and if the adjacent filament were not coated at all. If two adjacent filaments were each coated with IgG molecules so that in all instances both antigen-binding sites of a given antibody molecule had reacted with the same filament, the maximum interfilament distance possible with the antibody layers still in contact with each other would be 280 Å. These figures are based on the maximum spans of the antibody molecules. It seems likely that experimental observations will account for less

than these distances since the molecules may not occur fully extended and
further alterations in size probably occur during drying for electron microscopic observation.

If one considers the dimension of the flagellin subunits in the filament
(diameter is 40 to 50 Å) and the likelihood that flagellin possesses more
than one immunogenic group, it seems plausible that a given antibody
molecule might cover several antigenic determinants on the surface of the
filament. Moreover, for the following reasons, each filament may possess
several antigenic determinants (KAUFFMAN, 1950; NAKAYA *et al.*, 1952),
aside from the fact that each flagellin may have more than one antigenic
determinant. As is the case for filaments from cells of *B. pumilus,* each
filament may consist of more than one protein (OILER *et al.,* 1971).
Within a filament, identical flagellin molecules may exist in more than one
conformational state with different determinants exposed in the various
conformations. Furthermore, not only might each flagellin in the filament
induce the formation of specific antibodies, but also the interactions between several adjacent flagellins could give rise to determinants not found
in the individual subunits.

That more than one determinant exists per flagellin molecule is indicated by the observation that a single missense mutation can destroy several antigenic determinants (JOYS and STOCKER, 1966). Even more definitive is the fact that the structural gene for phase 1 flagellin (*H*1) in cells
of *Salmonella abortivoequina* can be divided into at least 5 antigen-deter-
mining groups (YAMAGUCHI and IINO, 1969). Thus this particular flagellin
contains at least 5 antigenic determinants.

With respect to the fourth possibility, intact filaments apparently induce the formation of specific antibodies not induced by soluble flagellin.
Soluble flagellin reacts with only 15 to 30% of the antibodies formed
against filaments (READ *et al.,* 1956; KOFFLER, 1957; READ, 1957; GRANT
and SIMON, 1968; SMITH and KOFFLER, 1971). ICHIKI and MARTINEZ
(1969) also have presented evidence that new antigenic determinants are
formed by the association of subunits in the filament. They defined four
classes of antibodies: (1) antifilament, immobilizing antibodies (react only
with filaments); (2) antifilament (react with both filaments and flagellin);
(3) antiflagellin, immobilizing (react with both filaments and flagellin);
and (4) antiflagellin (react only with flagellin). The antifilament, immobilizing antibodies, therefore, are induced by determinants that overlap
several subunits in the filament; consequently, these determinants are destroyed upon either acid or heat disintegration of the filament.

The complement fixation system used in Simon's laboratory also points
out immunologic differences between the soluble and polymerized forms
of flagellin (EMERSON and SIMON, 1971). Antisera were prepared against
both purified filaments and flagellin that had been coupled to methylated
albumin to prevent polymerization. When the homologous and heterologous systems were tested, the homologous but not the heterologous mixtures were found to fix complement. Regrettably, the results were not un-

equivocal, since only one rabbit among three injected with flagellin gave an antiserum that did not show complement fixation with filaments. Flagellin was able to inhibit complement fixation in the filament-antifilament system, although filaments did not inhibit the homologous flagellin system. Apparently flagellin did react with at least a portion of the antibodies directed against filaments, but the nature of the reaction was such that complement was not fixed.

The reaction of soluble flagellin with antiserum prepared against flagellin is typical of a precipitin reaction between antibody and a soluble antigen, in that regions of antibody excess, equivalence, and antigen excess occur. At equivalence, the antibody-nitrogen to antigen-nitrogen ratio is 12:7 (READ, 1957; SMITH and KOFFLER, 1971).

X. The Nature of the Hook

While an impressive body of knowledge has become available regarding the filamentous portion of the flagellum and its constituent protein, information on the basal structure and the hook is becoming available only slowly. No doubt this has been due largely to the fact that these components represent only a small fraction of the total organelle mass, and therefore are difficult to isolate in amounts sufficient for analysis. Aside from the fact that the hook differs from the filament in shape, fine structure, and relative stability to a variety of agents or conditions capable of bringing about the disintegration of the filament, LAWN (1967) observed that the hook might be specific in its antigenicity. Using antisera against whole cells absorbed with somatic antigens, he showed, electron microscopically, that anti-H sera react with filaments but not with hooks. Since then we have demonstrated this antigenic difference using purified antigens (OILER *et al.*, 1971, and unpublished observations). Reconstituted filaments of *B. pumilus* flagellin were used as antigens to stimulate the production of specific antibodies. A small amount of filaments broken off cell bodies by shearing, and still containing some hooks, was mixed with hooks purified in the manner described below and absorbed onto grids coated with carbon-stabilized collodion. These grids were floated in humid chambers for 10 minutes at 37°C or droplets of antiserum or control serum, then twice for 5 minutes on droplets of saline solution, and eventually twice on distilled water for 5 minutes. The grids were stained with 1% potassium phosphotungstate (PTA) at pH 6.8 to 7 and observed in a Philips —300 electron microscope. As is shown in Figure 7, the antifilament antibodies specifically coat the filament and leave naked the hook still attached to the filament and the purified hooks in the background. A complementary experiment was done by DIMMITT and SIMON (1970). They prepared antibodies against the hooks of *B. subtilis* strains 168 and 23 and found that antihook antibodies do not react with filaments. Interestingly, although the filaments of 168 and 23 are antigenically distinct, the hooks are not.

To study the nature of hooks further, we developed the following

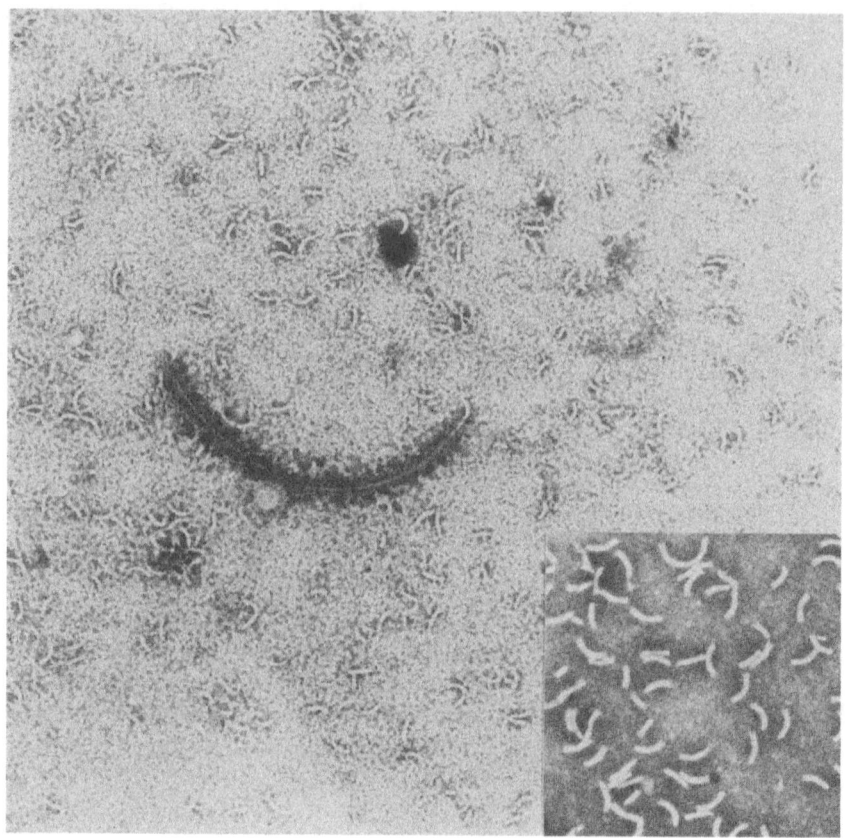

Fig. 7. Flagellar filaments and hooks in the presence of antifilament serum negatively stained with potassium phosphotungstate. 60,000 ✕. Inset: Micrograph of hooks purified as described in the text. Negatively stained with potassium phosphotungstate. 71,000 ✕. (J. Mitchen, H. Koffler, and R. W. Smith, unpublished observations).

method for purifying them (ABRAM *et al.*, 1967; MITCHEN, 1969; MITCHEN and KOFFLER, 1969; MITCHEN *et al.*, 1970a, b; MITCHEN, 1971). Flagella are broken off cells of *B. pumilus* suspended in phosphate buffer, pH 7, by shaking, and the cell bodies are separated from the flagella by fractional centrifugation. Forty percent of the hooks remain with the filaments, while the other 60% remain attached to the cells, as determined by electron microscopy. Further purification of hooks involves three critical steps. The first is the selective solubilization of filaments by treatment with HCl at pH 3 for 30 minutes at 37°C, a condition at which the hooks remain intact. Since native filaments are not completely dissolved by this treatment, either they are first fragmented by freezing and thawing, or the process is repeated until complete solubilization is achieved. The hooks, together with contaminant cell debris, are collected by centrifugation at 5000 ✕ g, 4°C, for 15 minutes. The next step involves treatment of the pellet by one fifth of the original volume of 4% Triton X-100, at 37°C for 60 minutes. This removes membrane and wall materials and also facilitates the resuspension of an otherwise unwieldy precipitate. Thirdly, further purification of the hooks is achieved by density gradient centrifugation in renografin (1.16 to 1.34 g per ml renografin-76, Squibb, in 0.1 N Tris-HCl, pH 8) for 8 hours at 200,000 × g and 15°C. To increase the sensitivity of the analysis, ^{14}C- or ^{3}H-labeled hooks were used isolated from cells grown on a defined medium containing biotin (2×10^{-7} M) ammonium sulfate (1.5×10^{-2} M), magnesium sulfate (10^{-3} M), calcium chloride (10^{-4} M), potassium phosphate buffer (0.15 M, pH 7) and U–^{14}C-labeled (22×10^{-2} M; 10 mc) or C$_6$–^{3}H-labeled (2.2×10^{-2} M; 20 mc) glucose as the sole source of carbon. The radioactive cells were mixed with ^{12}C-carrier cells before isolation of the hooks. In Figure 8 the appearance of two radioactive peaks upon density gradient centrifugation is shown. The portions containing radioactivity were pooled, diluted 5 times, were centrifuged at 100,000 × g for 1 hour at 4°C, and were examined electron microscopically. The hooks sediment under a single homogeneous peak when recentrifuged in a renografin gradient. Under these conditions the proteins of the filament do not enter more than a few fractions of the gradient. The inset in Figure 7 shows an electron micrograph of a preparation purified in this manner. As judged by the failure of purified hooks to react with specific antifilament antibodies, the preparation appears to be free from significant amounts of filament flagellin.

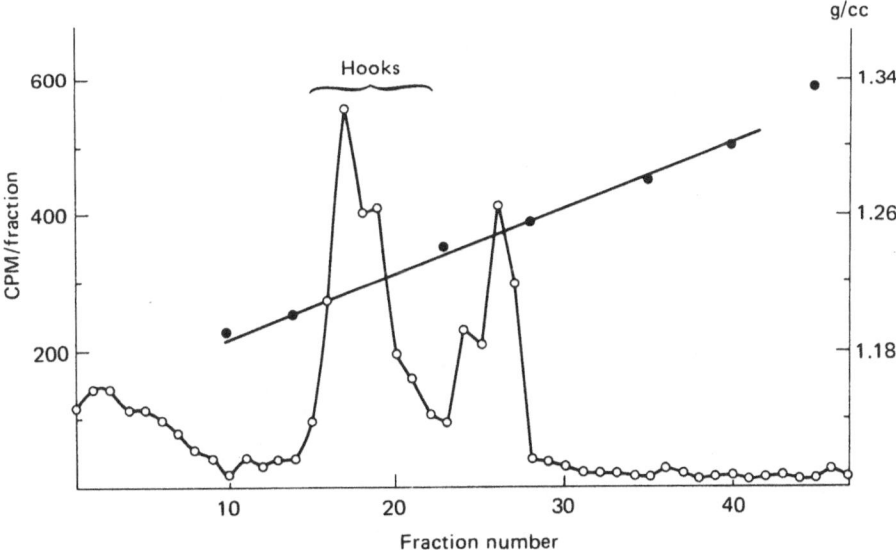

Fig. 8. Density gradient centrifugation of radioactive hooks through a gradient of renografin-76. The sample was run for 10 hours, 15°C, at 200,000 × g. The open circles indicate the location of radioactive material. The solid circles indicate the density as determined by refractive index measurements.

When radioactive hooks are precipitated with hot (90°C) 10% trichloracetic acid for 30 minutes, 90% of the counts become insoluble (70% of ^{14}C–RNAase used as a reference protein precipitate under identical conditions). This observation excludes several categories of compounds as likely constituents and suggests the protein nature of the hook.

Upon electrophoresis on 10% acrylamide sodium laurylsulfate (SLS) gels (0.05% per SLS, pH 7.2, 8 ma per tube, 2.5 hours), flagellar filaments pretreated with 1% SLS and 1% mercaptoethanol at 37°C for 1 hour give two bands—1 for flagellin A and the other for flagellin B. Hooks are not disintegrated by such a pretreatment and do not enter the gels. When hooks are layered on such gels, no flagellin bands can be demonstrated. Recently we have found phenolacetic acid acrylamide gel electro-

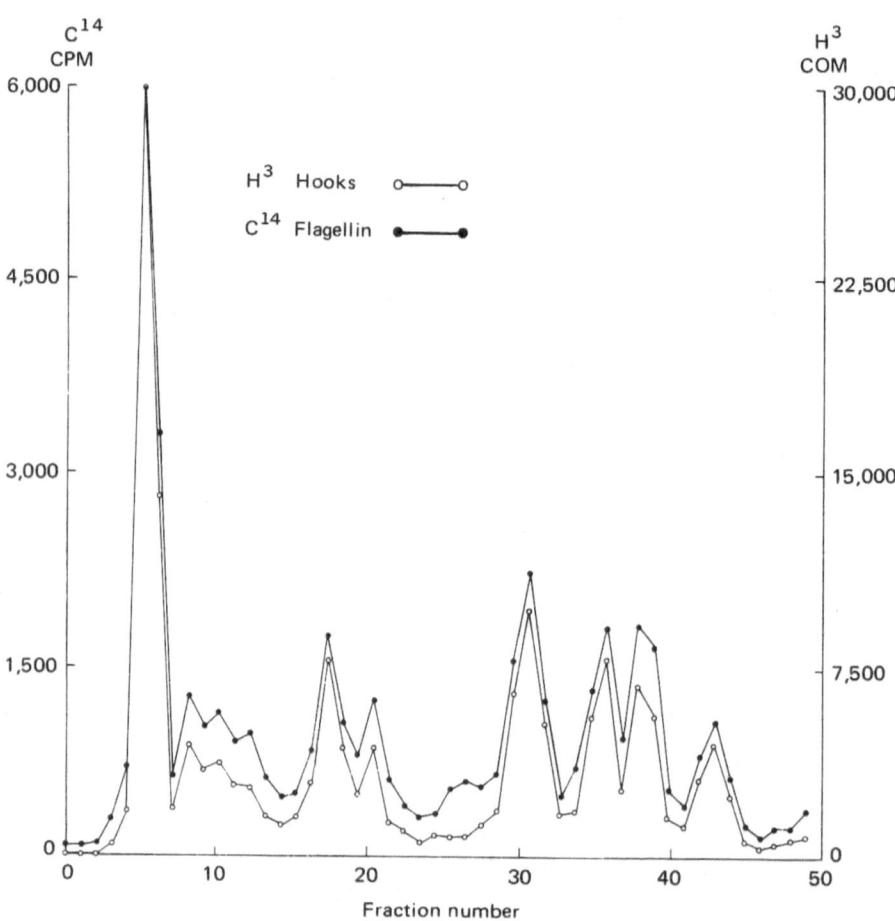

Fig. 9. Elution from a Dowex-50 type ion exchange resin of the first 50 fractions (1.5 ml volume) containing partial peptides of ^{3}H-hooks and ^{14}C-flagellin. Conditions are described in detail in the text.

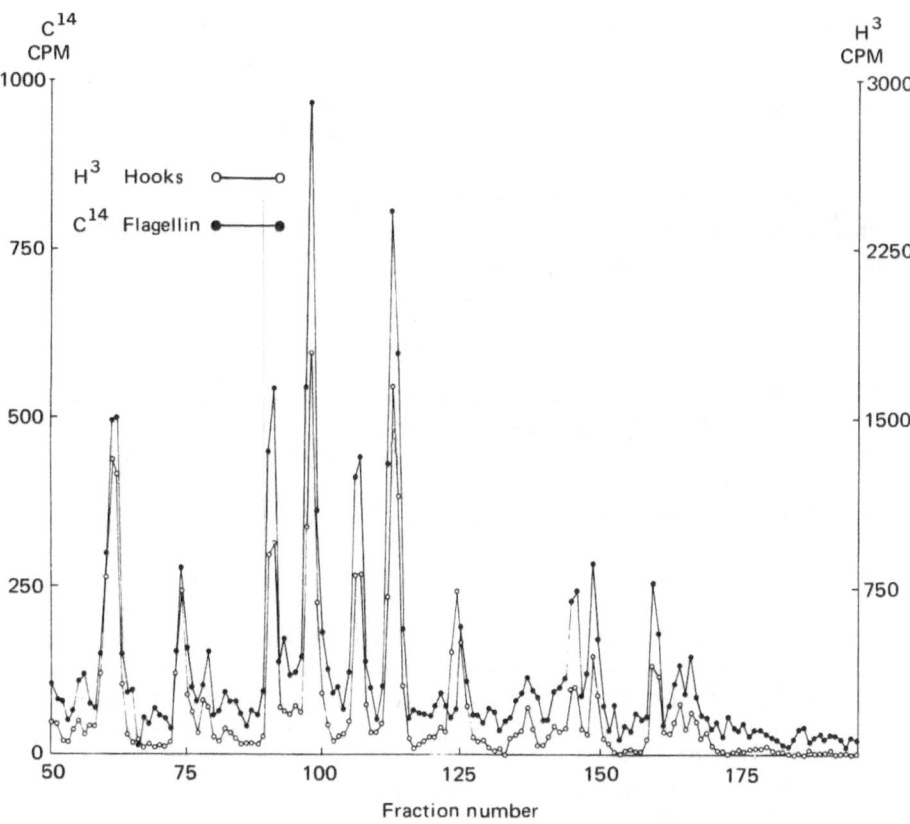

Fig. 10. Elution from a Dowex-50 type ion-exchange resin of fractions 50–200 (1.5 ml volume partial peptides) of ³H-hooks and ¹⁴C-flagellin after digestion with trypsin. Conditions are described in detail in the text.

phoresis (TAKAYAWA *et al.,* 1966) useful in comparing the proteins in the hook and filament, but it is yet too early to discuss results.

In order to learn more about the nature of the hook material, ¹⁴C-filaments, ¹⁴C-hooks, ³H-filaments, and ³H-hooks were isolated from cells grown as mentioned previously. The ¹⁴C- or ³H-hooks purified as described were mixed with ³H- or ¹⁴C-filaments, respectively, and after having been heated at 100°C and pH 2.5 for 30 minutes, the mixture was digested with trypsin (10 mg per gram protein), at pH 8, 37°C, for 10 hours. After removal of the insoluble material by centrifugation at 100,000 × g for 1 hour, the resulting partial peptides were separated on a Dowex-50 type ion-exchange column using the elution system of BENSON *et al.* (1960), and 1.5 ml portions were assayed for ¹⁴C- and ³H-radioactivity by differential counting in a scintillation spectrometer. Eighty-five percent of the hooks and 90% of the filaments are solubilized by trypsin under the conditions used. As seen in Figures 9 and 10, the partial peptides released from hooks and filaments appear to be almost identical. From $^3H/^{14}C$ ratios it appears that fraction 125 may contain one or more different peptides. The trypsin-resistant material from both hooks and filaments is highly insoluble and difficult to study. Nevertheless, we are making a serious effort to learn more about its identity through phenol-acetic acid acrylamide gel electrophoresis and other methods, since these materials may not only represent those portions of the flagellin molecule that are critical to flagellin-flagellin interactions, but also those that may include the differences in amino acid sequence between the filament flagellin and the hook protein.

References

Abram, D. (1968). Structural features at the sites of origin of flagella attached to cells and membrane fragments of *Proteus vulgaris*. *Bact. Proc.*, p. 30.

Abram, D., and H. Koffler (1963). Reconstitution of flagella-like fibers from flagellin. *Bact. Proc.*, p. 45.

—— (1964). *In vitro* formation of flagella-like filaments and other structures from flagellin. *J Molec. Biol.*, *9*:168–185.

Abram, D., H. Koffler, J. R. Mitchen, and A. E. Vatter (1967). Purification of flagellar hooks. *Bact. Proc.*, p. 39.

Abram, D., H. Koffler, and A. E. Vatter (1965). Basal structures and attachment of flagella in cells of *Proteus vulgaris*. *J. Bacteriol.*, *90*:1337–1354.

Abram, D., J. R. Mitchen, H. Koffler, and A. E. Vatter (1970). Differentiation within the bacterial flagellum and isolation of the proximal hook. *J. Bacteriol.*, *101*:250–261.

Abram, D., A. E. Vatter, and H. Koffler (1964a). Fine structure of bacterial flagella. *Proc. Biophys. Soc.*, WC5.

—— (1964b). Some specialized structures of bacterial flagella. *Bact. Proc.*, p. 25.

—— (1966). Attachment and structural features of flagella of certain bacilli. *J. Bacteriol.*, *91*:2045–2068.

Ada, G. L., G. J. V. Nossal, J. Pye, and A. Abbot (1963). Behavior of active bacterial antigens during the induction of the immune response. I. Properties of flagellar antigens from *Salmonella*. *Nature, Lond.*, *199*:1257–1259.

Ada, G. L., G. J. V. Nossal, J. Pye, and A. Abbot (1964). Antigens in immunity. I. Preparation and properties of flagellar antigens from *Salmonella adelaide*. *Aust. J. Exp. Biol. Med. Sci.*, *42*:267–282.

Ada, G. L., and C. R. Parish (1968). Low zone tolerance to bacterial flagellin in adult rats: a possible role for antigen localized in lymphoid follicles. *Proc. Nat. Acad. Sci. U.S.A.*, *61*:556–561.

Ada, G. L., and J. M. Williams (1966). Antigens in tissues. I. State of bacterial flagella in lymph nodes of rats injected with isotopically labelled flagella. *Immunology*, *10*:417–429.

Asakura, S. (1968). A kinetic study of *in vitro* polymerization of flagellin. *J. Molec. Biol.*, *35*:237–239.

Asakura, S., E. Eguchi, and T. Iino (1964). Reconstitution of bacterial flagella *in vitro*. *J. Mol. Biol.*, *10*:42–56.

—— (1968). Unidirectional growth of *Salmonella* flagella *in vitro*. *J. Mol. Biol.*, *35*:227–236.

Astbury, W. T. (1951). Flagella. *Scient. Am.*, *184*:21–24.

Austin, C. M., and G. J. V. Nossal (1966). Mechanism and induction of immunological tolerance. III. Cross-tolerance amongst flagellar antigens. *Aust. J. Exp. Biol. Med. Sci.*, *44*:341–354.

Balteanu, J. (1926). The receptor structure of *Vibrio cholerae* (*V. comma*) with observations on variations in cholera and cholera-like organisms. *J. Path. Bact.*, *29*:251–277.

Benson, J. V., Jr., R. T. Jones, J. McCormick, and J. A. Patterson (1966). Accelerated automatic chromatographic analysis of peptides on a spherical resin. *Anal. Biochem.*, *16*:91–106.

Bharier, M. A., and S. C. Rittenberg (1971). Immobilization effects of anticell and antiaxial filaments sera on *Treponema zuelzerae*. *J. Bacteriol.*, *105*:430–437.

Champness, J. N. (1971). X-ray and optical diffraction studies of bacterial flagella. *J. Mol. Biol.*, *56*:295–310.

Champness, J. N., and J. Lowy (1967). The structure of bacterial flagella. Symposium on Fibrous Proteins, Canberra. pp. 106–114.

Cohen-Bazier, G., and J. London (1967). Basal organelles of bacterial flagella. *J. Bacteriol.*, *94*:458–465.

Davis, B. J. (1964). Disc electrophoresis. I. Method and application to human serum proteins. *Ann. N.Y. Acad. Sci.*, *121*:404–427.

DePamphilis, M. L., and J. Adler (1971a). Attachment of flagellar basal bodies to the cell envelope: Specific attachment to the outer, lipopolysaccharide membrane and the cytoplasmic membrane. *J. Bacteriol.*, *105*:396–407.

—— (1971b). Fine structure and isolation of the hook-basal body complex of flagella from *Escherichia coli* and *Bacillus subtilis*. *J. Bacteriol.*, *105*:384–395.

—— (1971c). Purification of intact flagella from *Escherichia coli* and *Bacillus subtilis*. *J. Bacteriol.*, *105*:376–383.

Dimmitt, K., and M. Simon (1970). Antigenic nature of bacterial flagellar hook structures. *Infect. and Immunity*, *1*:212–213.

—— (1971). The purification and thermal stability of intact *Bacillus subtilis* flagella. *J. Bacteriol.*, *105*:369–375.

Di Pierro, J., and R. N. Doetsch (1967). Resumption of motility in bacteria agglutinated by flagellar antisera. *Bact. Proc.*, p. 40.

—— (1968). Enzymatic reversibility of flagellar immobilization. *Can. J. Microbiol.*, *14*:487–489.

Dorrington, K. J., and C. Tanford (1970). Molecular size and conformation of immunoglobulins. *Adv. Immunol.*, *12*:333–381.

Elek, S. V., K. Smith, and W. Highman (1964). The interaction of antigen and antibody in agglutination. A study by electron microscopy. *Immunology*, *7*:570–585.

Emerson, S. U., and M. I. Simon (1971). Variation in the primary structure of *Bacillus subtilis* flagellins. *J. Bacteriol.*, *106*:949–954.

Emerson, S. U., K. Tokuyosu, and M. Simon (1970). Polarity of bacterial flagellar elongation. *Science*, *169*:190.

Erlander, S. R., H. Koffler, and J. F. Foster (1960). Physical properties of flagellin from *Proteus vulgaris,* a study involving the application of the Archibald Sedimentation principle. *Arch. Biochem. Biophys., 90*:139–153.

Fitch, F. W., and J. Winebright (1962). Antibody formation in the rat. II. Agglutinin response to soluble flagellin from *Salmonella typhosa. J. Immunol., 89*:900–905.

Furness, G. (1958). The transfer of motility and tyrosine requirement to *Escherichia coli* strain B by recombination with *E. coli* strain K12. *J. Gen. Microbiol., 18*:782–786.

Gard, S. (1937). Ein Colistamm mit Salmonella-H-antigen, zugleich ein Beitrag zur Frage der Definition der Salmonellagruppe. *Z. Hyg. Infekt. Krankh., 120*:59–65.

Gard, S., and E. J. Erikson (1939). Studien uber Colistämme mit Salmonella-H-antigenen. *Z. Hyg. Infekt. Krankh., 122*:54–61.

Giesbrecht, P., K-E. Gillert, and S. Hofman (1964). Uber den electronmicroskopischen Nachweis von Antikörpen gegen Bakterien und ihre Organellen. *Zentbl. Bakt. Parasitkde,* Abt. I. Orig., *194*:503.

Glauert, A. M., D. Kerridge, and R. W. Horne (1963). The fine structure and mode of attachment of the sheathed flagellum of *Vibrio metchnikovii. J. Cell Biol., 18*:327–336.

Goto, S., J. Y. Homma, and S. Mudd (1967). Electron microscopy of the combination of antibodies with flagellar antigen and with a pyocine. *J. Bacteriol., 94*:751–758.

Grant, J. F., and M. Simon (1968). Use of radioactive antibodies for characterizing antigens and application to the study of flagella synthesis. *J. Bacteriol., 95*:81–86.

Greenbury, C. L., and D. H. Moore (1966). The mechanism of bacterial immobilization by anti-flagellar IgG antibody. *Immunology, 11*:617–625.

Hoeniger, J. F. M. (1965). Influence of pH on *Proteus* flagella. *J. Bacteriol., 90*:275–277.

Hoeniger, J. F. M., W. Van Iterson, and E. N. Van Zenten (1966). Basal bodies of bacterial flagella in *Proteus mirabilis. J. Cell Biol., 31*:603–618.

Hotani, H. (1971). Interconversion between flagella and P-filament *in vitro. J. Mol. Biol., 57*:575–587.

Ichiki, A. T., and R. J. Martinez (1968). Antigenic heterology between flagellin and flagella of *Bacillus subtilis. Bact. Proc.,* p. 30.

—— (1969). Antigenic heterology between flagellin and flagella of *Bacillus subtilis. J. Bacteriol., 98*:481–485.

Iino, T. (1969). Polarity of flagellar growth in *Salmonella. J. Gen. Microbiol., 56*:227–239.

Iino, T., and M. Mitani (1964). Mutation of flagellar antigen-1,2 in *Salmonella. Rep. Nat. Inst. Genet., Misima, 15*:98–99.

Joos, A. (1903). Untersuchungen uber die Verschiedenen Agglutinine des Typhusserums. *Zentbl. Bakt. Parasitkde.,* Abt. I Orig., *33*:762.

Joys, T. M., and B. A. D. Stocker (1963). Mutation and recombination of flagellar antigen i of *Salmonella typhimurium. Nature, Lond., 197*:413–414.

—— (1966). Isolation and serological analysis of mutant forms of flagellar antigen i of *Salmonella typhimurium. J. Gen. Microbiol.,* 44:121–138.

Katz, A. M., W. J. Dreyer, and C. B. Anfinson (1959). Peptide separation by two-dimensional chromatography and electrophoresis. *J. Biol. Chem.,* 234:2897–2900.

Kauffmann, F. (1950). The Diagnosis of *Salmonella* Types. Springfield: Charles C Thomas.

Kerridge, D., R. W. Horne, and A. M. Glauert (1962). Structural components of flagella from *Salmonella typhimurium. J. Mol. Biol.,* 4:227–238.

Klein, D., J. F. Foster, and H. Koffler (1967a). Structural conformation of flagellins in soluble and polymerized form. *Bact. Proc.,* p. 39.

—— (1967b). Conformational properties of flagellins and flagellar filaments from mesophilic bacteria. *Abst. Biophys. Soc.* WD9, p. 28.

Klein, D., J. F. Foster, and H. Koffler (1969). Changes in polarimetric parameters associated with the polymerization of flagellar filaments. *Biochem. Biophys. Res. Commun.,* 36:844–850.

Klein, D., M. Yaguchi, J. F. Foster, and H. Koffler (1968). Conformational transitions in flagellins. I. Hydrogen ion dependency. *J. Biol. Chem.,* 243:4931–4935.

Kobayashi, T., J. N. Rinker, and H. Koffler (1959). Purification and chemical properties of flagellin. *Arch. Biochem. Biophys.,* 84:242–262.

Koffler, H. (1957). Protoplasmic differences between mesophiles and thermophiles. *Bact. Rev.,* 21:227–240.

Koffler, H., and R. W. Smith. Unpublished results, 1971.

Lacey, B. W. (1961). Non-genetic variation of surface antigens in *Bordetella* and other micro-organisms, pp. 343–390. In *Microbial Reaction to Environment,* G. G. Meynell and H. Gooder (eds.), Symp. Soc. gen. Microbiol.

Lawn, A. M. (1967). Simple immunological labelling method for electron microscopy and its application to the study of filamentous appendages of bacteria. *Nature, Lond.,* 214:1151–1152.

Lederberg, J., and P. R. Edwards (1953). Serotypic recombination in *Salmonella. J. Immunol.,* 71:232–230.

Lederberg, J., and T. Iino (1956). Phase variation in *Salmonella. Genetics,* 41:743–757.

Lind, P. E. (1968). The immune responses of normal and tolerant rats to *Salmonella adelaide* flagellin in Freund's complete adjuvant. *Aust. J. Exp. Biol. Med. Sci.,* 46:179–188.

Lowy, J. (1965). Structure of the proximal ends of bacterial flagella. *J. Mol. Biol.,* 14:297–299.

Lowy, J., and J. Hanson (1964). Structure of bacterial flagella. *Nature, Lond.,* 202:538–540.

—— (1965). Electron microscope studies of bacterial flagella. *J. Mol. Biol.,* 11:293–313.

Lowy, J., and M. Spencer (1968). Structure and function of bacterial flagella. *Symp. Soc. Exp. Biol.,* 22:215–236.

Mäkelä, P. H., and G. J. V. Nossal (1961). Bacterial alherence: A method for detecting antibody production by single cells. *J. Immunol.,* 87:447–456.

Martinez, R. J., D. M. Brown, and A. N. Glazer (1967). The formation of bacterial flagella. III. Characterization of the subunits of the flagella of *Bacillus subtilis* and *Spirillum serpens*. *J. Mol. Biol., 28*:45–51.

McDowell, G. H., and A. K. Lascelles (1969). Local production of antibody by ovine mammary glands infused with *Salmonella* flagellar antigens. *Aust. J. Exp. Biol. Med. Sci., 47*:669–678.

McGroarty, E. J. (1971). Regulation of flagellar morphogenesis by temperature: The involvement of the bacterial cell surface in the synthesis of flagellin and the flagellum. Ph.D. Thesis, Purdue University, Lafayette, Indiana.

McGroarty, E. J., R. W. Smith, and H. Koffler. Unpublished observations, 1971.

Mitchen, J. R. (1969). The isolation of proximal hooks from *Bacillus pumilus* flagella. M.S. Thesis, Purdue University, Lafayette, Indiana.

—— (1971). The nature of flagellar hooks. Ph.D. Thesis, Purdue University, Lafayette, Indiana.

Mitchen, J. R., and H. Koffler (1969). Purification of flagellar hooks. *Bact. Proc.*, p. 29.

Mitchen, J. R., H. Koffler, and R. W. Smith (1970a). Nature of flagellar hooks. *Bact. Proc.*, p. 22.

—— (1970b). The nature of bacterial flagellar hooks. *J. Cell Biol., 47*:142a.

Moore, S., and W. H. Stein (1954). A modified ninhydrin reagent for the photometric determination of amino acids and related compounds. *J. Biol. Chem., 211*:907–913.

Nakaya, R., H. Uchida, and H. Fukumi (1952). Studies on the antigenic patterns of bacterial flagella. *Jap. J. Med. Sci. Biol., 5*:467–472.

Noelken, M. E., C. A. Nelson, C. E. Buckley, and C. Tanford (1965). Cross conformation of rabbit 7S γ-immunoglobulin and its papain-cleaved fragments. *J. Biol. Chem., 240*:218–224.

Nossal, G. J. V., G. L. Ada, and C. M. Austin (1964a). Antigens in immunity. II. Immunogenic properties of flagella, polymerized flagellin, and flagellin in the primary response. *Aust. J. Exp. Biol. Med. Sci., 42*:283–294.

—— (1964b). Antigens in immunity. IV. Cellular localization of [125]I- and ^{131}I-labelled flagella in lymph nodes. *Aust. J. Exp. Biol. Med. Sci., 42*:311–330.

—— (1965). Antigens in immunity. X. Induction of immunologic tolerance to *Salmonella adelaide* flagellin. *J. Immunol., 95*:665–672.

Nossal, G. J. V., and C. M. Austin (1966). Mechanism of induction of immunological tolerance. II. Simultaneous development of priming and tolerance. *Aust. J. Exp. Biol. Med. Sci., 44*:327–340.

Nossal, G. J. V., J. Mitchell, and W. McDonald (1963). Autoradiographic studies on the immune response. IV. Single cell studies on the primary response. *Aust. J. Exp. Biol. Med. Sci., 41*:423–435.

Nossal, G. J. V., K. D. Shortman, J. F. A. P. Miller, G. F. Mitchell, and J. S. Haskill (1967). The target cell in the induction of immunity and tolerance. Cold Spring Harb. Symp. Quant. Biol., *32*:369–379.

Oiler, L., F. Kocka, R. W. Smith, and H. Koffler (1971). The presence of flagellins A and B in the flagellum of *Bacillus pumilus* 101. *Bact. Proc.*, p. 27.

Oosawa, F., and S. Higashi (1967). Statistical thermodynamics of polymerization and polymorphism of protein. *Progr. Theor. Biol., 1*:79–164.

Ornstein, L. (1964). Disc electrophoresis. I. Background and theory. *Ann. N.Y. Acad. Sci., 121*:321–349.

Parish, C. R., R. Wistar, and G. L. Ada (1969). Cleavage of bacterial flagellin with cyanogen bromide. Antigenic properties of the protein fragments. *Biochem. J., 113*:501–506.

Read, K. S. (1957). Antigenic properties of bacterial flagella. Ph.D. Thesis, Purdue University, Lafayette, Indiana.

Read, K., M. Moskowitz, and H. Koffler (1956). Antibody-combining properties of bacterial flagella. *Fed. Proc., 15*:609.

Remsen, C. C., S. W. Watson, J. B. Waterbury, and H. E. Truper (1968). Fine structure of *Ectothiorhodospira mobilis* Pelsh. *J. Bacteriol., 95*:2374–2392.

Ritchie, A. E., and J. H. Bryner, Jr. (1969). Flagellar attachment in *Vibrio fetus. Bact. Proc.,* p. 29.

Ritchie, A. E., R. F. Keeler, and J. H. Bryner (1966). Anatomical features of *Vibrio fetus:* Electron microscopic survey. *J. Gen. Microbiol., 43*:427–438.

Shellam, E. R., and G. J. V. Nossal (1968). Mechanism of induction of immunological tolerance. IV. The effects of ultra-low doses of flagellin. *Immunology, 14*:273–284.

Smith, R. W., and H. Koffler (1971). Bacterial flagella. *Adv. Microbiol. Physiol., 6*:219–339.

Smith, R. W., L. R. Yarbrough, and H. Koffler (1968). Effects of chemical modifications on flagellin-flagellin interactions. *Abstr. Am. Soc. Cell Biol., 39*:127a.

Smith, T., and A. L. Reagh (1903). The non-identity of agglutinins acting upon the flagella and upon the body of bacteria. *J. Med. Res., 10*:89–101.

Stenesh, J., and H. Koffler. Unpublished observations, 1971.

Stocker, B. A. D. (1957). Methods of removing flagella from live bacteria: Effects on motility. *J. Path. Bact., 73*:314–315.

Sullivan, A. (1968). The preparative separation and study of the two electrophoretic protein components of *Bacillus pumilus* flagellin. M.S. Thesis, Purdue University, Lafayette, Indiana.

Sullivan, A., J. Bui, H. Suzuki, R. W. Smith, and H. Koffler (1969). Possible phase variation in *Bacillus pumilus. Bact. Proc.,* p. 30.

Suzuki, H., and H. Koffler (1969). *In vitro* synthesis of flagellin. *J. Cell Biol., 43*:143a.

—— (1970). *In vitro* enzymatic synthesis of flagellin. *Bact. Proc.,* p. 23.

Takayawa, K., D. H. MacLennan, A. Tazagoloff, and C. D. Stoner (1966). Studies on electron transfer system. LXVII. Polyacrylamide gel electrophoresis of mitochondrial electron transfer complex. *Arch. Biochem. Biophys., 114*:223.

Tauschel, H.-D. (1970). Der Geisselapparat von *Rhodopseudomonas palustris.* IV. Isolierung der Geissel und ihrer Komponenten. *Arch. Mikrobiologie., 74*:193–206.

Tauschel, H.-D., and G. Drews (1969). Der Geisselapparat von *Rhodopseudomonas palustris.* II. Entstehung und Feinstruktur der Geissel-Basalkorper. *Arch. Microbiologie, 66*:180–194.

Tomcsik, J., and J. B. Baumann-Grace (1956). Zellwandfreie Bakterienprotoplasten. *Verh. naturf, Ges.,* Basel, *67*:218–238.

Uratani, Y., S. Asakura, and K. Imahori (1971). A circular dichroism study of *Salmonella* flagellin: Evidence for conformational change on polymerization. *J. Mol. Biol.,* in press.

Vaituzis, Z., and R. N. Doetsch (1969). Relationship between cell wall, cytoplasmic membrane, and bacterial motility. *J. Bacteriol., 100*:512–521.

Van Iterson, W., J. F. M. Hoeniger, and E. N. Van Zanten (1966). Basal bodies of bacterial flagella in *Proteus mirabilis.* I. Electron microscopy of sectional material. *J. Cell Biol., 31*:585–601.

Vennes, J. W., and P. Gerhardt (1959). Antigenic analysis of cell structure isolated from *Bacillus megaterium. J. Bacteriol., 77*:581–592.

Wakabayashi, K., H. Hotani, and S. Asakura (1969). Polymerization of *Salmonella* flagellin in the presence of high concentrations of salts. *Biochim. Biophys. Acta, 175*:195–203.

Weibull, C. (1948). Some chemical and physico-chemical properties of *Proteus vulgaris. Biochim. Biophys. Acta, 2*:351–361.

—— (1949). Chemical and physicochemical properties of the flagella of *Proteus vulgaris* and *Bacillus subtilis:* A comparison. *Biochim. Biophys. Acta, 3*:378–382.

—— (1950a). Electrophoretic and titrimetric measurements on bacterial flagella. *Acta Chem. Scand., 4*:260–267.

—— (1950b). Investigations on bacterial flagella. *Acta Chem. Scand., 4*:268–276.

—— (1953). The free amino acid groups of the *Proteus* flagella protein. *Acta Chem. Scand., 7*:335–339.

Weinstein, D. (1959). The occurrence of flagellin within the cytoplasm of penicillin-induced spheroplasts of *Proteus vulgaris.* Ph.D. Thesis, Purdue University, Lafayette, Indiana.

Winebright, J., and F. W. Fitch (1962). Antibody formation in the rat. I. Agglutinin response to particulate flagella from *Salmonella typhosa. J. Immunol., 89*:891–899.

Yaguchi, M., J. F. Foster, and H. Koffler (1964). Configurational adaptability of structural proteins from bacteria. *Abstr. 6th Int. Congr. Biochem.,* New York, p. 189.

Yamaguchi, S., and T. Iino (1966). Genetic map of *H*1 gene in *Salmonella. Rep Nat. Inst. Genet., Misima, 17*:119–120.

—— (1969). Genetic determination of the antigenic specificity of flagellar protein in *Salmonella. J. Gen. Microbiol., 55*:59–74.

Yarbrough, L. R. (1971). The likely participation of tyrosine residues in intra- and intermolecular interactions in flagellin from cells of *Bacillus stearothermophilus.* Ph.D. Thesis, Purdue University, Lafayette, Indiana.

Yarbrough, L. R., H. Koffler, and J. F. Foster (1969). Conformation of flagellin from a thermophilic bacillus. *Biophys. J., 9*:A-107.

Yarbrough, L. R., H. Koffler, W. J. Ray, and J. F. Foster (1971a). Flagellin-flagellin interactions in the bacterial flagellum. *Abstr. Biophys. Soc.,* p. 126a.

—— (1971b). Hydrophobic residues in flagellin-flagellin interactions. *Fed. Proc.,* p. 1303.

SUMMARY COMMENTS ON CELLULAR ANTIGENS OF GRAM-NEGATIVE BACTERIA

STEPHAN E. MERGENHAGEN

Laboratory of Microbiology and Immunology, National Institute of Dental Research, National Institutes of Health, Bethesda, Maryland

I. Ultrastructure of the Gram-Negative Cell

For several decades it has been realized that the gram-negative bacterial cell contains a number of complex antigens. With the advent of electron microscopy the complex nature of the cell surface could be visualized and correlated with the chemical and antigenic nature of each of the cell surface membranes. Figure 1 illustrates, in thin section, the ultrastructure of *Veillonella alcalescens,* a gram-negative anaerobic coccus. In this section, three separate structural entities surrounding the protoplasm of the cell are observed and are referred to as the outer membrane, the solid membrane, and the plasma membrane. The outer membrane appears as two dense layers and is considered to be a unit membrane. This membrane consists of lipopolysaccharide (endotoxin, O antigen) and lipoprotein. The solid membrane contains the mucopeptide (peptidoglycan), which is the substrate for lysozyme. The plasma membrane encloses the cytoplasm and consists mainly of lipid and protein. If one uses some imagination, one might also envision ribosomes and strands of DNA within the cytoplasm. The only structure which is obviously lacking in this illustration is the flagellum, which is attached to and arises from the cell surface layers on certain strains of bacteria and provides the cell with additional antigenic specificities. The reader is referred to the excellent discussion of the flagellum by Dr. Koffler (KOFFLER and SMITH, 1972). Following this brief introduction on the ultrastructure of the gram-negative cell, I would now like to comment on certain of the individual contributions before entertaining certain thoughts and relating experiments from our laboratory concerned with the mechanism of action of endotoxins.

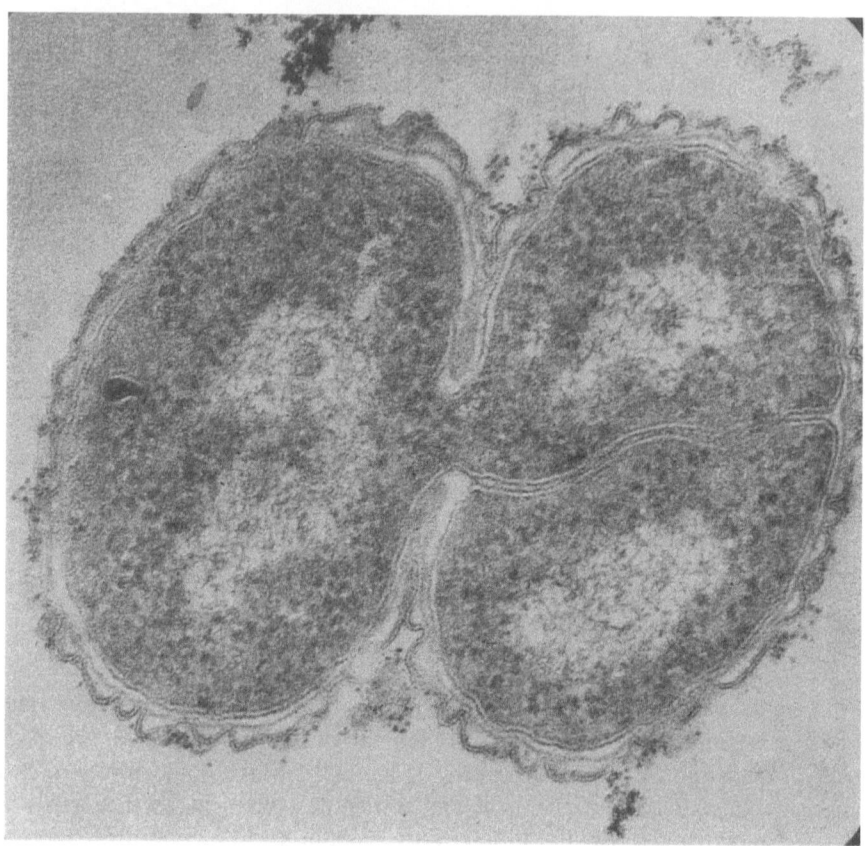

Fig. 1. Three separate structures surround the Gram-negative cell (*Veillonella alcalescens*): the outer membrane, solid membrane, and the plasma membrane. The outer membrane (lipopolysaccharide, lipoprotein), a unit membrane, appears extremely convoluted, whereas the solid membrane (mucopeptide) tightly follows the periphery of the protoplasmic constituents. 120,000 ×.

II. RNA as a Potential Vaccine against Gram-Negative Infections

Dr. Berry and his coworkers have recently evaluated subcellular immunizing antigens from *Salmonella typhimurium* (BERRY and VENNEMAN, 1972). In this elegant work, Dr. Berry explained how he was provoked into these experiments following the initial exciting observations of the Youmans', who showed that a ribosomal vaccine from *Mycobacterium tuberculosis* provided significant protection against tuberculous challenge in mice (YOUMANS and YOUMANS, 1972). In his experiments, Dr. Berry prepared ribosomal and RNA fractions from salmonellae and found that both preparations afforded mice significant protection against experimental infection with a virulent strain of *S. typhimurium*. Protection was assessed by the ability of the mice to survive the challenge and also by the inability of the microorganisms to proliferate within the tissues of the vaccinated host. The contributions of humoral and/or cellular immunity in these experiments was also studied. Interestingly enough, passively transferred serum from mice vaccinated with RNA failed to protect animals against infection, while the passive transfer of peritoneal cells from vaccinated animals to normal recipients transferred a high degree of immunity. It is not known whether cells other than macrophages are involved in this protection; however, Dr. Berry commented that lymphocytes from vaccinated mice failed to elaborate macrophage inhibition factor when exposed to ribosomes or RNA *in vitro*. These experiments indicate that "activated" macrophages may be an extremely important determinant in immunity to gram-negative infections. This conclusion is in essential agreement with the work of the Youmans. These experiments further indicate that a number of our currently employed vaccines may have to be re-evaluated because of our previous concern with the potential role of humoral factors in protection against infections with such organisms as *Salmonella typhosa, Vibrio cholera, Bordetella pertussis,* and others. It may well turn out that purified RNA from these organisms will stimulate a more long-lasting immunity in the human against infection. Certainly additional work in this area is needed in order for us to fully assess the possible role of contamination of the RNA preparation with other components of the cell which could lead to the formation of a "superantigen." This seems to be an unlikely possibility, however, because the contaminant would have to be extremely potent in very minute concentrations.

III. Heterogenetic Antigens of Enterobacteriaceae

Heterogenetic antigens have long been known to be involved in the immune response of the host against antigenically unrelated genera or species of bacteria. In discussing one of these antigens, the so-called common antigen (CA) of enterobacteriaceae, Dr. Neter showed that because of its association with lipopolysaccharide the host does not develop an immune response to this antigen (NETER and WHANG, 1972). Thus the lipopolysaccharide is masking the determinants of the CA. Dr. Neter and his associates have studied the immune response to this antigen by employing the very sensitive passive hemagglutination procedure. When the CA is separated from the O antigen by virtue of its being ethanol-soluble, it becomes fully immunogenic in the rabbit. If, however, the CA is first combined with lipopolysaccharide and then injected, the blocking effect of the lipopolysaccharide can be demonstrated. This novel type of immunosuppression is provocative from a number of standpoints. For instance, are there other antigens on the surface of the gram-negative cell which could elicit some form of immunologic priming (cellular or humoral) but whose detection is difficult because of the blocking effects of associated molecules? Future experiments should be directed at a more complete chemical and physical analysis of the CA and its role in natural and acquired immunity to related and unrelated gram-negative pathogens. Indeed, Dr. Neter presented data which showed that rabbits immunized with *Salmonella* CA are significantly protected against histological evidence of pylonephritis produced with *Proteus* organisms.

IV. Chemistry of Lipopolysaccharides and Biologically Active Centers

A great deal of the panel discussion in this session centered around the chemistry and biological activities of endotoxic lipopolysaccharides.* This is understandable, since these microbial products have stimulated a great deal of research during the past four decades and also because the Conference Chairman, Dr. Nowotny, unintentionally influenced these discussions because of his personal interests and notable contributions in this area. Dr. Wheat reviewed rather thoroughly the chemistry of lipopolysaccharides and indicated that we are now on the threshold of defining endotoxin as a homogeneous chemical entity with a more complete recognition of its component parts (BARROW and WHEAT, 1972). In recent studies, the possible significance of the polysaccharide components of endotoxic lipopolysaccharides could be ruled out by using mutant strains of *Salmonella* which synthesize a cell wall endotoxin free of O-polysaccharide and heptose. Therefore, polysaccharide-free lipid A can be extracted from such strains without the application of drastic hydrolytic procedures. In this regard, Dr. Nowotny and coworkers have recently identified a chromatographically homogeneous fraction derived from the lipid moiety of endotoxin from *S. minnesota* R 595 (NOWOTNY, 1971). The preparation was lethal for chick embryos, and other biological parameters showed similar results. Analyses of the fraction showed that it consists of a high percentage of β-hydroxymyristic and similar acids. Hexosamine-P could be detected in the preparation. Thus the biologically active centers in the endotoxin molecule are being better defined.

V. Effectors of the Endotoxin Response

A. Nucleic Acids

In discussion on the biological activities of endotoxin, it is now becoming more evident that endotoxins, like immune complexes, rely upon host-effector systems to express certain of their biological activities. In the case of the well-known adjuvant effect of lipopolysaccharides, it now appears that this activity may be mediated by nucleic acids which are released from cells damaged by the endotoxin. This hypothesis was supported in work

* For a more complete review of this subject, the reader is referred to several recent articles (LÜDERITZ et al., 1968; NOWOTNY, 1969; NETER, 1969; MERGENHAGEN et al., 1969).

by Dr. Johnson, who explained how nucleic acids *per se* increase antibody formation and how synthetic polynucleotides act similarly. The latter appear to cause an increase in the rate of cell division of the thymus-derived lymphocytes. These experiments have far-reaching implications in defining more precisely the nature of the immune response and how it may be regulated by nonspecific agents.

B. Complement

It has long been known that endotoxin produces severe disturbances in microcirculation, an effect that resembles those produced with pharmacologically active amines (*e.g.,* histamine). Dr. Urbaschek related here how he has employed the mesenteric blood vessels of the guinea pig to study the influence of endotoxin on microcirculation (URBASCHEK, 1967, 1971). These disturbances are among the first observable alterations caused by endotoxins. The phenomena include slowdown of the bloodstream, degranulation of mast cells, wall adhesion of granulocytes, thrombocyte aggregation, formation of microthrombi, rouleaux-formation of erythrocytes, swelling of endothelial and periendothelial cells, and an increase in permeability of the vessel walls, which leads to microbleeding.

Dr. Urbaschek documented his studies with superb photomicrographs. Scanning electron microscopy in their laboratory revealed bridges of fibrin connecting thrombocytes to other cells, leading to the formation of thrombi and to changes in the plasticity of erythrocytes. All these alterations in the capillary bed are comparable to those after biogenic amines, the only difference being the time at which the disturbances begin. Whereas the blood flow begins to slow down about 10 minutes after endotoxin (at which time tissue mast cells are in the process of degranulation), it slows down immediately in response to histamine or serotonin. The combination of threshold doses of endotoxin and histamine or serotonin leads to the severe alterations of the microcirculation described above. Dr. Urbaschek and associates found that the release of biogenic amines after endotoxin application plays an important role in the initial stage of endotoxemia.

Furthermore, Dr. Urbaschek's laboratories, in cooperation with Dr. Nowotny and associates, found that chemically detoxified endotoxins, called endotoxoids, can prevent all these microcirculatory disturbances if applied 24 hours before endotoxin. In addition, it was found that a nonspecific, endotoxin tolerance is induced by a single injection of endotoxoid. This was shown in animal experiments as well as in human volunteers (URBASCHEK, 1967; URBASCHEK and NOWOTNY, 1968; FRITSCH *et al.,* 1968). A glycolipid preparation isolated from a rough mutant of *Salmonella minnesota,* was found to be nontoxic in normal mice and guinea pigs, but capable of inducing nonspecific tolerance towards lethal doses of endotoxins. The effect of this glycolipid on microcirculation was found to be significantly milder than toxic endotoxin.

Since endotoxins are potent activators of the complement system, the

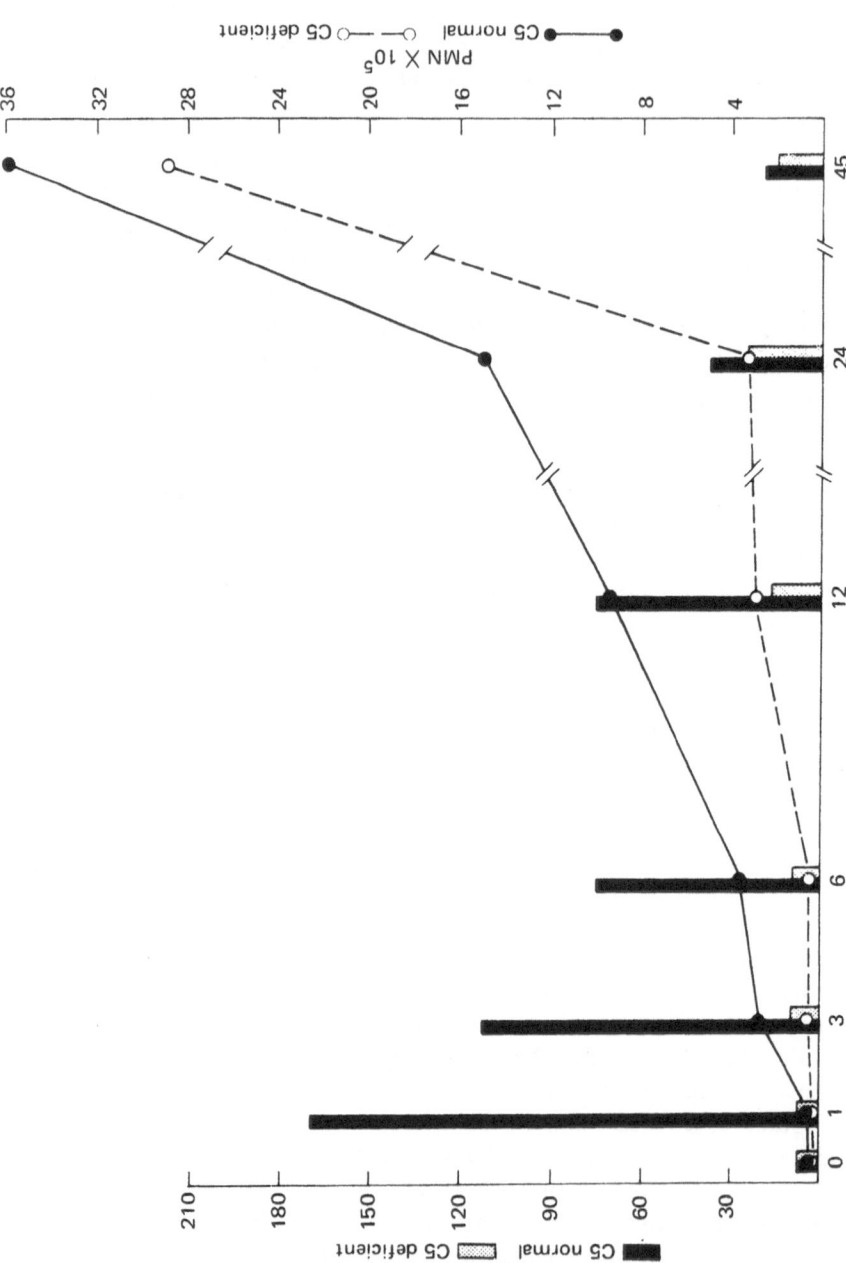

Fig. 2. Accumulation of polymorphonuclear leukocytes (PMNs) in the peritoneal cavity of C5-deficient and C5-normal mice treated with endotoxin. Also shown is the chemotactic activity for mouse PMNs in the peritoneal fluids at various times after endotoxin administration. Mice were injected with 250 μg of *Salmonella typhosa* endotoxin, and the peritoneal fluids were removed at varying times thereafter. Each point represents the mean PMN count and chemotactic activity of the pooled exudates of 6 mice. The standard deviation of the means of the C5-deficient and C5-normal mice at hours 3, 6, and 24 did not overlap, and the curves differ significantly ($P < 0.01$).

question was raised as to whether these observed effects might be due to the formation of anaphylatoxin from complement since anaphylatoxin effects capillary permeability and releases histamine from mast cells. Complement is also required for the immune adherence and degranulation by endotoxin of platelets, and this could account for certain of the observed effects in the capillary bed. In addition to the effects of endotoxin on microcirculation, another cardinal event in the endotoxin-induced acute inflammatory response is the accumulation of polymorphonuclear leukocytes (PMNs) at the inflammatory site. Our laboratory has shown that during the activation of guinea pig complement by endotoxin, a polypeptide (C5a) is liberated from the fifth component of complement, which is anaphylatoxic and chemotactic *in vitro* for rabbit PMNs and mononuclear leukocytes. Simlar results are obtained using human complement and human peripheral leukocytes. Dr. Snyderman has related our recent *in vivo* experiments which suggest an important role for C5 in the early accumulation of PMNs in inflammatory exudates (SNYDERMAN *et al.,* 1971). C5-deficient and C5-normal mice were injected intraperitoneally with endotoxin. Mice were sacrificed at various times after endotoxin administration and peritoneal exudates were removed and the total number of leukocytes in the exudates was determined. Supernatants of peritoneal exudates from the endotoxin-treated mice were tested *in vitro* in modified Boyden chambers for chemotactic activity for peripheral mouse leukocytes. A significant increase in the accumulated PMNs was noted in C5-normal mice as early as 3 hours after endotoxin administration, and the number of cells increased linearly during a 24-hour period (Figure 2). In contrast, mice deficient in C5 did not have a significantly increased number of PMNs until 48 hours, and at no time in the 24-hour period did the numbers of PMNs approach those found in C5-normal mice. In addition, in C5-normal mice, marked chemotactic activity for PMNs was detected within 1 hour after endotoxin injection (Figure 2). The activity diminished greatly over a 24-hour period but was detectable as late as 24 hours after endotoxin. In contrast, supernatants of C5-deficient mouse exudates contained no detectable chemotactic activity for at least 6 hours, and at 12 and 24 hours contained barely detectable levels of activity.

It is now becoming clear that the complement and clotting systems are important humoral mediators of endotoxin-induced inflammation. However, evidence detailing a specific function related to the mediation of endotoxin inflammation *in vivo* to either of these systems or component parts is incomplete. The data reported by Dr. Snyderman from our laboratory suggest that C5 plays an important role in the early phases of leukocyte accumulation in response to endotoxin. The accumulation of PMNs in inflammatory sites is of major significance to the host for at least two reasons. First, the PMN is involved in host defense against microbial invasion; and second, in certain inflammatory disease states, the PMN appears to be responsible for a large degree of host-tissue destruction.

In concluding my remarks, I would like to state that I think this Conference has been highly successful. It has introduced each of us to scientists in other fields who encounter similar problems when working with cellular antigens from mammalian as well as from bacterial cells. I anticipate the time when another conference will take place on a similar subject and when the free exchange of ideas and approaches to problems will again be desirable.

References

Barrow, R., and R. W. Wheat (1972). Amino sugars of *Pseudomonas aeruginosa* ATCC 7700 lipopolysaccharide. This volume, p. 22.

Berry, L. J., and M. R. Venneman (1972). Immune response of mice to subcellular vaccines of *Salmonella typhimurium*. This volume, p. 3.

Fritsch, H., H.-J. Krecke, B. Becker, B. Urbaschek, and A. Nowotny (1968). Zur Erzeugung einer Endotoxintoleranz durch detoxifiziertes Endotoxin (Endotoxoid) beim Menschen. *Verh. Dtsch. Ges. Inn. Med., 74:*1151–1154.

Lüderitz, O., K. Jann, and R. Wheat (1968). Somatic and capsular antigens of Gram-negative bacteria. In *Comprehensive Biochemistry,* 26A, Amsterdam/London/New York: Elsevier.

Koffler, H., and R. W. Smith (1972). The nature of flagellar antigens. This volume, p. 31.

Mergenhagen, S. E., R. Snyderman, H. Gewurz, and H. S. Shin (1969). Significance of complement to the mechanism of action of endotoxin. *Curr. Topics Microbiol. Immunol., 50:*37–77.

Neter, E. (1969). Endotoxins and the immune response. *Curr. Topics Microbiol. Immunol., 47:*82–124.

Neter, E., and H. T. Whang (1972). The common antigen of Gram-negative enteric bacteria. This volume, p. 14.

Nowotny, A. (1969). Molecular aspects of endotoxic reactions. *Bacteriol. Rev., 33:*72–98.

—— (1971). Relationship of structure and biological activity of bacterial endotoxins. *Naturwissenchaften,* in press.

Snyderman, R., J. K. Phillips, and S. E. Mergenhagen (1971). Biological activity of complement *in vivo:* Role of C5 in the accumulation of polymorphonuclear leukocytes in inflammatory exudates. J. Exp. Med., *134:*1131–1143.

Urbaschek, B. (1967). *Zur Frage des Wirkungssmechanismus bakterieller Endotoxine und seiner Beeinflussung.* Habilitationsschrift, Medizinische Fakultät der Universität Heidelberg.

—— (1971). The effects of endotoxins in the microcirculation. In *Microbial Toxins,* pp. 261–275, vol. V. New York: Academic Press.

Urbaschek, B., and A. Nowotny (1968). Endotoxin tolerance induced by detoxified endotoxin (endotoxoid). *Proc. Soc. Exp. Biol. Med., 127:*650–652.

Youmans, G. P., and A. S. Youmans (1972). Immunizing antigens of mycobacteria. This volume, p. 95.

Part 2

Gram-Positive Bacterial Antigens

NATURE OF THE STREPTOCOCCAL AND MYOCARDIAL ANTIGENS INVOLVED IN THE IMMUNOLOGIC CROSS-REACTION BETWEEN GROUP A STREPTOCOCCUS AND HEART

Melvin H. Kaplan*

Department of Medicine, Case-Western Reserve University School of Medicine, and Cleveland Metropolitan General Hospital, Cleveland, Ohio

I. Introduction

The antigenic cross-reaction between mammalian myocardium and the group A streptococcus has been substantiated by several different investigative groups. The available evidence as to the nature of the respective antigens involved is limited and, in some respects, controversial. In this paper, the nature of these shared antigens of streptococcus and myocardium, and their relation to other defined antigens of these structures will be reviewed.

II. Nature of the Streptococcal Antigen(s) Cross-Reactive with Myocardium

The initial reports of KAPLAN and MEYESERIAN (1962b) and KAPLAN (1963) demonstrated the cross-reaction of rabbit antisera to group A, type 5, streptococcal cells and cell extracts with human heart and skeletal muscle fibers by immunofluorescence and complement fixation. As illustrated in Figure 1, the immunofluorescent staining was localized mainly to sarcolemma and subsarcolemmal sarcoplasm of these structures, and also to smooth muscle of vessel walls and of endocardium. As assayed by inhibition of the immunofluorescent reaction of such rabbit cross-reactive antisera, the responsible antigen was found associated with the streptococ-

* Research Career Awardee, USPHS (K6-HE-4576). Supported by USPHS grant H-3726.

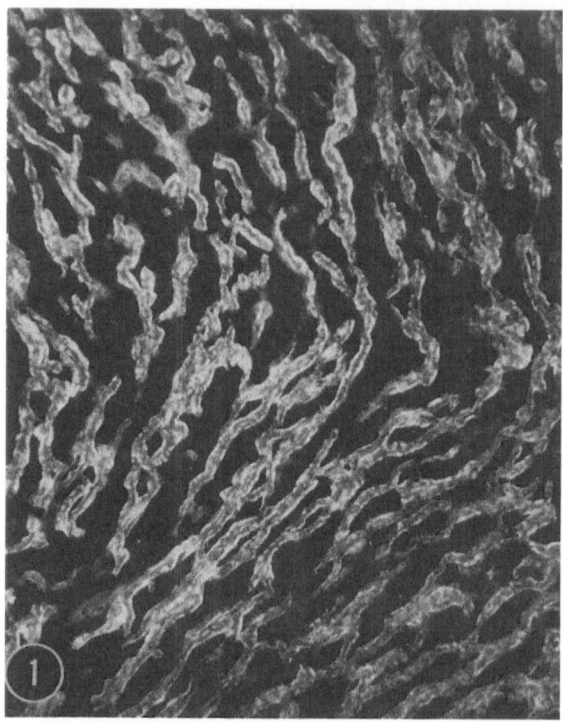

Fig. 1. Immunofluorescent reaction with human heart of rabbit antiserum (635) to type 5 streptococcal acid extract preparation. Cross-reaction involves sarcolemma and subsarcolemmal structures. This immunofluorescent reaction was inhibited selectively by absorption with extracts of streptococcal strains of the homologous serotype. × 125. (From KAPLAN, 1969b.)

cal cell-wall fraction, and could be released by cell-wall lytic enzymes derived from *Streptomyces albus* and also by the group C streptococcal phage-associated lysin (KAPLAN, 1967). Absorption of these antisera with streptococcal cell membrane preparations resulted in little or no inhibition of immunofluorescent reaction. In contrast, absorption was effective with small amounts of cell wall preparations, which had been treated with sodium lauryl sulfate in order to solubilize membrane contaminants and were shown to contain less than 10% contamination by cell membranes based on glucose analysis. The absorbing activity of cell-wall preparations could therefore not be attributed to contamination by cell membranes. The antigen was solubilized from cells or cell walls by hot HCl acid extraction by the Lancefield procedure, precipitated between 0.3 to 0.7 saturation ammonium sulfate, and partially purified by DEAE-cellulose and CM-cellulose chromatography. In all separation procedures as well as in quantitative inhibition studies, the active antigen was found closely associated with M protein, the virulence factor of the group A streptococcus. What was of particular interest was the finding that the immunofluorescent reaction with heart of these type 5 antisera was absorbed by streptococcal extracts from strains of the homologous serotype, and only partially or not at all by streptococcal extracts of other serotypes or groups. Purified preparations of the cross-reactive antigen derived from acid extracts of cell walls was found to be protein in nature and to contain little or no acetylhexosamine, rhamnose, glucose, or other carbohydrate. The antigen was highly susceptible to digestion by trypsin, pepsin, and chymotrypsin.

ZABRISKIE and FREIMER (1966) reported that rabbit antisera to groups A and C streptococci reacted with human cardiac and skeletal muscle and with smooth muscle of vessel walls by the immunofluorescent technique. They noted that antisera to a number of group A streptococcal strains of different serotype were reactive without evidence of type specificity. Antisera to isolated group A streptococcal membranes showed similar cross-reactivity, and could be absorbed by membranes of groups A, C, and G streptococci. They attributed the cross-reactive antigen to a constituent of streptococcal cell membranes. LYAMPERT *et al.* (1966) and DANILOVA (1966) reported a cross-reaction between rabbit antisera to types 5 and 29 streptococci and cardiac myofibers of mammalian heart. The immunofluorescent reaction of the type 5 antiserum was inhibited by absorption with streptococcal cells of the homologous serotype, but not by streptococci of heterologous serotype. Specificity of the type 29 antiserum was not described. NAKHLA and GLYNN (1967) reported that antisera to 9 strains of group A streptococci of different serotypes exhibited cross-reaction with cardiac and skeletal muscle. The serologic reaction was absorbed by cells and cell extracts of both homologous and heterologous types.

In further studies from this laboratory (KAPLAN, 1967), it was observed that rabbit antisera to type 12 and to type 19 streptococci gave a typical immunofluorescent reaction with cardiac and skeletal muscle fibers and smooth muscle, as shown in Figure 2; however, as determined by specific absorption tests, these antisera differed in their serologic specificity. With several of these antisera, immunofluorescent reaction could be inhibited following absorption by streptococcal extracts only of the homologous serotype. With other antisera, the reaction was absorbed by streptococcal extracts of all group A strains tested, independent of serotype. Analysis of the specificity of these two groups of antisera indicated that in the case of the first group with serotype specificity, the absorbing antigen was associated with the cell-wall fraction, was released from cell walls by phage-associated lysin, and was consistently associated with the homologous type M protein in chromatographic fractions of acid extracts. The nature of this association of absorbing antigen to M protein was examined in immunoabsorption experiments (KAPLAN, 1969a). Type-specific anti-M serum of the homologous Lancefield type was added in excess to partially purified preparations of the absorbing antigen to precipitate the M protein

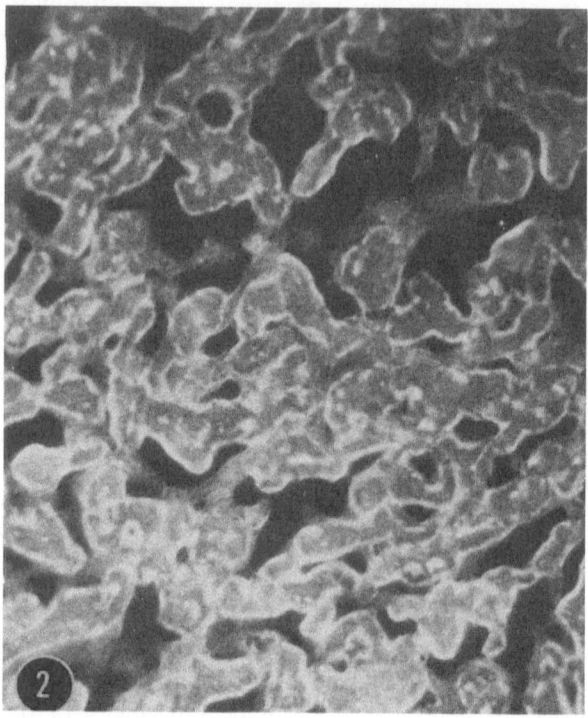

Fig. 2. Immunofluorescent reaction with human heart of rabbit antiserum (2036) to type 19 streptococci. Reaction involves sarcolemma and subsarcolemmal structures. The immunofluorescent reaction in this case was inhibited following absorption with extracts of group A streptococcal strains independent of serotype. × 312.

present in these preparations. The supernatant was shown to be free not only of M protein, but it was also devoid of immunofluorescent-absorbing activity. The latter was presumably also removed by treatment with anti-M serum. The M-anti-M immune precipitate was then washed repeatedly and extracted with hot HCl acid according to the Lancefield procedure in order to recover the M protein dissociated from antibody. This extract contained not only M protein but was also effective in inhibiting the immunofluorescent reaction of homologous type cross-reactive antiserum. This immunoabsorption experiment was carried out with cross-reactive antigen preparations from types 19, 12, and 5 streptococci, employing anti-M sera of the homologous type. The results in all three cases were as described above. It was concluded from these data that the streptococcal cross-reactive antigen was closely linked to M protein antigen and that both of these latter antigenic determinants were expressed in the specificity of the resulting cross-reactive antisera in close analogy to carrier-hapten specificity.

As was noted above, other cross-reactive antisera to type 12 and 19 streptococci were not specific for serotype but could be absorbed by extracts of group A strains independent of serotype. The reaction of certain of these antisera was found related to cell membranes, as reported by ZABRISKIE and FREIMER (1966).

It is as yet not possible to reconcile completely the data from the different investigative groups as to the identity of the streptococcal antigen or antigens. The immunochemical basis of these cross-reactions remains to be more fully clarified. In particular, data are not yet available as to the chemical linkage of cross-reactive antigen to other streptococcal constituents and the nature of the antigenic determinants present in the linkage region. Further, the question has not yet been approached whether the cross-reactive antigen may exhibit antigenic variation or polymorphism analogous to the behavior of M protein.

Sera from patients with rheumatic fever and other streptococcal disease have also been shown to contain cross-reactive antibody (KAPLAN and SVEC, 1964; ZABRISKIE, 1967). Such sera may exhibit immunofluorescent reaction with sarcolemmal-subsarcolemmal constituents of cardiac and skeletal muscle fibers and less frequently with smooth muscle. This reaction may be abolished following absorption with streptococcal preparations. Data by the above groups of investigators have ascribed the absorbing antigen to cell walls or to cell membranes. Evidence has also been obtained that these sera may exhibit varying specificity in absorption tests (KAPLAN, 1969a). It is probable that in the case of human sera, the determinants of specificity are even more complex than in the case of rabbit antistreptococcal sera. Purification of the responsible antigens derived from both streptococcus and myocardium will be a prerequisite to final resolution of the problem of serologic specificity.

III. Nature of the Myocardial Antigen Cross-Reactive with Group A Streptococci

Immunofluorescent staining by rabbit antistreptococcal sera indicated, as noted above, that the antigen in human heart cross-reactive with group A streptococci was distributed in sarcolemmal-subsarcolemmal sites of cardiac myofibers and skeletal muscle and in smooth muscle. Reaction with smooth muscle, however, was found to vary with specimens of heart tissue from different individuals, since only 14 of 40 different human heart specimens showed reactivity of smooth muscle of vessels with rabbit cross-reactive antiserum (KAPLAN, 1963). The antigen was present also in heart and skeletal muscle of rabbit, rat, mouse, dog, guinea pig, and rhesus and cyomologous monkeys (KAPLAN and MEYESERIAN, 1962; NAKHLA and GLYNN, 1967; LYAMPERT et al., 1966; DANILOVA, 1966; KAPLAN, unpublished observations). Absorption tests indicated that the antigen was not present in saline extracts of heart but was associated with the sedimentable fraction of heart homogenates. This antigen could be solubilized by acid extraction of washed heart homogenates (KAPLAN et al., 1967).

Comparable data have been obtained also with sera containing cross-reactive antibodies from patients with rheumatic fever and other streptococcal disease, including glomerulonephritis and uncomplicated streptococcal infection. As was noted above, these sera were reactive with sarcolemma-subsarcolemma of cardiac myofibers and skeletal muscle, and less frequently with smooth muscle. In addition, KAPLAN and SVEC (1964) demonstrated a precipitin reaction in agar diffusion plates between sera of patients with rheumatic fever and other streptococcal disease and partially purified preparations of streptococcal cross-reactive antigen. This precipitin reaction was abolished after absorption of the sera with washed human heart homogenate but not after absorption with washed homogenates of other organs, including skeletal muscle, liver, kidney, lung, spleen, uterus, or aorta. These data suggested that the precipitating activity of these sera was directed to a streptococcal antigen which was immunologically related to an organ-specific antigen of heart. These serologic data based on immunofluorescent and precipitin reactions pointed to the possibility that the cross-reactive relationship between streptococcus and heart involved antigens with specificity for heart only, as well as for heart and skeletal muscle.

These preceding data on the cross-reactive relationship of streptococcus and heart have stimulated interest in the direct analysis of the antigenic composition of heart and, in particular, in the definition of cardiac antigens with organ-specific or organ-restricted properties. The available information on the antigenic composition of heart tissue has been reviewed by ESPINOSA et al. (1969) and will be summarized here in brief.

In early studies, BAILEY and RAFFEL (1941) demonstrated heat-stable water-soluble antigens in both heart and skeletal muscle, which were not present in other organs. HENLE et al. (1941) described antigenic activity characteristic of heart tissue only in a sedimentable particulate fraction of saline extracts of bovine and mouse heart. KAPLAN and MEYESERIAN (1962a) also described a particulate organ-specific antigen of heart which they localized by immunofluorescence to sarcoplasmic inclusions localized between myofibrils. In addition, they demonstrated presence of an antigen common to heart and skeletal muscle in several different species which was localized to sarcolemmal-subsarcolemmal sites. GERY and DAVIES (1961) and GERY et al. (1964) demonstrated by passive hemagglutination an antigen specific for heart in saline extracts of heart; however, this antigen was not sedimentable on high-speed centrifugation. KUSHNER and KAPLAN (1967) demonstrated the presence in human heart of a soluble antigen common to heart and skeletal muscle by immunodiffusion and immunoelectrophoretic analysis, which was identified as myoglobin. A second antigen common to heart and kidney was found present in primate species. HALBERT et al. (1968) reported that rabbits immunized with rabbit heart extract exhibited specificity for at least two and possibly three soluble antigens of heart with strict organ specificity.

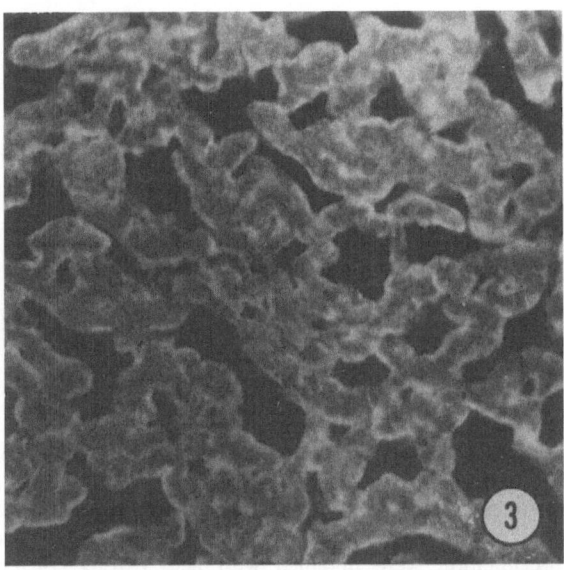

Fig. 3. Reaction with human heart of antiserum to heart-specific antigen derived from a DEAE-cellulose fraction from acid extract of human heart. Antiserum was absorbed with plasma and skeletal muscle extracts. Staining is predominantly of sarcolemma and subsarcolemmal sarcoplasm. × 312. (From ESPINOSA and KAPLAN, 1971.)

In studies directed to the antigen(s) restricted to heart and skeletal muscle which were localized to sarcolemmal-subsarcolemmal sites (cf. Figure 3), ESPINOSA and KAPLAN (1968) reported that human cardiac sarcolemmal antigens could be solubilized in dilute acid, *i.e.,* 0.05 *M* citric acid, pH 2.5. Such acid extracts of heart and of other organs were analyzed by immunodiffusion and immunofluorescence using rabbit antisera to acid extracts of human myocardium. As determined by immunodiffusion, two sarcolemmal antigens common to heart and skeletal muscle were differentiated. One of these antigens was found widely distributed among mammalian species; the other was found in primate species only. These antigens were not related to myoglobin or other saline-soluble components of heart.

In an extension of their studies, ESPINOSA and KAPLAN (1970) separated acid extracts of human heart tissue into two fractions by chromatography on DEAE-cellulose, and prepared hyperimmune rabbit antisera against these fractions for use in antigenic analysis by double diffusion and immunoelectrophoresis. Combined immunodiffusion and specific absorption tests demonstrated at least six antigens common to cardiac and skeletal muscle in acid extracts of myocardium (Figure 4). Of these six antigens, four were found present in several different mammalian species, including dog, rat, beef, and rabbit, while two showed specificity for primates,

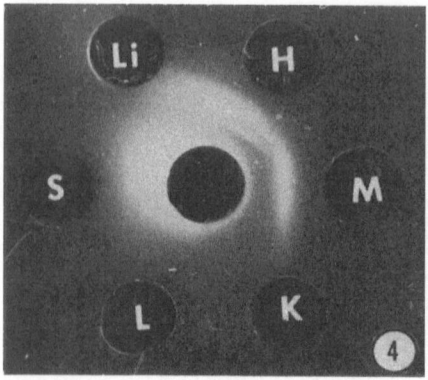 

Fig. 4. Central well contains antiserum to acid extract of human heart homogenate (Fraction II). Antiserum is absorbed with human plasma and kidney acid extract. Peripheral wells contain acid extracts of human organs at 1% concentration: H, heart; M, skeletal muscle; K, kidney; L, lung; S, spleen; and Li, liver. From four to six lines of precipitation were obtained with acid extracts of heart and skeletal muscle. (From ESPINOSA and KAPLAN, 1970.)

Fig. 5. Central well as in Figure 4, except that the antiserum was additionally absorbed with acid extract of skeletal muscle. Only a single line of precipitation specific for heart is demonstrable. (From ESPINOSA and KAPLAN, 1970.)

i.e., man and rhesus monkey. These antigens varied in susceptibility to different proteolytic enzymes.

ESPINOSA and KAPLAN (1970, 1971) also described a heart-specific antigen in both acid and saline extracts of human myocardium (Figure 5). The antigen exhibited both species-specific and species-cross-reactive antigenic properties. It was highly susceptible to proteolytic enzyme digestion. It had an electrophoretic mobility between serum α_1 and β_2-globulins, was relatively thermostable to boiling temperature, and was precipitated at 30 to 70% saturated ammonium sulfate and 20 to 70% ethanol concentration. The sedimentation coefficient by sucrose gradient ultracentrifugation was 1.8S to 2.0S. The molecular weight was estimated at between 20,000 to 23,500. Immunofluorescent staining of heart tissue with antisera specific for heart-specific antigen showed a reaction related to structures in sarcolemma-subsarcolemmal sites (Figure 3).

Thus the sarcolemmal immunofluorescent staining pattern for the antigens restricted to heart and skeletal muscle and for heart-specific antigen was similar to that obtained with rabbit antisera to group A streptococci cross-reactive with heart tissue, and with that obtained with sera of patients with streptococcal disease (cf. Figures 1, 2, and 3). It will be recalled also that sera of patients with streptococcal disease frequently exhibited a precipitin reaction with streptococcal cross-reactive antigen preparation which exhibited strict organ-specificity for heart, as demonstrated by absorption tests with organ homogenates (KAPLAN and SVEC, 1964). Further, it was shown that acid extracts of human heart were capable of absorbing completely the immunofluorescent reaction of cross-reactive antibody from sera of patients with rheumatic fever as well as well as from rabbit cross-reactive antisera (KAPLAN *et al.,* 1967).

These observations raise the following questions. Which of the several sarcolemmal antigens common to heart and skeletal muscle is related to the cross-reactive antigen demonstrated in heart and skeletal muscle with rabbit antistreptococcal sera? Is there a relationship between the organ-specific antigen of heart and the cross-reaction demonstrated with sera of rheumatic fever patients by the precipitin-absorption techniques?

These questions are under intensive investigation at present. Combined immunopathologic, serologic, and epidemiologic studies have provided support for the concept that autoantibodies to heart induced by streptococcal infection are associated with rheumatic fever, and may play a pathogenic role in the myocardial involvement in rheumatic fever (KAPLAN, 1969c; ZABRISKIE, 1967). Evaluation of the precise role which this cross-reactive relationship plays in the rheumatic process will depend in large measure on the definition of the responsible antigenic determinants and on the biologic properties and specificities of the antibodies thereto.

References

Bailey, G. H., and S. Raffel (1941). Organ specificity of tissues of the dog and man as shown by passive anaphylaxis in guinea pigs. *J. Exp. Med., 73*:617–628.

Danilova, T. A. (1966). An immunofluorescent study of the antigen of group A streptococcus and heart tissue from human and experimental animals. *Fed. Proc.,* 25 (Transl. Suppl.):1099–1102.

Espinosa, E., and M. H. Kaplan (1968). Antigenic analysis of human heart tissue: Identification of antigens with specificity restricted to heart and skeletal muscle in acid extracts of myocardium. *J. Immunol., 100*:1020–1031.

—— (1970). Antigenic analysis of human heart tissue: Antigens with restricted organ distribution in acid extracts of human myocardium. *J. Immunol., 105*:416–425.

—— (1971). Antigenic analysis of human heart tissue: Further characterization of an organ-specific antigen of heart tissue. *J. Immunol., 106*:611–615.

Espinosa, E., I. Kushner, and M. H. Kaplan (1969). Antigenic composition of heart tissue. *Am. J. Cardiol., 24*:508–513.

Gery, I., and A. M. Davies (1961). Organ specificity of the heart. I. Animal immunization with heterologous heart. *J. Immunol., 87*:351–361.

Gery, I., A. M. Davies, and E. Lazarov (1964). Immunity and immunotolerance to bovine heart antigens in the rat: Heterogeneity of the serological response. *Immunology, 7*:183–192.

Halbert, S. P., S. E. Holm, and A. Thompson (1968). Cardiac autoantibodies.

I. Immunodiffusion analysis of multiple responses evoked homologously and heterologously. *J. Exp. Med., 127*:613–631.

Henle, W., L. A. Chambers, and V. Groupe (1941). The serological specificity of particulate components derived from various normal mammalian organs. *J. Exp. Med., 74*:495–510.

Kaplan, M. H. (1963). Immunologic relation of streptococcal and tissue antigens. I. Properties of an antigen in certain strains of group A streptococci exhibiting an immunologic cross-reaction with human heart tissue. *J. Immunol., 90*:595–606.

—— (1967). Multiple nature of the cross-reactive relationship between antigens of group A streptococci and mammalian tissue, pp. 48–60. In *Cross-Reacting Antigens and Neoantigens*, J. J. Trentin (ed.), Baltimore: Williams and Wilkins.

—— (1969a). Cross-reaction of group A streptococci and heart tissue: Varying serologic specificity of cross-reactive antisera and relation to carrier-hapten specificity. *Transpl. Proc., 1*:976–980.

—— (1969b). Autoimmunity to heart and its relation to heart disease. *Prog. Allergy, 13*:408–429.

—— (1969c). The cross-reaction of group A streptococci with heart tissue and its relation to induced autoimmunity in rheumatic fever. *Bull. Rheum. Dis., 19*:560–567.

Kaplan, M. H., E. Espinosa, and J. D. Frengley (1967). Properties of solubilized cross-reactive antigens of streptococcal cell walls and mammalian heart and heart valves. *Fed. Proc., 26*:701.

Kaplan, M. H., and M. Meyeserian (1962a). Immunologic studies of heart tissue. V. Antigens related to heart tissue revealed by cross-reaction of rabbit antisera to heterologous heart. *J. Immunol., 88*:450–461.

—— (1962b). An immunological cross-reaction between group A streptococcal cells and human heart tissue. *Lancet, i*:706–710.

Kaplan, M. H., and K. H. Svec (1964). Immunologic relation of streptococcal and tissue antigens. III. Presence in human sera of streptococcal antibody cross-reactive with heart tissue: Association with streptococcal infection, rheumatic fever, and glomerulonephritis. *J. Exp. Med., 119*:651–666.

Kushner, I., and M. H. Kaplan (1967). Antigenic analysis of extracts of human heart tissue: Cardiac antigens with limited distribution in other organs. *J. Immunol., 99*:526–533.

Lyampert, I. M., O. I. Vvedenskaya, and T. A. Danilova (1966). Study on streptococcus group A antigens common with heart tissue elements. *Immunology, 11*:313–320.

Nakhla, L. S., and L. E. Glynn (1967). Studies on the antigen in β-haemolytic streptococci that cross-reacts with an antigen in human myocardium. *Immunology, 13*:209–218.

Zabriskie, J. B. (1967). Mimetic relationships between group A streptococci and mammalian tissues. *Adv. Immunol., 7*:147–188.

Zabriskie, J. B., and E. H. Freimer (1966). An immunological relationship between the group A streptococcus and mammalian muscle. *J. Exp. Med., 124*:661–678.

POLYSACCHARIDES OF HEMOLYTIC STREPTOCOCCI*

RICHARD M. KRAUSE

The Rockefeller University, New York, New York

I. Introduction

Beginning with World War I and the concomitant mass epidemics of streptococcal disease, interest has focused on the antigens of hemolytic streptococci. Lancefield and others began the systematic study to uncover the antigenic mosaic of these organisms. One of the highlights of this sustained preoccupation was the recognition by LANCEFIELD (1940) that the hemolytic streptococci could be divided into distinct serologic groups on the basis of somatic carbohydrate antigens. From the work of McCARTY (1969) and others, it became clear that these somatic antigens were major components of the streptococcal cell wall. The major subcellular components of the streptococcus are shown in Figure 1. Both groups A and C streptococci possess an outermost capsule consisting of hyaluronic acid. The cell wall contains several important protein antigens including those designated M, T and R, but discussion of these is beyond the scope of this paper. Streptococcal cell walls, freed of all protein antigens by proteolytic enzyme treatment, contain two major elements: a group-specific carbohydrate antigen and the peptidoglycan matrix.

Two major areas of investigation are considered here. The first is concerned with the antigenic and immunochemical properties of the group-specific carbohydrates of groups A, A-variant, and C streptococci. The second area of investigation is concerned with the immunochemical properties of streptococcal peptidoglycan.

* This work was supported by National Institutes of Health grant AI-08429 and by a grant-in-aid from the American Heart Association. The work was also sponsored by the Commission on Streptococcal and Staphylococcal Diseases of the Armed Forces Epidemiological Board, and was supported in part by the U.S. Army Medical Research and Development Command, under research contract No. DADA 17-67-C-7043.

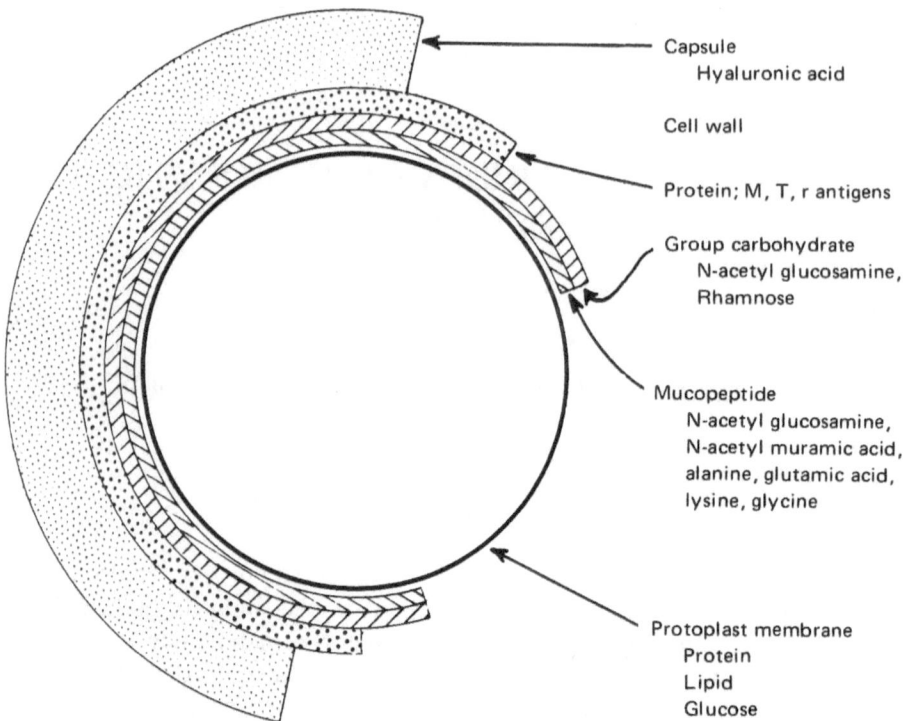

Fig. 1. Schematic cross-section of Group A hemolytic streptococci.

II. Group-Specific Carbohydrate

The carbohydrates of groups A, A-variant, and C streptococci have been studied in detail (McCARTY, 1971). The antigen extracted from the cell wall by chemical means is undoubtedly heterogeneous with respect to size. The group A antigen is depicted in Figure 2. Certainly a portion of the extracted material has a molecular weight in the range of 10,000. A mole of group A antigen possesses 17 moles of N-acetyl-glucosamine and 35 moles of rhamnose. Eleven moles of N-acetyl-glucosamine can be selectively stripped off with β-N-acetyl-glucosaminidase, leaving a residual polymer which is predominantly rhamnose. McCARTY (1970) has shown that the terminal β-N-acetylglucosaminide residues are the immunodominant determinants of specificity.

A companion carbohydrate to this one of group A has been isolated from group C streptococci (KRAUSE, 1963). The evidence suggests that group C carbohydrate has a similar branched rhamnose polymer but terminal α-N-acetyl-galactosaminide residues confer specificity. The potential antigenicity of the rhamnose moiety in the case of group A and group C carbohydrates is masked by the terminal amino sugar residues. A mutant of group A, termed group A-variant, however, lacks terminal amino sugar

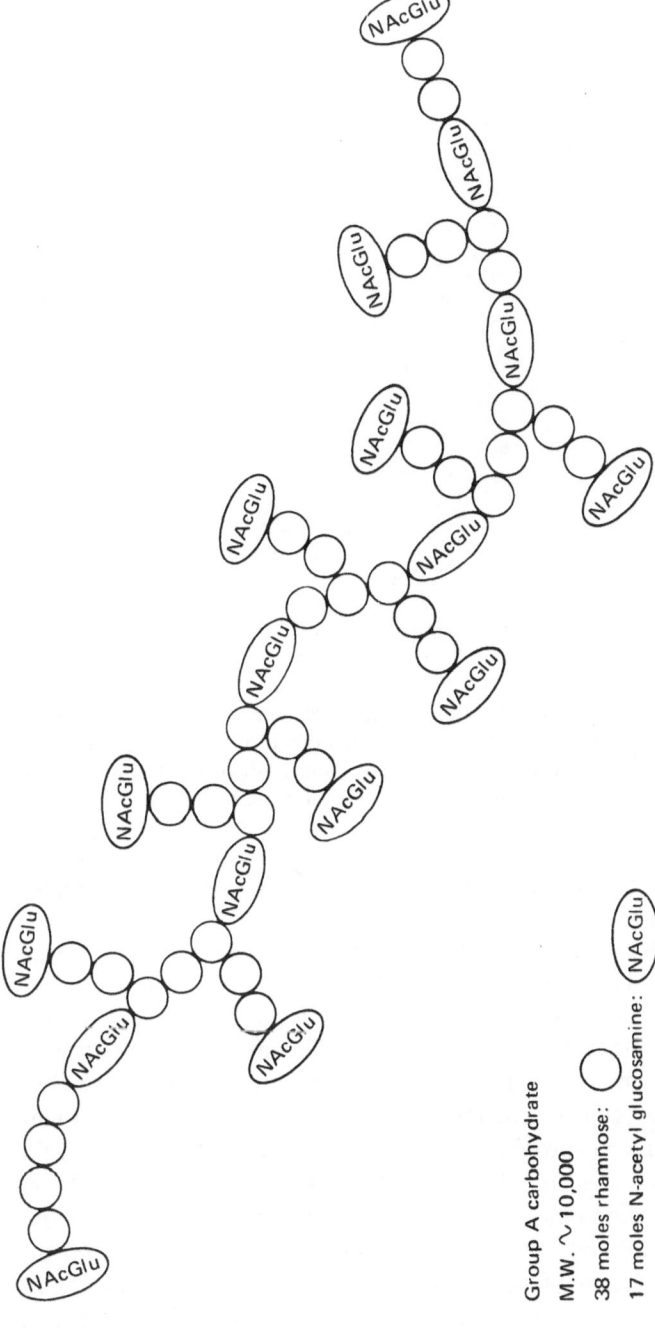

Group A carbohydrate

M.W. ∿ 10,000

38 moles rhamnose: ◯

17 moles N-acetyl glucosamine: (NAcGlu)

Fig. 2. Schematic diagram of the chemical structure of the Group A carbohydrate.

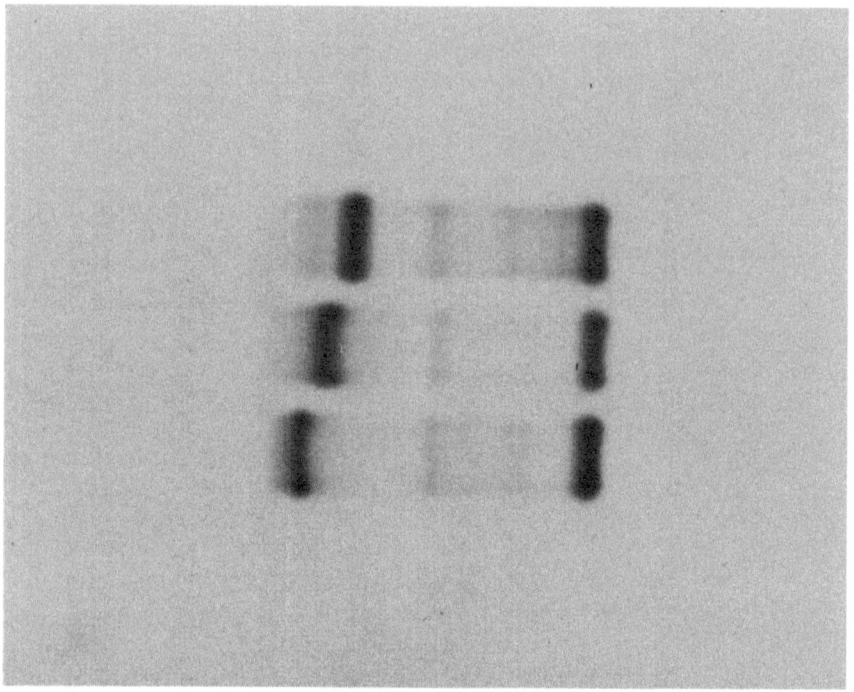

Fig. 3. Microzone electrophoretic patterns of antisera from three rabbits after intravenous immunization with Group C hemolytic streptococci. The monodisperse gammaglobulin components are to the left. Almost all of the gammaglobulin in each component is precipitating antibody to the Group C carbohydrate.

residues and, in this instance, the antigenicity resides in the rhamnose polymer itself (McCARTY, 1969).

Although this information of the immunochemistry of the streptococcal carbohydrate is not recent, it has been reviewed briefly here because of certain unusual features of the immune response to these antigens. It is a remarkable fact, for example, that antibodies to this family of antigens may have a molecular uniformity so striking that they resemble in this respect the myeloma proteins (KRAUSE, 1970). Depicted in Figure 3 are the microzone electrophoretic patterns of antisera of three rabbits immunized intravenously with group C streptococcal vaccine. The monodisperse gammaglobulin components in these sera contain the bulk of the antibodies. Antibody concentrations range between 30 and 60 mg/ml. It is not the purpose of this report to review this subject in detail. Suffice it to say that the dogma of conventional wisdom has held that the immune response to a single antigen is crowded with heterogeneous antibodies. For this reason, progress on the molecular structure of the immunoglobulin stems primarily from an examination of the homogeneous myeloma proteins and not from an examination of antibodies. It would now appear, however, that in selected cases, rabbit antibodies to certain microbial carbohydrates are sufficiently uniform so that sequence studies can be undertaken.

Evidence to support the view the antibodies to streptococcal carbohydrate have molecular uniformity is as follows. These antibodies migrate as one distinct component on microzone electrophoresis. The isolated light chains migrate as one distinct component on polyacrylamide gel electrophoresis. N-terminal analysis reveals only 1 amino acid alternative at the N-terminal position of the light chains. Data from sequence analysis, which employed the Beckman protein sequencer model 1890, indicated that there is a single amino acid sequence through the first 30 positions of the N-terminus of the light chains (HOOD *et al.*, 1970). With the means at hand to procure antibodies sufficiently uniform for complete sequence analysis, it should prove feasible to examine the topography of antigen-binding sites and to study the structure-function relationship between antigens and antibodies.

III. Peptidoglycan

In the remainder of this paper the immunochemistry of streptococcal peptidoglycan will be considered. Depicted in Figure 4 is an electron micrograph of cell wall-like structures devoid of all antigen materials and consisting solely of streptococcal peptidoglycan (KRAUSE and MCCARTY, 1961). These structures were obtained by extracting the cell walls with hot formamide at 180°C.

Recent evidence from a number of sources indicates that the cell wall peptidoglycan of gram-positive cocci are potentially antigenic (KARAKAWA and KRAUSE, 1966; HISATSUNE *et al.*, 1967; GOTSCHLICH and LIU, 1967). The studies of KARAKAWA and KRAUSE (1966) have shown that the peptide moiety of peptidoglycan is a major antigenic component. The hexosamine polymer is also antigenic (KARAKAWA *et al.*, 1967). Each of these components of peptidoglycan gives rise to distinct antibodies. The most recent studies of SCHLEIFER and KRAUSE (1971a, b) have focused on the antigenicity of the pentapeptide of streptococcal peptidoglycan. The pentapeptide has been synthesized by the solid phase synthesis technique and employed in quantitative precipitin inhibition tests. Such information, combined with inhibition data achieved with other peptides with other amino acid sequences, has indicated that the C-terminal D-ala-D-alanine is the immunodominant determinant of the pentapeptide of the peptidoglycan.

Because of the similar structure of the peptidoglycans of all bacteria, it comes as no surprise that there is extensive immunologic cross-reactivity among the microbial peptidoglycans and peptidoglycan antibodies (KARAKAWA *et al.*, 1968).

The interest of biochemists and microbiologists has centered on peptidoglycan, because this substance is the single most characteristic feature

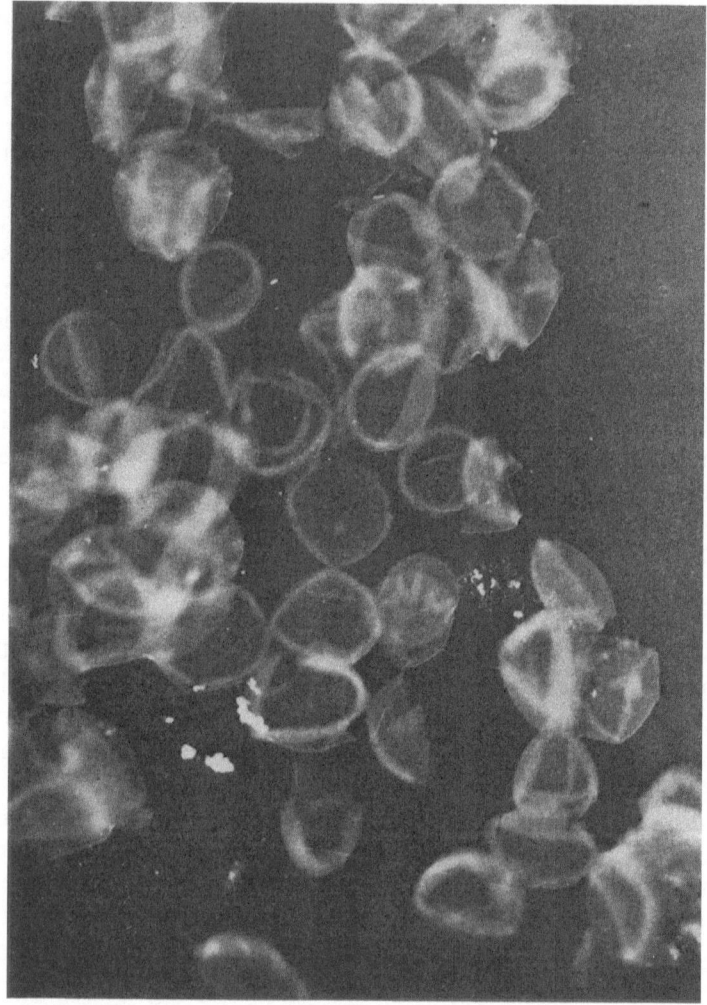

Fig. 4. Electronmicrograph of the hot formamide residue of Group C streptococcal cell walls. These structures consist entirely of peptidoglycan.

of all bacteria and because it is such an essential element for the structural integrity of microorganisms. But there are other reasons for interest in peptidoglycan. It has been shown that peptidoglycan and the endotoxins of gram-negative bacteria have biological properties in common, even though they do not have similar chemical properties (ROTTA *et al.,* 1965; ABDULLA *et al.,* 1966; OHANIAN and SCHWAB, 1967; OHANIAN *et al.,* 1969; ROTTA and BEDNAR, 1969). These biological properties are summarized in Table 1. Peptidoglycan produces fever in the rabbit and may initiate shock and death. A necrotic skin lesion develops in rabbits follow-

Table 1. Biologic and Toxic Properties of the Peptidoglycan

---

1. Antigenic   a. peptide moiety
              b. hexosamine polymer
              c. cross-link peptide bridge
2. Induces fever in the rabbit
3. Enhances nonspecific resistance in mice
4. Produces dermal necrosis in rabbits
5. Prepares and provokes the local Shwartzman reaction in rabbits
6. Intravenous injection causes carditis in rabbits
7. Lyses red blood cells and thrombocytes

---

ing local injection of peptidoglycan. Peptidoglycan will prepare and provoke the localized Shwartzman reaction. One of the most dramatic toxic properties of streptococcal peptidoglycan is the extensive carditis which occurs within three days after intravenous injection into the rabbit. The basis of this reaction remains a mystery. It is possible that these toxic manifestations are mediated through immunologic mechanisms due to the antigenicity of peptidoglycans. It is conceivable as well that the toxicity of peptidoglycan is due to a direct toxic reaction. It is provocative to note, for example, that peptidoglycan from some bacteria will lyse red blood cells and thrombocytes. While the mechanism of this lytic activity is obscure, the toxic manifestations observed after the administration of peptidoglycan to experimental animals may be related to the lytic effect of this substance for cellular membranes.

IV. Summary

The studies of numerous investigators on the cell wall antigen of hemolytic streptococci, and the immune response to them, have progressed on a broad front during the past several years. Certain features of the immune response to these carbohydrates were wholly unanticipated. It has been observed, for example, that the antibodies to the carbohydrate antigens produced in the rabbit may be as homogeneous as the myeloma proteins.

The immunochemistry of the bacterial mucopeptides is a recent advance and the fine structure of the antigenic determinants is now being explored in detail. This class of substances appears to have a peculiar toxicology. The basis for the *in vitro* and *in vivo* toxicity of mucopeptide has yet to be determined.

References

Abdulla, E. M., and J. H. Schwab (1966). Biological properties of streptococcal cell wall particles. III. Dermonecrotic reaction to cell wall mucopeptides. *J. Bacteriol., 91*:374–383.

Gotschlich, E. C., and T. Liu (1967). Structural and immunological studies on the pneumococcal C polysaccharide. *J. Biol. Chem., 242*:463–470.

Hisatsune, K., S. J. DeCourey, and S. Mudd (1967). Studies on the carbohydrate-peptide fraction of centrifugal supernatant of *Staphylococcus aureus* cultures. *Biochem., 6*:586.

Hood, L., K. Eichmann, H. Lackland, R. M. Krause, and J. J. Ohms (1970). Rabbit antibody light chains and gene evolution. *Nature, Lond., 228*:1040–1044.

Karakawa, W. W., D. G. Bruan, H. Lackland, and R. M. Krause (1968). Immunochemical studies on the cross-reactivity between streptococcal and staphylococcal mucopeptide. *J. Exp. Med., 128*:325–340.

Karakawa, W. W., and R. M. Krause (1966). Studies on the immunochemistry of streptococcal mucopeptide. *J. Exp. Med., 124*:155–171.

Karakawa, W. W., H. Lackland, and R. M. Krause (1967). Antigenic properties of the hexosamine polymer of streptococcal mucopeptide. *J. Immunol., 99*:1178–1182.

Krause, R. M. (1963). Symposium on relationship of structure of microorganisms to their immunological properties. IV. Antigenic and biochemical composition of hemolytic streptococcal cell walls. *Bacteriol. Rev., 27*:369–380.

—— (1970). Factors controlling the occurrence of antibodies with uniform properties. *Fed. Proc., 29*:59–65.

Krause, R. M., and M. McCarty (1961). Studies on the chemical structure of the streptococcal cell wall. I. The identification of a mucopeptide in the cell walls of groups A and A-variant streptococci. *J. Exp. Med., 114*:127–140.

Lancefield, R. C. (1940). The specific relationship of cell composition to biological activity of hemolytic streptococci. The Harvey Lectures, Series 36, p. 251.

McCarty, M. (1971). The streptococcal cell wall. The Harvey Lectures, Series 65, pp. 73–96.

Ohanian, S. H., and J. H. Schwab (1967). Persistence of group A streptococcal cell walls related to chronic inflammation of rabbit dermal connective tissue. *J. Exp. Med., 125*:1137–1148.

Ohanian, S. H., J. H. Schwab, and W. J. Cromartie (1969). Relation of rheumatic-like cardiac lesions of the mouse to localization of group A streptococcal cell walls. *J. Exp. Med., 129*:37–49.

Rotta, J., and B. Bednar (1965). Biological properties of cell wall mucopeptide of hemolytic streptococci. *J. Exp. Med., 130*:31–47.

Rotta, J., T. J. Prendergast, W. W. Karakawa, C. K. Harmon, and R. M. Krause (1965). Enhanced resistance to streptococcal infection induced in mice by cell wall mucopeptide. *J. Exp. Med., 122*:877–890.

Schleifer, K. H., and R. M. Krause (1971a). The immunochemistry of peptidoglycan. I. The immunodominant site of the peptide subunit and the contribution of each of the amino acids to the binding properties of the peptides. *J. Biol. Chem., 25*:986–993.

—— (1971b). The immunochemistry of peptidoglycan. II. Separation and characterization of antibodies to the glycan and to the peptide subunit. *Eur. J. Biochem., 19*:471–478.

IMMUNIZING ANTIGENS OF
MYCOBACTERIA*

GUY P. YOUMANS and ANNE S. YOUMANS

Department of Microbiology, Northwestern University
Medical School, Chicago, Illinois

I. Introduction

It has been recognized for years that when animals, including man, are infected with tubercle bacilli they will become much more resistant to reinfection. Furthermore, animals, if vaccinated with viable cells of attenuated mycobacterial strains, will acquire a much greater resistance to infection. This actively acquired immune state is characterized not only by an increased capacity of phagocytic cells (macrophages) to kill tubercle bacilli, but also by the development of the ability of these cells to inhibit the intracellular multiplication of the parasite. In fact, intracellular bacteriostasis is the *major* manifestation of acquired immunity to tuberculosis, and tubercle bacilli may remain viable but nonmultiplying *in vivo* for years. It is important to emphasize that the maintenance of intracellular bacteriostasis is not due to the activity of a conventional circulating antibody (RAFFEL, 1955). It appears to be mediated solely by cells (SUTER and RAMSIER, 1964; PATTERSON and YOUMANS, 1970), although involvement of a cell-bound antibody has not been ruled out.

It is also essential, before discussing the nature of mycobacterial immunogens, to point out that it has recently been clearly shown that the cellular acquired immune response in tuberculosis to viable attenuated mycobacterial cells involves at least three mechanisms. These are (1) a stimulation of the reticuloendothelial system, which is usually referred to as "macrophage activation"; (2) the rapid accumulation and proliferation of macrophages (the granulomatous response) (YOUMANS and YOUMANS, 1964); and (3) a lymphocyte-mediated immune response of unknown nature (PATTERSON and YOUMANS, 1970). These protective responses and

* This investigation was supported by Public Health Service research grant AI-01636 from the National Institute of Allergy and Infectious Diseases, The Canal Zone Tuberculosis Association, and The Tuberculosis Institute of Chicago and Cook County.

certain responses we consider nonprotective are listed in Table 1. Reference to tuberculin hypersensitivity is made again later in this paper. The increased resistance due to macrophage activation and the granulomatous response usually is not great, nor long-lasting, and it is nonspecific. The lymphocyte-mediated immunity, on the other hand, is specific, long-lasting, and of greater potency.

Table 1. Host Responses to Mycobacterial Components

Protective: (1) macrophage activation; (2) granulomatous response; (3) specific lymphocyte mediate immuntiy.
Nonprotective: (1) tuberculin hypersensitivity; (2) circulating antibody (IgM, IgG, etc.); (3) cytophylic antibody?

Finally, we must recognize that these multiple mechanisms of increased resistance to tuberculous infection may be activated by different components of the mycobacterial cell (YOUMANS and YOUMANS, 1969a). It will be the purpose of this paper to describe certain of these components and to define which immune mechanisms each stimulates. Our major purpose, however, will be to provide information on the nature of the material found in the tubercle bacillus which we have found to be responsible for the induction of the specific, long-lasting, lymphocyte-mediated immune mechanism.

II. The Immunizing Capacity of Viable and Killed Mycobacterial Cells

In order to understand the reasons for our approach to the problem of the isolation of the major immunogen from mycobacterial cells it is essential to know that when using whole cells for immunization, they must be viable in order to induce an appreciable degree of immunity. Viable cells are from 100 to 1000 times more effective as immunizing agents (YOUMANS and YOUMANS, 1969b). The usual explanation for this phenomenon is that the viable cells multiply *in vivo* and produce adequate amounts of immunogen. This explanation does not apply to our experimental situation in which the attenuated H37Ra strain is used to vaccinate CF-1 strain mice. These avirulent mycobacterial cells do not multiply in this mouse strain; instead they are rapidly killed (SEVER and YOUMANS, 1957).

If the difference between the immunizing capacity of viable and killed attenuated mycobacterial cells cannot be accounted for on a quantitative basis, then there must be a qualitative difference, such as the presence of a potent but labile immunizing substance in viable cells. This, as will be seen, we have been able to demonstrate by experiment. However, a number of other substances from the tubercle bacillus are immunogenic in the hands of some investigators (Table 2), particularly preparations in which the predominant component appears to be a polysaccharide, such as the trypsin-extracted immunogen of CROWLE (1969) or lipopolysaccharide. The immunizing capacity of such preparations, however, usually does not exceed that of killed cells. Probably many of these substances owe their

Table 2. Mycobacterial Products which Induce Increased Resistance to Infection with Virulent Tubercle Bacilli

---

(1) Miscellaneous (polysaccharides, lipopolysaccharides, Wax-D, etc.); (2) trypsin-extracted immunogen (probably a polysaccharide); (3) mycobacterial oil-cell wall vaccine; (4) cord factor (Trehalose, 6, 6'Dimycolate); (5) ribosomal and RNA preparations.

---

activity either to macrophage activation or to the granulomatous response, or both.

III. Mycobacterial Cell Walls

We have shown that mycobacterial cell walls have only a weak capacity to induce increased resistance to tuberculous infection in mice when the mice are vaccinated intraperitoneally and challenged intravenously. Drs. Larson and Ribi and coworkers, however, have shown that mycobacterial cell walls under certain conditions can induce an appreciable degree of increased resistance to tuberculous infection in mice and in monkeys (LARSON *et al.*, 1963; RIBI and LARSON, 1964; ANACKER *et al.*, 1967; BREHMER *et al.*, 1968). This work has recently been reviewed by KANAI (1967) and SMITH *et al.* (1968).

These investigators found, when using pulmonary challenge of mice with aerosolized virulent tubercle bacilli, that an immune response to viable or heat-killed BCG cells could only be detected when the route of vaccination was intravenous. Cell walls prepared from BCG and suspended in water were inactive. Cell walls prepared from cells disrupted by passing through a pressure cell while suspended in mineral oil, or cell walls prepared from cells disrupted in water and then dried and resuspended in mineral oil, however, were immunogenic, but again only if administered by the intravenous route. On the other hand, these same substances were shown to be immunogenic when vaccination was by the subcutaneous or intraperitoneal route, provided the challenge injection with virulent mycobacteria was given intravenously.

These apparently contradictory results find a ready explanation in findings of BARCLAY *et al.* (1967), who reported that the intravenous injection of mycobacterial oil-cell wall vaccine induced an intense granulomatous response in the lungs of mice. It is well known that the intravenous injection of viable, or killed mycobacterial cells into mice or rabbits will induce a pulmonary granulomatous response (YOUMANS and YOUMANS, 1964). Broken mycobacterial cells, or cell components isolated by centrifugation from broken cells, such as intracellular fluid, ribosomes and/or other cellular particulates, have little capacity to do so. With the initiation of the granulomatous response, there is an increase in resistance to tuberculous infection which, in turn, is directly proportional to the magnitude of the granulomatous response. It has been shown that mineral oil increases the pulmonary granulomatous response produced by whole cells and that the macrophages which accumulate are metabolically more active and have a greater capacity to destroy mycobacterial cells (LEAKE *et al.,* 1966). Therefore, the granulomatous response can account completely for the increased resistance noted following intravenous vaccination of mice with oil-cell wall preparations, whether challenge is by the pulmonary or intravenous route. There is a strong possibility that the granulomatous response obtained following the intravenous injection of oil-cell wall preparations may be due to the "cord factor" present in the cell walls.

IV. The "Cord Factor"

BEKIERKUNST (1968) and BEKIERKUNST *et al.* (1969) have recently provided the information that "cord factor" (trehalose-6,6'-dimycolate) is the major substance in mycobacterial cells responsible for the induction of the granulomatous response. The name "cord factor" has been given to this substance because at one time it was thought to be responsible for the manner in which virulent mycobacterial cells stuck together and grew in cords. Now it is recognized that many mycobacteria, including some saprophytes, may contain some "cord factor." Dr. Bekierkunst has shown that "cord factor," when dissolved in mineral oil and injected intravenously into mice in the form of an oil in water emulsion, will induce as good a granulomatous response as intact mycobacterial cells. The mineral oil is essential, as is injection by the intravenous route; these findings are in keeping with what is observed when mycobacterial oil-cell wall vaccine is used. Animals in which a granulomatous response had been induced with "cord factor" were not only more resistant to infection with virulent mycobacteria but to a variety of other pathogenic bacteria as well. In other words, as would be expected, the increased resistance to infection conferred by the granulomatous response is nonspecific.

V. Mycobacterial Ribosomal and RNA Preparations

We have spent quite a few years testing the hypothesis that the greater immunizing capacity of viable cells might reside in their possession of a highly labile but potent immunizing substance (YOUMANS and YOUMANS, 1964, 1969a, b, c, 1970). Because of technical problems and the prolonged period required to assay immunizing activity, progress has been slow, but has been crowned with complete success. Our many results can only briefly be summarized here.

Viable cells of the H37Ra strain were employed in our studies. These cells were ruptured mechanically by passing them through a French pressure cell. This avoided the harsh chemical extraction procedures customarily employed. The ruptured cell mass originally was fractionated by differential centrifugation into mycobacterial cell protoplasm, cell walls, and internal structures such as cytoplasmic membranes, particles, ribosomes, etc. Each of these was tested to determine its capacity to immunize mice to subsequent intravenous challenge with virulent tubercle bacilli. The first and very surprising finding was that most of the immunizing material was found in the fraction containing the intracellular components. This we called particulate fraction. The cytoplasmic fluid was inactive and the

Table 3. Response of CF-1 Mice to Immunization with Mycobacterial Intracellular Particles and Cell Walls

Cell walls				Particulate fraction			
Amount injected Mg*	Number of mice	Number of S-30 mice	Per-cent of S-30 mice†	Amount injected Mg	Number of mice	Number of S-30 mice	Per-cent of S-30 mice†
40	32	3	9	40	16	7	44
30	10	0	0	30	84	49	58
20	30	1	3	20	20	8	40
10	30	5	17	10	97	59	61
4	18	2	11	5	40	16	40
2	39	3	8	2.5	20	1	5
				0	139	7	5
H37Ra	193	131	68				

* Moist weight.

† S-30 mice = percent of mice which survived >30 days. Modified from KANAI and YOUMANS, *J. Bact. 80*:607–614, 1960.

cell walls had only slight immunizing activity. Table 3 shows the immunizing activity of cell walls and particulate fraction. The particulate material was found to be very unstable, and it was difficult to obtain reproducible results; therefore, early studies were directed toward obtaining a more stable immunogenic particulate fraction. Also, efforts were made to increase the immunizing potency and to determine the nature of the immunizing ingredient. However, the last was not feasible until the stability of the particulate fraction was such that it could be subjected to greater manipulation.

It was found that the immunogenic activity was decreased or abolished by even a moderate increase in temperature, by a pH other than 7.0, and by surface active agents. Also, using 0.44 M sucrose buffer for breakage of the cells instead of 0.25 M, gave more active preparations, as did the addition of 3×10^{-2} M $MgCl_2$ to the sucrose solution. A comparison of the potencies of the preparations shown in Table 4 with those shown in

Table 4. Effect of 0.44 M Sucrose and Magnesium Ions on the Immunogenicity of Particulate Fraction

Immunizing preparation	Amount injected Mg*	Number of mice	Number of S-30 mice	Percent of S-30 mice†
Sucrose (0.25 M)	20.0	271	102	38
	5.0	224	44	20
Sucrose (0.44 M)	20.0	300	156	52
	5.0	309	117	38
	1.0	114	13	11
Sucrose (0.44 M) plus MgCl₂ (3×10^{-2} M)	20.0	349	245	70
	5.0	136	63	46
	1.0	114	40	35
H37Ra cells	1.0	580	324	56
None	—	595	39	7

* Moist weight.

† S-30 mice = percent of mice which survived >30 days. Modified from YOUMANS and YOUMANS, *J. Bact.* 87:1346–1350, 1964.

Table 3 reveals the very substantial increase in activity obtained when all procedures during preparation were carried out at 0 to 4°C, at a neutral pH, and in 0.44 M sucrose containing 3×10^{-2} M $MgCl_2$. Apparently, greater stability had been achieved since there was less loss of immunizing activity during preparation.

We wished to localize the immunogenic component further; therefore, this more stable and active particulate fraction was resuspended in sucrose magnesium chloride buffer and fractionated by differential centrifugation according to the procedure indicated in Table 5. The pellets from each centrifugation were resuspended and injected intraperitoneally into mice to determine their capacity to immunize against tuberculous infection. From the data shown in Table 5 it appeared that all of the immunizing activity was located in the larger particles which sedimented at

Table 5. Effect of Differential Centrifugation on the Immunogenicity
of Particulate Fraction

Immunizing preparation	Amount injected Mg*	Number of mice	Number of S-30 mice	Percent of S-30 mice†
Particulate fraction				
(i) 144,700 × g	20.0	262	175	67
(ii) 144,700 × g, pellet suspended and recentrifuged at				
56,550 × g	20.0	198	124	63
100,000 × g	20.0	106	11	10
144,700 × g	20.0	135	13	10
H37Ra cells	1.0	258	139	54
None	—	264	8	3

* Moist weight.

† S-30 mice = percent of mice which survived >30 days. Modified from
YOUMANS and YOUMANS, *J. Bact. 88*:1030–1037, 1964.

55,550 × g. Experiments also showed that treatments such as freezing,
followed by rapid thawing, or sonic oscillation, destroyed the immunogenicity of both the particulate fraction and the 55,550 × g fraction, as
did treatment of these preparations with sodium dodecyl sulfate (SDS)
or sodium deoxycholate. All of these treatments would disrupt membranous
structures or organelles. Therefore, the integrity of some structure seemed
necessary for immunogenicity.

We tested the effect of Freund's incomplete adjuvant on the immunogenicity of the particulate fraction, the 55,550 × g fraction, and the
144,700 × g ribosomal fraction obtained by differential centrifugation of
the particulate fraction. Table 6 shows that the ribosomal fraction, which
is only slightly immunogenic in the largest dose if injected alone, when
incorporated into incomplete adjuvant was fully as immunogenic as the
particulate fraction from which it was derived. Of particular importance
is that the immunogenicity of the particulate fraction itself and the
55,550 × g fraction was not increased by the adjuvant, showing that the
Freund's incomplete adjuvant was not merely promoting the activity of
a substance in particulate fraction which was immunogenic only in this
adjuvant.

The most probable explanation for this phenomenon is that the membranous material present in the particulate fraction and the 55,550 × g
fraction acted as an adjuvant, or protective substance, for the ribosomes,
which are known to occur both free and membrane-bound in the cell. The
free ribosomes would be concentrated in the 144,700 × g fraction, whereas
the membrane-bound ribosomes would occur in the 55,550 × g fraction.

Table 6. Immunogenic Activity of the Particulate Fraction and a Ribosomal Fraction, with and without Freund's Incomplete Adjuvant

Immunizing preparation	Freund's adjuvant added	Amount injected Mg*	Number of mice	Number of S-30 mice	Percent of S-30 mice†
Particulate	−	20	126	90	71
fraction	+	20	94	65	69
	−	1	112	40	36
	+	1	84	33	39
Ribosomal fraction	−	20	117	45	39
(144,700 × g)	+	20	120	81	68
	−	1	29	3	10
	+	1	52	18	35
H37Ra cells		1	115	93	81
None		0	118	4	3

* Moist weight.

† S-30 mice = percent of mice which survived >30 days. Modified from YOUMANS and YOUMANS, *J. Bact. 89*:1291–1298, 1965.

Further experience confirmed this, for if the particulate fraction, or the 55,550 × g fraction was treated with SDS or sodium deoxycholate, immunogenicity was lost, but could be fully restored by the use of Freund's incomplete adjuvant. The same restoration of full immunogenicity by incomplete adjuvant was seen following the destruction of immunizing power of the same fractions by the physical agents previously mentioned. Electron photomicrographs of these ribosomes have shown that the particles are approximately 20 mμ in diameter (YOUMANS and YOUMANS, 1969a).

These results established that some component of the ribosomal subfraction of the particulate fraction comprised the labile immunogen. The results also pointed the way to a greatly simplified and more efficient method for the preparation of immunogenic ribosomal fractions. The particulate fraction was prepared as usual and then treated with an equal volume of 0.5% SDS. This removed membranous material, much of the protein, and also inhibited the activity of endogenous ribonuclease (RNAase). This solution as recentrifuged for 3 hours at 144,700 × g to sediment both free and formerly membrane-bound ribosomes. Finally, the ribosomal fraction was collected, homogenized, and resuspended in 0.01 M phosphate buffer, pH 7.0, containing 10^{-4} M MgCl$_2$. Such preparations were 100 times more immunogenic than ribosomal material obtained only by differential centrifugation (Table 7). These ribosomes were analyzed chemically and found to contain 67% RNA and 33% protein. No lipopolysaccharides, DNA or sugars other than pentose were detected chemically. These ribosomes were characterized in 5 to 20% sucrose gradients using 10^{-1} M MgCl$_2$ and typical 70S particles were obtained (Figure 1).

Table 7. A Comparison of Immunogenic Activities of Ribosomal Fractions
Prepared by Differential Centrifugation and by Using
Sodium Dodecyl Sulfate

Immunizing preparation	Amount injected Mg*	Number of mice	Number of S-30 mice	Percent of S-30 mice†
Ribosomes prepared by	20.0	120	81	68
differential centrifugation	1.0	52	18	35
Ribosomes prepared by	1.0	175	134	77
using SDS	0.1	174	86	49
	0.01	175	53	30
H37Ra cells	1.0	115	93	81
Controls	—	118	4	3

* Moist weight.

† S-30 = percent of mice which survived >30 days. Modified from YOUMANS
and YOUMANS, *J. Bact. 91*:2139–2145, 1966.

These broke down into 50S and 30S subunits when the magnesium ion
concentration was lowered to 10^{-4} *M*.

Mycobacterial RNA has been prepared from ribosomal fractions by
using a modification of the method of CRESTFIELD *et al.* (1955), in which
the RNA is precipitated by ethanol. Chemically, again only RNA and pro-

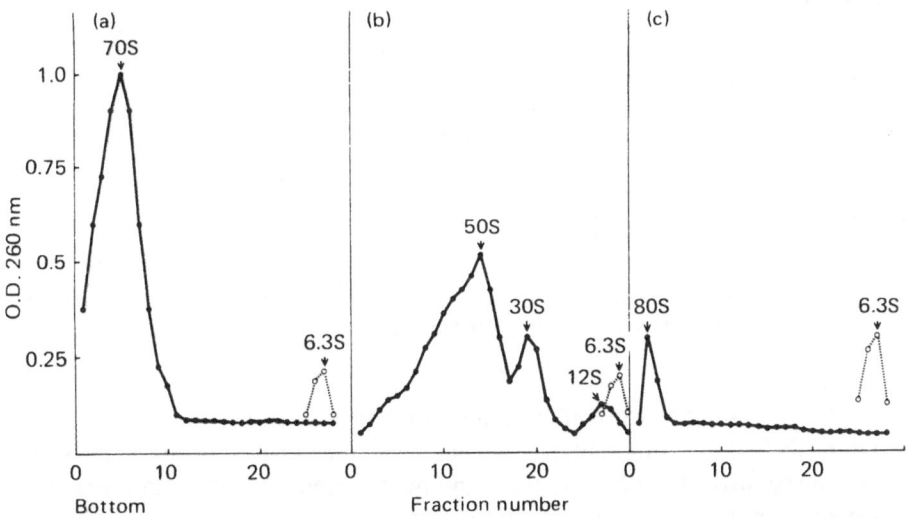

Fig. 1. Sedimentation patterns of (a) 70S mycobacterial ribosomes, sedimented in
a 5 to 20% sucrose gradient containing 10^{-1} *M* MgCl₂; (b) 70S ribosomes, dialyzed
against 10^{-4} *M* MgCl₂ overnight, and sedimented in a 5 to 20% sucrose gradient
containing 10^{-4} *M* MgCl₂; and (c) 80S mammalian brain ribosomes sedimented in a
5 to 20% sucrose gradient containing 10^{-2} *M* MgCl₂. Alkaline phosphatase marker, 0.
Reprinted from *Infect. Immun.* 1970, 2:659–668.

tein were present; no DNA or polysaccharide could be detected. Protein consisted of approximately 20 to 40% of the RNA present. The UV absorption peak was 258 mμ and the 260:280 ratio was approximately 2. Sucrose gradient determinations generally revealed only one peak, and the S value of this peak was approximately 16. If careful attention is paid to detail in order to avoid degradation during preparation, the immunogenic activity of this RNA is similar to that obtained with viable H37Ra cells. As little as 0.5 μg of RNA will protect 50 to 60% of vaccinated mice and, as with the ribosomal fraction, at times as little as 0.05 μg of RNA will be immunogenic. It is important to emphasize here that careful quantitative assessments of immunizing potency revealed that on a comparative nucleic acid weight basis these ribosomal, and RNA protein preparations, were as effective as immunizing agents as intact viable H37Ra cells. These comparisons are shown in Table 8.

Table 8. Immunogenic Activity of Mycobacterial Ribosomal and RNA Preparations and Viable H37Ra Cells

Immunizing preparation	RNA injected, μg	Number of mice	Number of S-30 mice	Percent of S-30 mice*	Range between experiments
Ribosomal fraction	50.0	334	191	87	76–100
	5.0	303	194	64	47–86
	0.5	252	123	49	21–86
RNA preparation	50.0	239	193	81	73–89
	5.0	234	171	73	64–90
	0.5	157	87	55	47–78
H37Ra cells	50.0	272	229	84	77–100
	5.0	236	201	85	72–97
	0.5	239	156	65	60–87
Controls		270	42	16	

* S-30 mice = percent of mice which survived >30 days. Modified from YOUMANS and YOUMANS, *J. Bact*, *99*:42–50, 1969.

At this moment we cannot state with certainty whether the mycobacterial RNA is the immunogen or whether some other substance, possibly complexed with the RNA, is the immunizing agent. We can state that the protein present in these preparations is not involved, because the elimination of protein with proteolytic enzymes does not affect immunizing activity. In addition, treatment of the RNA preparations with large amounts of RNAase appreciably reduces immunizing activity, and this loss is accompanied by partial degradation of RNA. Furthermore, complete degradation of the RNA with weak alkali completely destroys the capacity to immunize. There is also a high correlation between nativeness (lack of degradation) of the RNA and potency as an immunogen (YOUMANS and YOUMANS, 1969c). Finally, for maximum activity, the RNA apparently must have a complex secondary or tertiary structure, or exist in a complete or partial double-stranded form (YOUMANS and YOUMANS, 1970).

The manner in which these mycobacterial RNA preparations induce immunity to infection with *Mycobacterium tuberculosis* is unknown. Nucleotides are known to be potent stimulators of immune responses. We have found that these mycobacterial RNA preparations are as effective potentiators of the antibody response in mice to bovine gamma globulin as poly (A:U). Mycobacterial RNA is not granulomatogenic. Whether it may produce an unusual degree of macrophage activation has not been determined. In this connection it is important to point out that the immunity produced by these RNA preparations is specific. At least, no significant increase in resistance to infection with *Listeria monocytogenes* or *Klebsiella pneumoniae* can be detected following vaccination with mycobacterial RNA.

Of particular importance to current concepts of the nature of antimicrobial cellular immunity is that extensive experiments using both mice and guinea pigs have failed to reveal any evidence of tuberculin hypersensitivity following injection of mycobacterial ribosomal and RNA preparations. Thus these findings show that high degrees of immunity to tuberculous infection can be produced without the concomitant production of detectable tuberculin hypersensitivity. This supports the work of RAFFEL (1950), who showed that a high degree of tuberculin hypersensitivity can be produced in guinea pigs by the injection of Wax D and tuberculoprotein without any increase in resistance to tuberculous infection. These results indicate that tuberculin hypersensitivity does not play a large role in acquired cellular immunity to tuberculosis.

It is of more than passing interest to point out that immunogenic ribosomal and RNA preparations also have been isolated from *Salmonella typhimurium* by other investigators (VENNEMAN et al., 1968; VENNEMAN and BIGLEY, 1969; VENNEMAN *et al.,* 1970), from *Staphylococcus aureus* by WINSTON and BERRY (1970a), from *Pseudomonas aeruginosa* by WINSTON and BERRY (1970b), and from *Diplococcus pneumoniae* by THOMPSON and SNYDER (1971). Mycobacteria, therefore, are not unique in this respect.

Since the immunizing capacity of viable attenuated mycobacterial cells resides in the mycobacterial RNA preparations, this raises the question as to how animals immunized with RNA bring about the intracellular inhibition of multiplication of virulent tubercle bacilli. We have obtained information bearing on this point from the use of macrophage tissue cultures infected with virulent tubercle bacilli. This tissue culture system provides a means whereby the interaction between cell and parasite can be directly observed, and the effect of various conditions or substances on the host-parasite interaction can more easily be measured. Using such a tissue culture system we have shown that, in the absence of streptomycin in the tissue culture fluid, virulent tubercle bacilli will grow just as rapidly within macrophages obtained from mice immunized with viable attenuated mycobacterial cells as within macrophages obtained from nonimmunized animals. This is illustrated in Figure 2, (PATTERSON and YOUMANS 1970a). However, when splenic lymphocytes from immunized animals are added to the infected macrophages, marked inhibition of intracellular multiplication of the parasite occurs. This is illustrated in Figure 3, (PATTERSON and YOUMANS 1970b). Similar inhibition of intracellular multiplication will occur when the supernatant fluid, from immune lymphocyte cultures which

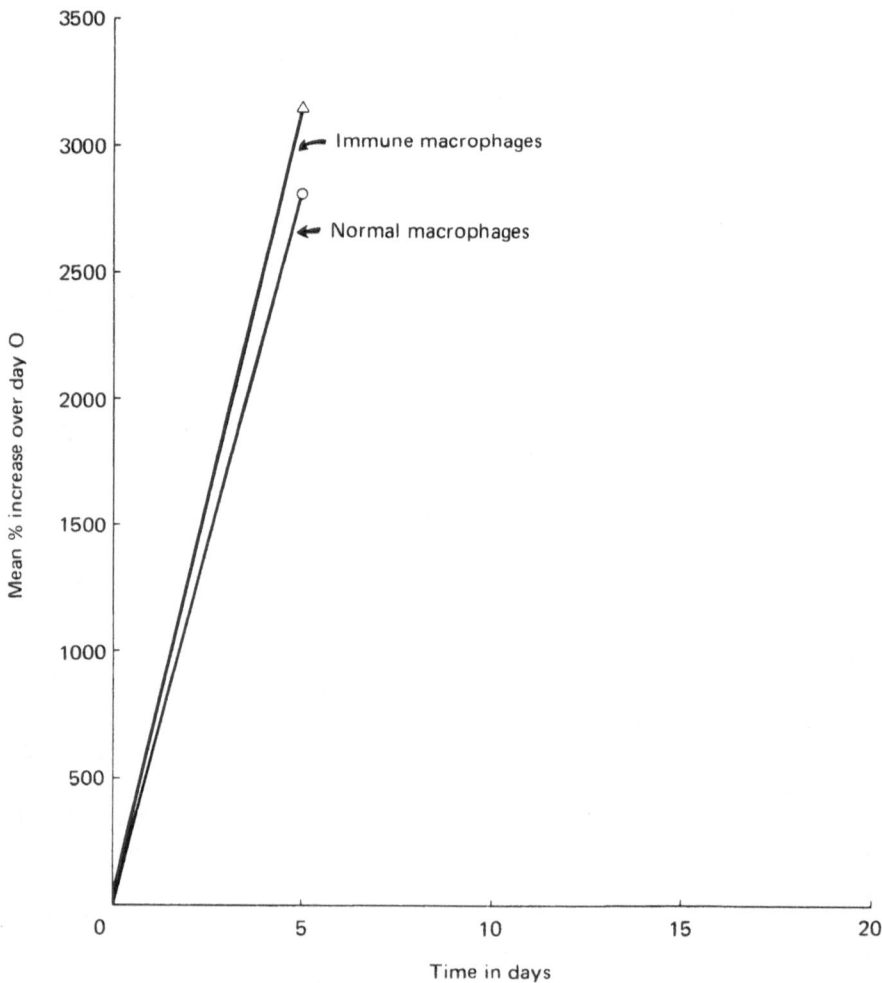

Fig. 2. Mean percentage increases of virulent H37Rv cells within normal and "immune" macrophages. Modified from *Infect. Immun.* 1970, *1*:30–40.

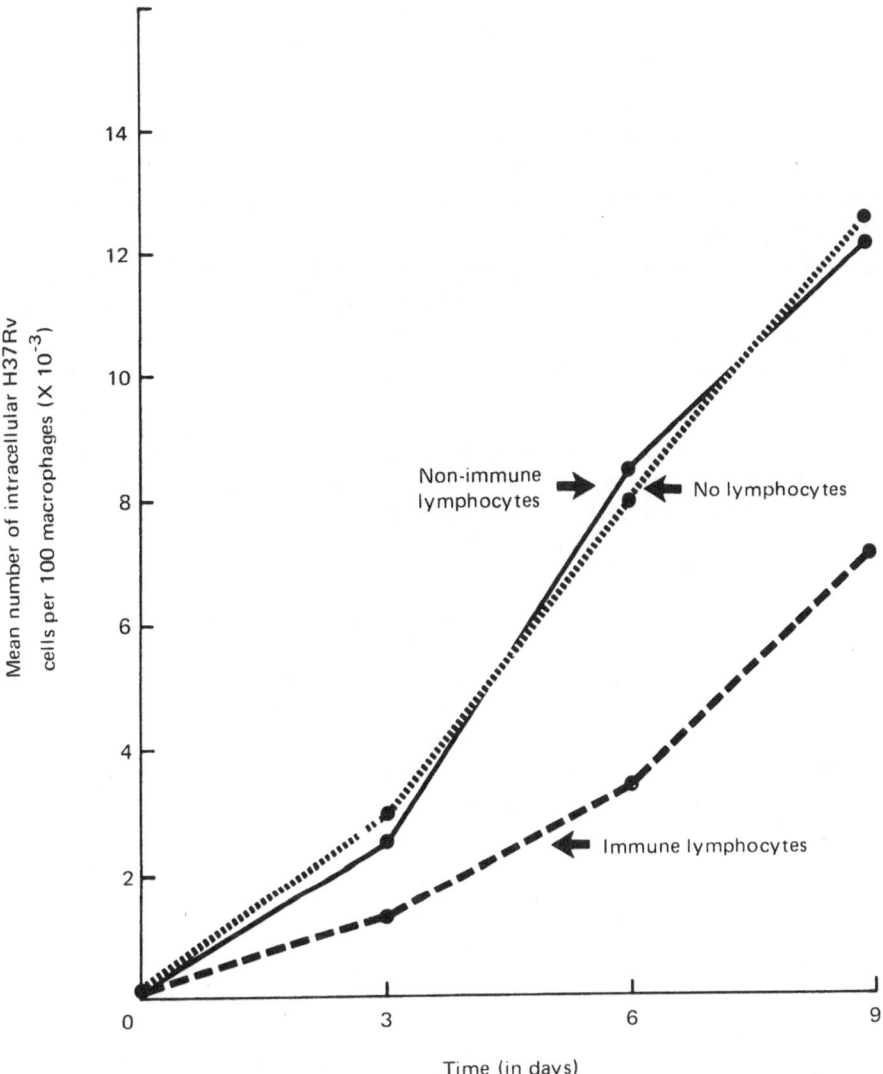

Fig. 3. Growth of *M. tuberculosis* (H37Rv) within normal macrophages exposed to splenic lymphocytes from immunized and nonimmunized mice. Reprinted from *Infect. Immun.* 1970, *1*:600–603.

have been stimulated with mycobacterial cells or mycobacterial products, is added to infected macrophages. Thus there is no "immune" macrophage; rather, it is the "immune" lymphocyte that is the affector cell.

Using this tissue culture system, information also has been obtained upon the nature of the acquired immune response engendered by some of the mycobacterial immunizing agents to which we have already referred. We have postulated that the major immunizing component of viable atten-

uated mycobacterial cells resides in the labile RNA material we have isolated. We have also suggested that the low-grade immune response obtained with killed mycobacterial cells or cell wall constituents resides
mainly in their capacity to induce the granulomatous response and to produce macrophage activation. Therefore, since the major immune mechanism, activated by viable attenuated cells, appears to be lymphocyte-
mediated, splenic lymphocytes obtained from animals immunized with our
RNA preparations should be effective inhibitors of intracellular mycobacterial proliferation in our macrophage tissue culture system, whereas lymphocytes from animals immunized with heat-killed cells (labile antigen destroyed) should not be active. This has been confirmed by experiment,
and the data are shown in Table 9. Only splenic lymphocytes obtained

Table 9. Percent Reduction in the Numbers of H37Rv Cells within
Macrophages Treated with Lymphocytes From Immunized Animals
and the Survival of Comparably Immunized Animals With
Virulent Tubercle Bacilli

Number of lymphocytes	Source of Lymphocytes			
	Non-immunized mice	Heat-killed H37Ra immunized mice	Viable H37Ra immunized mice	RNA immunized mice
10^7	0	17	not done	40
5×10^7	26	20	55	70
10^8	36	27	not done	72
In vivo challenge	23*	30*	100*	79*

* Percent of mice surviving 30 days following challenge with H37Rv. Modified
from PATTERSON and YOUMANS, Infection and Immunity, *1*:600–603, 1970.

from animals immunized with viable H37Ra cells or immunized with RNA
preparations were effective. Included also are data showing the high
immune response induced by injecting these same viable H37Ra cells and
RNA preparations into mice, and the low immune response following the
injection of killed cells (PATTERSON and YOUMANS, 1970b). Thus the *in
vitro* tissue culture data correlate well with that obtained by challenge of
animals with virulent tubercle bacilli.

The above conclusions, derived from the evidence given in this review,
provide a great deal of clarification for the whole area of acquired cellular
immunity to tuberculosis and, by analogy, cellular immunity to many other
facultative and obligate intracellular parasites. The realization that mycobacterial cells contain a number of immunizing substances provides the
basis for defining and measuring the potency of each substance individ-

ually. The realization that there are several immune responses not only should help to resolve discrepancies in results of different investigators but should permit investigators to examine more carefully the nature of each mechanism. It should permit an assessment of the contribution of each to the total immune response obtained when either viable cells are used for immunization or when the immune state is reached after recovery from actual infection.

The realization that the major specific immunogen is a labile intracellular substance will permit attention to be focused on this agent until its exact nature and mode of action are defined. The dissociation of tuberculin hypersensitivity from acquired immunity to infection eliminates the necessity for providing an explanation for the paradox whereby an acute, destructive, inflammatory process, such as tuberculin hypersensitivity, can be responsible for a protective effect that results in the intracellular inhibition of multiplication of tubercle bacilli. Many of these factors have been

Table 10. Protective Mechanisms Induced by Various Immunizing Preparations

Immunizing preparations	Macrophage activation	Pulmonary granulomatous response	Specific lymphocyte mediated immunity
1. Killed cells	+	+	0
2. Living cells	+	+	+++
3. Miscellaneous products	+	+	0
4. Trypsin-extracted immunogen	?	?	?
5. Oil-cell wall vaccine	+	+	0
6. Cord factor	+	+	0
7. Ribosomal and RNA preparations	?	0	+++

covered in greater detail in a recent review by YOUMANS and YOUMANS (1969a).

Finally, the demonstration that the lymphocyte rather than the macrophage is the affector cell for the major mechanism of acquired immunity in tuberculosis, brings the phenomenon of acquired cellular immunity in this disease into the mainstream of modern immunology. The relationship between the soluble lymphocyte substance described here to other known lymphocyte products will be of intense interest. Hopefully, since the role of the lymphocyte can be assessed in tissue culture, we will now be able much more rapidly to gain an insight into the basic mechanism responsible for the intracellular bacteriostasis so characteristic of acquired immunity in this disease.

In summary, Table 9 will show the mycobacterial immunizing substances already mentioned, and the immune responses we feel are stimulated by each.

References

Anacker, R., W. Barclay, W. Brehmer, C. L. Larson, and E. Ribi (1967). Duration of immunity to tuberculosis in mice vaccinated intravenously with oil-treated cell walls of *Mycobacterium bovis* strain BCG. *J. Immunol.,* 98:1265–1273.

Barclay, W., R. Anacker, W. Brehmer, and E. Ribi (1967). Effects of oil-treated mycobacterial cell walls on the organs of mice. *J. Bacteriol.,* 94:1736–1745.

Bekierkunst, A. (1968). Acute granulomatous response produced in mice by trehalose-6,6'-dimycolate. *J. Bacteriol.,* 96:958–961.

Bekierkunst, A., I. S. Levij, E. Yarkoni, E. Vikas, A. Adam, and E. Lederer (1969). Granuloma formation induced in mice by chemically defined mycobacterial fractions. *J. Bacteriol.,* 100:95–102.

Brehmer, W., R. Anacker, and E. Ribi (1968). Immunogenicity of cell walls from various mycobacteria against air-borne tuberculosis in mice. *J. Bacteriol.,* 95:2000–2004.

Crestfield, A. M., K. C. Smith, and F. W. Allen (1955). The preparation and characterization of ribonucleic acids from yeast. *J. Biol. Chem.,* 216:185–193.

Crowle, A. J. (1969). Immunogen extracted from tubercle bacilli with trypsin. *Z. Immunitatsforsch. Allerg. Klin. Immunol.,* 137:71–79.

Kanai, K. (1967). Acquired resistance to tuberculous infection in experimental model. *Jap. J. Med. Sci. Biol.,* 20:21–72.

Larson, C. L., E. Ribi, W. C. Wicht, R. H. List, and G. Good (1963). Resistance to tuberculosis in mice immunized with BCG disrupted in oil. *Nature, Lond.,* 198:1214–1215.

Leake, E. S., D. Gonzalez-Ojeda, and Q. N. Myrvik (1966). Digestive vacuole formation in alveolar macrophages after phagocytosis of *Mycobacterium smegmatis in vivo. J. Reticuloendoth. Soc.,* 3:83–100.

Patterson, R. J., and G. P. Youmans (1970a). Multiplication of *Mycobacterium tuberculosis* within normal and immune mouse macrophages cultivated with and without streptomycin. *Infect. and Immunity, 1*:30–40.

Patterson, R. J., and G. P. Youmans (1970b). Demonstration in tissue culture of lymphocyte-mediated immunity to tuberculosis. *Infect. and Immunity, 1*:600–603.

Raffel, S. (1950). Chemical factors involved in the induction of infectious allergy. *Experientia,* 6:410–419.

—— (1955). The mechanism involved in acquired immunity in tuberculosis. In *Ciba Foundation Symposium on Experimental Tuberculosis.* Boston: Little, Brown & Co.

Ribi, E., and C. Larson (1964). Immunological properties of cell wall versus protoplasm from mycobacteria. *Zbl. Bakt. Abt.* 1, *194*:673–685.

Sever, J. L., and G. P. Youmans (1957). Enumeration of viable tubercle bacilli from the organs of nonimmunized and immunized mice. *Am. Rev. Tuberc.,* 76:616–635.

Smith, D. W., A. A. Grover, and E. Wiegeshaus (1968). Nonliving immunogenic substances of mycobacteria. *Adv. Tuberc. Res., 16*:191–227.

Suter, E., and H. Ramsier (1964). Cellular reactions in infection, pp. 117–173. In *Advances in Immunology*, vol. 4, F. J. Dixon, Jr., and J. H. Humphrey (eds.), New York: Academic Press.

Thompson, H. C. W., and I. S. Snyder (1971). Protection against pneumococcal infection by a ribosomal preparation. *Infect. and Immunity, 3*:16–35.

Venneman, M. R., and N. J. Bigley (1969). Isolation and partial characterization of an immunogenic moiety obtained from *Salmonella typhimurium*. *J. Bacteriol., 100*:140–148.

Venneman, M. R., N. J. Bigley, and L. J. Berry (1970). Immunogenicity of ribonucleic acid preparations obtained from *Salmonella typhimurium*. *Infect. and Immunity, 1*:574–582.

Venneman, M. R., N. J. Bigley, C. L. Klun, A. Zachary, and M. C. Dodd (1968). The induction of cellular immunity in experimental salmonellosis. *Fed. Proc., 27*:429.

Winston, S. H., and L. J. Berry (1970a). Immunity induced by ribosomal extracts from *Staphylococcus aureus*. *J. Reticuloendoth. Soc., 8*:66–73.

—— (1970b). Antibacterial immunity induced by ribosomal vaccines. *J. Reticuloendoth. Soc., 8*:13–24.

Youmans, G. P., and A. S. Youmans (1964). An acute pulmonary granulomatous response in mice produced by mycobacterial cells and its relation to increased resistance and increased susceptibility to experimental tuberculous infection. *J. Infect. Dis., 114*:135–151.

—— (1969a). Recent studies on acquired immunity in tuberculosis, pp. 130–178. In *Current Topics in Microbiology and Immunology*, vol. 48. New York: Springer-Verlag.

—— (1969b). Immunizing capacity of viable and killed attenuated mycobacterial cells against experimental tuberculosis infection. *J. Bacteriol., 97*:107–113.

—— (1969c). Factors affecting immunogenic activity of mycobacterial ribosomal and ribonucleic acid preparations. *J. Bacteriol., 99*:42–50.

—— (1970). Immunogenic mycobacterial ribosomal and ribonucleic acid preparations: chemical and physical characteristics. *Infect. and Immunity, 2*:659–668.

GRAM-POSITIVE BACTERIAL ANTIGENS

G. Ivánovics

Medical University, Szegea, Hungary

One can safely say that the breakthrough on cellular antigens of Gram-positive bacteria occurred through the classical work of the Rockefeller Institute's teams in the 1920s. I am very pleased to welcome Dr. M. Heidelberger, one of the pioneers in this outstanding work, as a participant in this Conference.

The chemical characterization of pneumococcus antigens pointed a new way to the biochemical investigation of bacteria. This impetus soon resulted in the isolation and characterization of several distinct immunospecific bacterial substances.

I think it is appropriate to refer in my summary to the valuable comments which Professor Heidelberger made during the panel discussion.

His comments concerned the sometimes tenuous relationship between type-specificity and group-specificity. For example, it was found (HIGGINBOTHAM *et al.*, 1970) that the quite rigorous type-specific capsular polysaccharide of pneumococcal type IV readily loses its 9% of pyruvic acid, bound as a ketal, and is thereby converted, without loss of sugars, into a substance which precipitates a large proportion of serum antibodies to the group-specific pneumococcal C-polysaccharide, regardless of type. The depyruvylated type IV substance even precipitates C-reactive protein. The only component in common between C and depyruvylated IV is N-acetyl-D-galactosamine. The remarkable thing is that removal of a nonsugar residue, that of pyruvic acid, presumably from the D-galactose of the type IV substance, can greatly reduce the homologous type-specificity and unmask a massive group-specificity. Dr. Heidelberger and coworkers are still trying to elucidate the reasons for this astounding reversal.

When we talk of cellular antigens of bacteria, we are usually inclined to refer only to the different structural components of the bacterial surface. Most of these antigens are strictly integrated into the cell wall; the others may only form a dense or a loose layer on the bacterial surface. This latter may only appear as a result of some special environmental condition. This fact indicates their nonessential role in the viability of bacteria, yet there arose considerable interest as to their significance in the virulence. The cell wall with a basically identical architecture in different Gram-positive species has become the center of interest.

The immunochemistry of the peptidoglycan has been studied by SCHLEIFER and KRAUSE (1971a, b). These authors produced streptococcus group A variant antiserum, and found that this reacted with sonicated peptidoglycan preparations of *Staphylococcus epidermis, Lactobacillus acidophilus, Corynebacterium poinsettiae,* and *Micrococcus lysodeikticus.* Certain Gram-positive bacterial peptidoglycans carried the immunodeterminant groups in the pentapeptide portion, while others reacted with the antisera through their glycan moiety. This was due to differences on the pentapeptide moiety's C terminal amino acid sequence. The antibodies to the pentapeptide were isolated by immunoadsorption. Their capacity to agglutinate whole cells was investigated. It was found that the antibodies can agglutinate only group A variant of streptococci but not the cells of groups A or C. This has been explained by these authors on the basis of availability of the peptidoglycan's peptide moiety on the surface of the cells of group A variant, while the same peptide is efficiently masked in groups A and C or in staphylococci. More details about the immunochemistry of the peptidoglycan, together with discussion of their recently found endotoxin-like biological properties can be found in the paper presented at this Conference by Dr. Krause (this volume, p. 87).

Dr. Shockman wished to comment at this Conference on a few aspects of cell-surface antigens of Gram-positive bacteria. His comments were limited to those compounds that are known to be part of the cell wall structure and covalently linked to the rigid heteropolymer, the peptidoglycan. He said that the evidence seemed to be rather clear that the so-called "accessory wall polymers," such as the teichoic acids, or neutral polysaccharides of many species, such as the group polysaccharide of the group A streptococci (GHUYSEN 1968), were covalently linked to the peptidoglycan. This differed from the situation in Gram-negative species where the lipopolysaccharide appeared to lie outside the peptidoglycan layer and not to be covalently linked to it.

Also, there appears to be little evidence that the cell wall of Gram-positive bacteria is a layered structure with the peptidoglycan as the innermost layer, as depicted in Dr. Krause's presentation. HUGHES (1971) used antibody to peptidoglycan to study wall growth of *Bacillus licheniformis* by means of the immunofluorescence technique. Thus in addition to the other wall polymers in this species, at least portions of peptidoglycan polymer were close enough to the exterior surface of the cell to react with antibody. Thus the surface (and antigens) that the bacterial cell presented to its environment was one that was a mosaic which was complex, dynamic, and alterable.

One should also say a few words about the teichoic acid antigens of Gram-positive bacteria, which have been studied by Baddily and associates (SHABAROVA *et al.,* 1962). Shabarova and associates obtained 21 type-specific substances of pneumococci and examined the preparations for their constituent polyols. It has been found that 8 of them contain ribitol and anhydroribitol and 7 had glycerol as the main hydrolytic product. Continued investigation of those specific substances which were rich in polyols, characteristic constituents of teichoic acids, revealed that ribitol or glycerol teichoic acids form the type-specific polysaccharides of several pneumococci (VENKATA RAO et al., 1966a, b; KENNEDY et al., 1969; ROBERTS et al., 1963; VENKATA RAO et al., 1969; CHITTENDEN et al., 1968; DIXON et al., 1966).

In speaking of bacterial antigens, we sometimes also forget to mention the numerous metabolic products of bacteria formed on the surface of cytoplasmic membrane. These macromolecules with enzymic activity—the so-called exoenzymes such as alkaline phosphatase, coagulase, hyaluronidase, etc.—are produced, like most of the immunospecific structural elements of bacteria, under control of the chromosome.

Important antigenic materials derived from bacterial cells, such as bacterial toxins, are not referred to above.

The review of Gram-positive bacterial antigens would be incomplete without mentioning those frequently highly antigenic proteins produced by several strains. Some of the Gram-positive toxins, on the other hand, are hardly antigenic if injected without adjuvants. The fact that several of the highly toxic proteins can be converted to still-antigenic but nontoxic derivatives (toxoids) provided preparations suitable for detailed immunological studies (GLENNY and HOPKINS, 1923; HARRISON, 1963). These toxoids are capable of eliciting immune reactions, and the antibodies produced react and neutralize the corresponding toxic parent proteins. Details of immunization schedules, choice of the proper animals, and purification of the immunoglobulins obtained have recently been reviewed by OAKLEY (1969).

It should be kept in mind that the production of bacterial toxins is apparently not governed by bacterial chromosomes, but by a prophage. Other extrachromosomal elements besides prophage—plasmids or episomes—determine the production of certain specific macromolecules with antigenic activity. In the case of Gram-negative bacteria, these macromolecules may be integrated into the cell structure like sex pilus, the production of which is governed by the F-factor episome. The somatic antigen may be altered by lysogenic conversion; a well-known example is *Salmonella anatum,* which is converted to another antigenic species by infection with certain phages (UETAKE et al., 1958).

In addition to the toxin production of Gram-positive bacteria, some examples of macromolecular synthesis controlled by certain extrachromosomal elements should also be considered. Penicillinase production by *Staphylococcus aureus* associated with the presence of a plasmid in the cytoplasm has been extensively studied (NOVICK, 1963; RICHMOND,

1965). An antibacterial protein substance (megacin) is produced *de novo* by certain strains of *Bacillus megaterium* after induction by agents which disturb DNA synthesis (IVÁNOVICS and ALFÖLDI, 1954). This newly formed protein is distinct in its antigenic specificity from those of bacterial cell constituents. Megacin was found to be identical with phospholipase A (OZAKI *et al.*, 1966), and the immunospecificity of an individual megacin varied with the strain which produced it. It was found recently (Gaál and Ivánovics, in preparation) that phospholipase A production is also inducible by mitomycin C in some strains of *Bacillus cereus*. The genetic control of these immunospecific phospholipases is also related to an extrachromosomal genetic element (IVÁNOVICS, 1965). I think that these soluble proteins liberated by *de novo* synthesis from certain bacterial cells after induction should also be kept on record among the antigenic substances of Gram-positive bacteria.

As far as I know, there is no clear evidence as to the antigenic structure of the bacterial cell wall which is related to a plasmid. There are, however, some observations which point to a probably existence of plasmids which may be involved in controlling the specificity of cell wall antigens in *B. megaterium*. The great variety of immunospecificity in individual strains of *B. megaterium* is amazing (IVÁNOVICS, 1955; IVÁNOVICS *et al.*, 1957). A proportion of strains of this organism isolated from different natural sources ("wild type" strains) exhibited at cultivation a high segregation rate of cells differing from the original ones by their antigenic structure. The segregants are common in their antigenic specificity (IVÁNOVICS *et al.*, 1957). This "dissociation" of bacterial cultures can hardly be explained by mutation, since the high rate of segregation, as well as the common immunospecificity of the segregants contradict a chromosomal mutation. At the time of observation (IVÁNOVICS, 1955), the plasmid concept was in its embryonic stage. Therefore no explanation could have been given for the phenomenon. Retrospectively, our observations suggest the involvement of a plasmid as a determining factor in the antigenic structure of *B. megaterium*.

The aspect of immunologists studying bacterial antigens is gradually overlapping that of bacterial geneticists. The integration of these two disciplines may lead to answers to several issues related to the phylogenesis of bacteria.

References

Chittenden, G. J. F., W. K. Roberts, J. G. Buchanan, and J. Baddily (1968). The specific substance from *Pneumococcus* type 34 (41). *Biochem. J., 109*:597–602.

Dixon, J. R., J. G. Buchanan, and J. Baddily (1966). The specific substance from *Pneumococcus* type 34: the configuration of the glycosidic linkages. *Biochem. J., 100*:507–511.

Ghuysen, J.-M. (1968). Use of bacteriolytic enzymes in determination of wall structure and their role in cell metabolism. *Bacteriol. Rev.,* *32*:425–464.

Glenny, A. T., and B. A. Hopkins (1923). Diphtheria toxoid as an immunising agent. *Brit. J. Exp. Path., 4*:283–288.

Harrison, K. J. (1963). The preparation and properties of staphylocoagulase toxoid. *J. Path. Bacteriol., 85*:341–348.

Higginbotham, J. D., M. Heidelberger, and E. C. Gotschlich (1970). Degrada· tion of a pneumococcal type-specific polysaccharide with exposure of group-specificity. *Proc. Nat. Acad. Sci. U.S.A., 67*:138–142.

Hughes, R. C., and E. Stokes (1971). Cell wall growth in *Bacillus licheniformis* followed by immunofluorescence with mucopeptide-specific antiserum. *J. Bacteriol., 106*:694–696.

Ivánovics, G. (1955). Dissociation of *B. megaterium* associated with the change of the cell wall antigenic structure. *Acta microbiol. Acad. Sci. Hung., 3*:135.

—— (1965). Megacin and megacin-like substances. *Zentbl. Bakt.* Orig. I., *196*:318–329.

Ivánovics, G., and L. Alföldi (1954). A new antibacterial principle: Megacin. *Nature, Lond., 174*:465.

Ivánovics, G., L. Alföldi, and A. Szell (1957). Serological types of *Bacillus megaterium* and their sensitivity to phages. *Acta microbiol. Acad. Sci. Hung., 4*:333–351.

Kennedy, D. A., J. G. Buchanan, and J. Baddily (1969). The type-specific substance from *Pneumococcus* type 11A (43). *Biochem. J., 115*:37–45.

Krause, R. M. (1972). Polysaccharides of hemolytic streptococci. This volume, pp. 87–94.

Novick, R. P. (1963). Analysis by transduction of mutations affecting penicil- linase formation in *Staphylococcus aureus. J. Gen. Microbiol., 33*:121–136.

Oakley, C. L. (1969). Immunology of bacterial protein toxins, pp. 389–400. In *Microbial Toxins,* vol. I, S. J. Ajl, S. Kadis and T. C. Montie (eds.), New York/London: Academic Press.

Ozaki, M. Y., H. Yigashi, T. Saito, T. An, and T. Amano (1966). Identity of megacin A with phospholipase A. *Biken. J., 9*:201–213.

Richmond, M. H. (1965). Penicillinase plasmids in *Staphylococcus aureus. Brit. Med. Bull., 21*:260–263.

Roberts, W. K., J. G. Buchanan, and J. Baddily (1963). The specific substance from *Pneumococcus* type 34 (41): the structure of a phosphorus-free repeating unit. *Biochem. J., 88*:1–7.

Schleifer, K. H., and R. M. Krause (1971a). The immunochemistry of peptido- glycan. I. The immunodominant site of the peptide subunit and the con- tribution of each of the amino acids to the binding properties of the peptides. *J. Biol. Chem., 25*:986–993.

—— (1971b). The immunochemistry of peptidoglycan. II. Separation and characterization of antibodies to the glycan and to the peptide subunit. *Eur. J. Biochem., 19*:471–478.

Shabarova, Z. A., J. G. Buchanan, and J. Baddily (1962). The composition of pneumococcus type-specific substances containing phosphorus. *Biochim. Biophys. Acta, 57*:146–148.

Uetake, H., S. E. Luria, and J. W. Burrous (1958). Conversion of somatic antigens in *Salmonella* by phage infection leading to lysis or lysogeny. *Virology*, 5:68–91.

Venkata Rao, E., J. G. Buchanan, and J. Baddily (1966a). The type-specific substance from *Pneumococcus* type 10A (34): structure of the dephosphorylated repeating unit. *Biochem. J., 100*:801–810.

—— (1966b). The type-specific substance from *Pneumococcus* type 10A (34): the phosphodiester linkages. *Biochem. J., 100*:811–814.

Venkata Rao, E., M. J. Watson, J. G. Buchanan, and J. Baddily (1969). The type-specific substance from *Pneumococcus* type 29. *Biochem. J., 111*:547–556.

Part 3

Erythrocyte Antigens

ISOLATION OF THE BLOOD GROUP A RECEPTOR COMPONENT FROM HUMAN ERYTHROCYTE MEMBRANE

Roy F. Davis and Seymour Bakerman

Department of Pathology, Medical College of Virginia Richmond, Virginia

I. Introduction

The isolation and characterization of erythrocyte membrane blood group receptors have received considerable attention in the past few years. This followed the detailed analysis of soluble ABO blood group substances found in various secretions (Marcus, 1969; Watkins, 1966; Kabat, 1956). Yamakawa and Suzuki (1952) isolated a glycolipid from equine red blood cells which was shown to have A, B, and O blood group activity corresponding to the group of the red cells from which it was obtained. Numerous other investigators have isolated antigenically active glycolipids from various membrane sources (Koscielak and Zakrzweski, 1960; Hakomori and Jeanloz, 1961; Hakomori and Strycharz, 1968; Koscielak, 1963). A and B blood group activity has also been shown to be present in glycoprotein components derived from human erythrocyte membranes (Whittemore et al., 1969), but as yet these blood group substances have not been purified. Recently, Springer (1969) succeeded in isolating and purifying a highly active MM, MN, and NN glycoprotein from erythrocyte membranes.

It has been demonstrated that erythrocyte membranes are composed of at least 20 components (Schneiderman and Junga, 1968; Limber et al., 1970) as shown by disc gel electrophoresis. In this paper we describe the isolation and purification of one of these components from human red cell membranes, having A blood group activity. This component may be termed the A blood group receptor component.

II. Materials and Methods

The procedures used for the assay of protein, phospholipid, cholesterol, neutral sugars and sialic acids and for molecular sieve chromatography

(P-300) were those described by BAKERMAN and WASEMILLER (1967). Glucosamine and galactosamine were detected on the amino acid analyzer on the same column that was used to separate the acidic and neutral amino acids. Qualitative separation of carbohydrates was performed using thin-layer chromatography. For the qualitative identification of the lipid classes, the membrane or membrane fractions were extracted with chloroform: methanol, 2:1 (WAY and HANAHAN, 1964). The separation of the lipids was achieved using thin-layer chromatography following the method of SKIPSKI *et al.* (1962).

A. Polyacrylamide Disc Gel Electrophoresis

Disc gel electrophoresis of the whole membrane and subsequent fractions was carried out using one of the two different procedures depending upon the solubility of the membrane fraction.

The whole membrane was solubilized in phenol-urea-acetic acid and water (2:1.2:1:1/W:W:V:V) and electrophoresed for two hours at 5 mA per tube (LIMBER *et al.,* 1970). For those fractions which were soluble in aqueous solution, disc gel electrophoresis was performed, using the method of DAVIS and ORNSTEIN (1959).

All of the gels were fixed in 13% trichloroacetic acid, stained in 0.1% (W/V) Coomassie brilliant blue in 10% trichloroacetic acid. The gels were destained overnight in 10% trichloroacetic acid.

B. Isoelectric Focusing

Isoelectric focusing was carried out in the LKB 8121 column having a total volume of 110 ml.

The dense electrode solution (cathode), which was layered at the bottom of the column, was prepared by mixing 0.2 g sodium hydroxide and 12.0 g sucrose dissolved in 14.0 ml distilled water. A density gradient was layered on top of the cathode solution. The density gradient consisted of two solutions: a "light" solution made up of distilled water, 20 ml of sample, 0.7 ml pH 3 to 10 40% Ampholine, and a "heavy" solution consisting of 28 g surose in 31.5 ml of water and 1.8 ml Ampholine. The final Ampholine concentration was 1%. The anodic solution, 1% sulphuric acid, was carefully layered on the surface of the density gradient. The electrodes were connected and then the voltage was applied. The voltage was regulated so that the maximum power at the start of the run did not exceed 3 watts. During the run the voltage was increased to compensate for the increasing resistance along the column. The entire run was carried out at 4°C using a refrigerated circulating waterbath. The average run time for the pH 3 to 10 range of the Ampholine was 48 hours.

Following separation of the membrane components on the isoelectric-focusing column, the column was emptied from the bottom using an LKB 10200 Perpex pump with a flow rate of 1 ml per minute. The effluent was monitored at 280 mμ, and fractions were collected in a Gilson fraction collector. The pH of each fraction was measured on a Beckman expanded scale pH meter.

Having established the approximate pI of the proteins on the wide-range Ampholine, the electro-focusing was repeated using a narrow-range Ampholine in order to obtain greater separation and a more accurate pI of the proteins.

C. Hemagglutination Inhibition

The activity of A antigen on the membrane and its fractions was measured using a microhemagglutination inhibition titration. Anti-A and the positive control of group-specific AB substance was obtained from Ortho Diagnostics.

D. Immunization of Rabbits

New Zealand Giant white rabbits were injected subcutaneously at weekly intervals for 4 to 6 weeks. The samples to be injected were prepared by prior dialysis against isotonic saline, and then an aliquot was emulsified with an equal volume of Freund's complete adjuvant. At the end of the fourth week, a sample of blood was removed from the animals and the antibody titer was determined. If antibody was present, the animals were sacrificed, exsanguinated and the collected blood was centrifuged at 2000 rpm for 15 minutes; the serum was removed and frozen.

Agar gel diffusion studies were carried out using the technique described by OUCHTERLONY (1958) while the immunoelectrophoresis technique was that of GRABAR and WILLIAMS (1955).

III. Experimental Procedure and Results

A. Membrane Preparation

Red cells were obtained through the courtesy of the Medical College of Virginia Blood Bank. The cells were typed for their ABO blood group using Ortho Diagnostic's antisera and were also checked for the presence of bound antibodies by the direct Coombs' test. Hemoglobin-free membrane was prepared using the method described by LIMBER et al. (1970).

An aliquot of the membrane preparation was dialyzed against 0.85% sodium chloride for 48 hours prior to running hemagglutination inhibition tests, and was shown to retain its antigenic activity.

B. Fractionation of Erythrocyte Membrane

The initial separation of membrane components was carried out on the basis of their solubility (ROSENBERG and GUIDOTTI, 1969). See Figure 1.

The distribution of protein in the fractions is shown in Table 1. Aliquots of these fractions were run on acrylamide gel electrophoresis, and following dialysis against isotonic sodium chloride, were tested for anti-

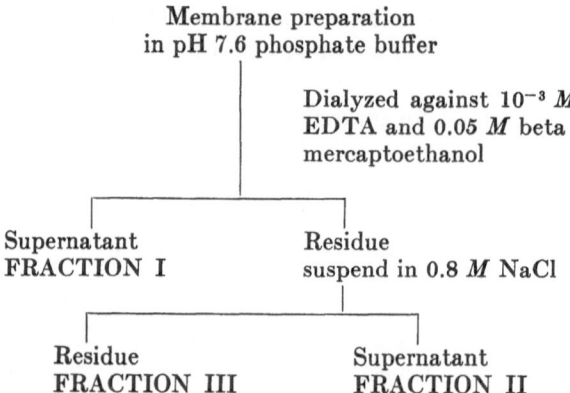

Fig. 1. Flow diagram for fractionation of erythrocyte membrane developed by ROSENBERG and GUIDOTTI (1969)

Table 1. This Table Shows the Distribution of the Protein as a Percentage of the Total Membrane Protein in the Three Fractions Obtained using Differential Solubility. (See Figure 1.)

Fraction	Percent of total membrane protein
I	11 ± 2
II	26 ± 5
III	63 ± 9

Table 2. This Table Compares the Activity per mgm of Protein of the Whole Membrane and Fractions I, II, and III which were Obtained from Differential Solubility. (See Figure 1.)

Antigen	Activity/mgm
Whole membrane	3.5
Fraction I	6.0
Fraction II	16.4
Fraction III	3.4

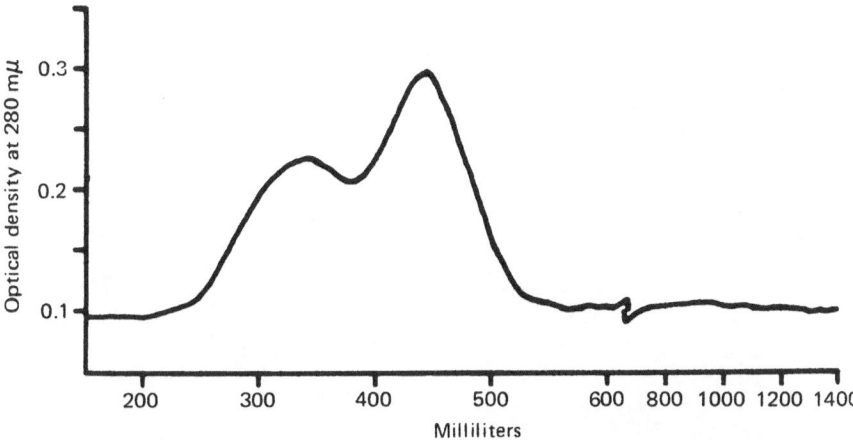

Fig. 2. Gel filtration pattern on P300 of Fraction II of human erythrocyte membrane which was fractionated according to the schemata in Figure 1. There is a broad peak at the void volume followed closely by a second peak V/Vo = 1.4.

genic activity by hemagglutination inhibition. From the hemagglutination studies, it was shown that fraction II contained the highest antigenic activity per mgm of protein compared to the other two fractions and the whole membrane (Table 2). The gel electrophoresis studies showed this fraction to be heterogeneous, and it was further purified by column chromatography. A 50-ml sample of fraction II was applied to the Bio-gel P-300 column, which was eluted and monitored at 280 mμ. The column chromatography of fraction II resulted in its separation into two peaks, fraction IIa and IIb (Figure 2). The fractions under these peaks were pooled and concentrated. An aliquot of each was dialyzed against isotonic saline, and hemagglutination inhibition was run on both samples. The activity per mgm of protein is tabulated in Table 3.

Table 3. Hemagglutination Inhibition
Activity of Fractions IIa and IIb
which were Obtained from a
Bio-Gel P-300 Column

Antigen	Activity/mgm
Fraction IIa	25.6
Fraction IIb	3.2

Fraction IIa, which exhibited an antigenic activity of 25.6 per mgm of membrane protein, was run on polyacrylamide gel electrophoresis (Figure 3) using the method of DAVIS and ORNSTEIN (1959). Two bands were demonstrated following staining with Coomassie blue.

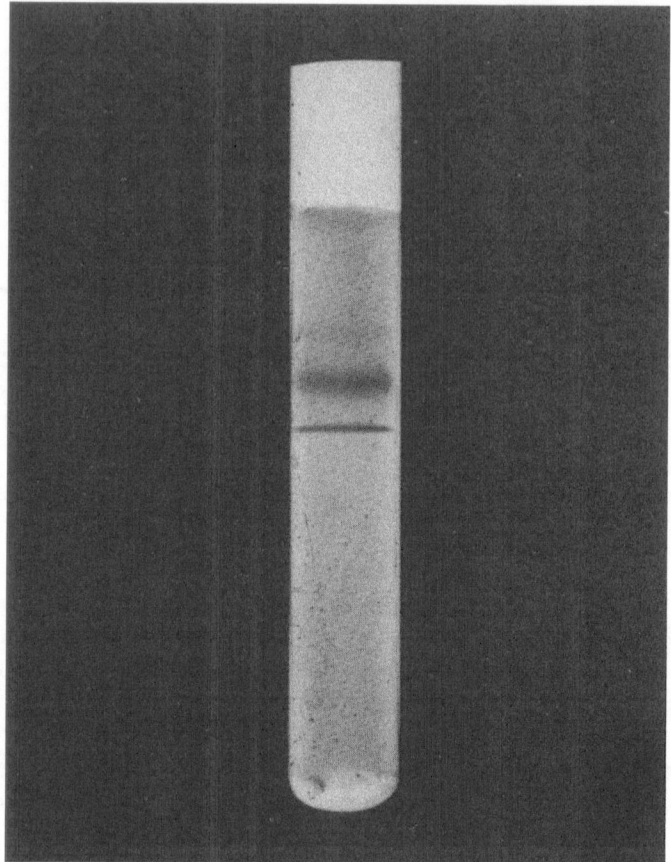

Fig. 3. Gel electrophoresis of Fraction IIa shows the sharp band of the tracking dye followed by a broad dense band and finally a light sharp band.

Final purification of fraction IIa was carried out using isoelectric focusing. The LKB isoelectric-focusing column was set up as described in the analytical procedures section. Following electro focusing on the wide-range Ampholine, pH 3 to 10, for 48 hours, the column was emptied; 1.0 ml fractions were collected, and their pH was determined. Two prominent peaks were obtained—the main one at pH 6.1 ± 0.2 and a minor one at pH 7.8 ± 0.2. The fractions under these peaks were extensively dialyzed against saline and were then tested for antigenic activity. The fractions under the peak at pH 6.1 showed inhibiting activity against anti-A, while the other fraction showed no activity. The isoelectric focusing was repeated using a pH 5 to 8 range Ampholine under the same conditions as described above. Figure 4 shows the elution profile obtained. The tubes under the major peak were combined and dialyzed exhaustively against 0.8 *M* NaCl for 5 days.

The component, under the major peak, will be referred to as the antigen receptor component.

C. Chemical Characterization of the Antigen Receptor Component

The component contained 0.059 mgm of phospholipid per mgm of protein. Thin-layer chromatography of a chloroform-methanol extract demonstrated the presence of the following lipids: phosphatidyl ethanolamine, phosphatidyl choline, and sphingomyelin.

The carbohydrate composition of the antigen carrier component was shown to be galactose, glucosamine, galactosamine, and fucose.

D. Immunological Characterization of the Antigen Carrier Component

The antigen carrier component was used for immunization of rabbits. Titration of the rabbit antisera against human group A, B, and O erythrocytes demonstrated agglutination of the group A red cells to a titer of 1:128 (Table 4). Group B red cells were hemolyzed in concentrated antisera, but no agglutination of the B or O cells was detected.

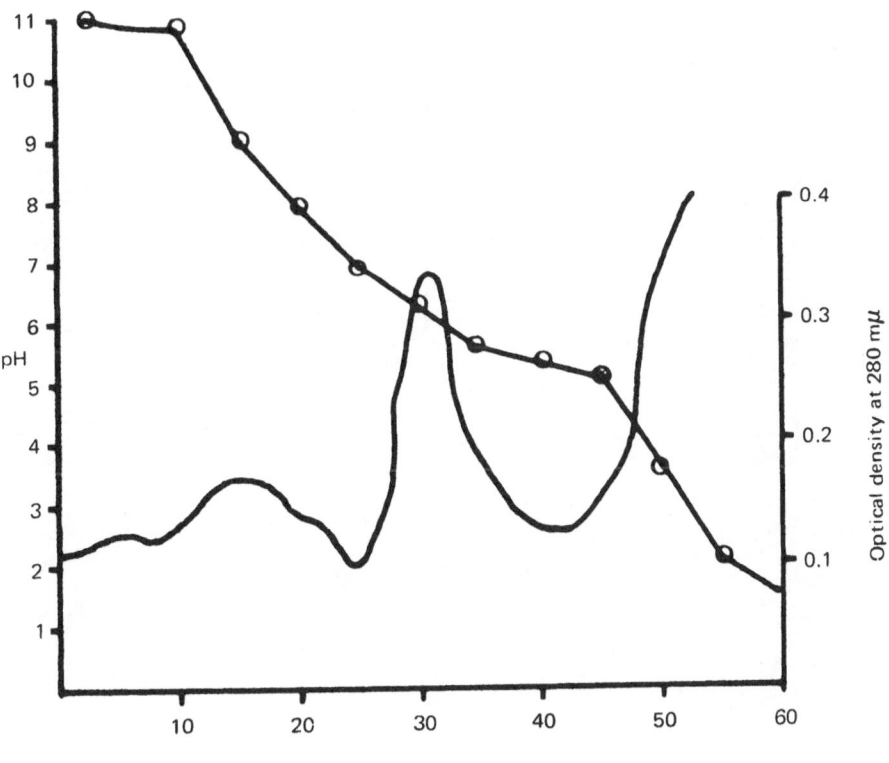

Fig. 4. Isoelectric focusing of Fraction IIa run on a narrow range pH 5 to 8 ampholine gradient for 48 hours at a maximum voltage of 500 V. 0 ——————— 0 pH gradient, ——————— absorbance at 280 mu.

Table 4. This Table Shows the
Hemagglutination Titer of the
Antibody Produced Following
the Immunization of Rabbits
with the Antigen Carrier
Component

Red cells (blood type)	Agglutination titer
A	1:128
B	1:1 (hemolysis)
O	zero

The results of immunodiffusion studies are shown in Figure 5. A single line of identity was obtained when the antisera produced in the rabbit (center well) was diffused against two different preparations of the antigen (antigen carrier component). Figure 6 shows that identity was also obtained when the antisera was diffused against fraction IIa, whole membrane

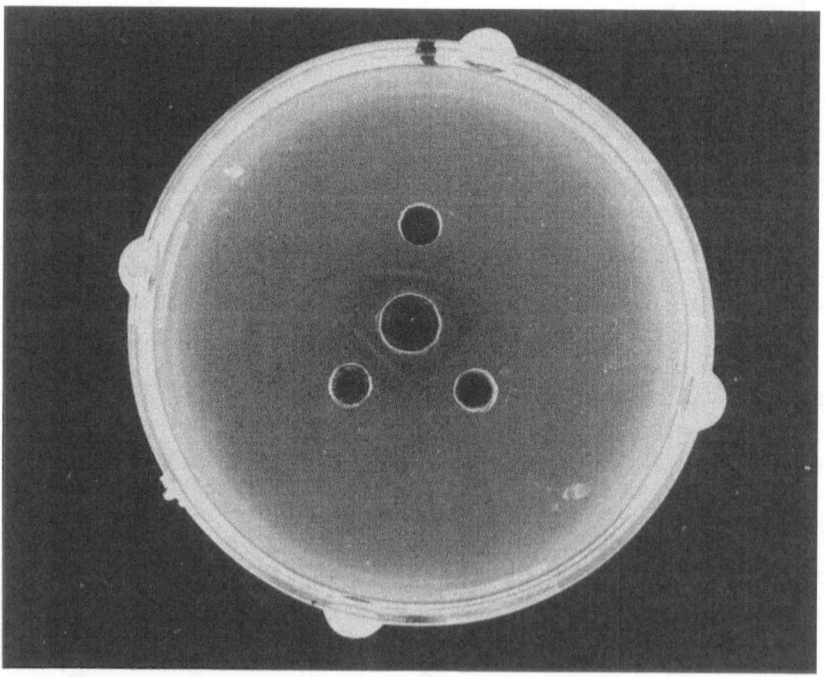

Fig. 5. A single precepitin band was produced following diffusion of two preparations designated A and B of the membrane component against the antisera (center well) to preparation A. (Upper well preparation A, right hand well preparation B, left hand well preparation A.)

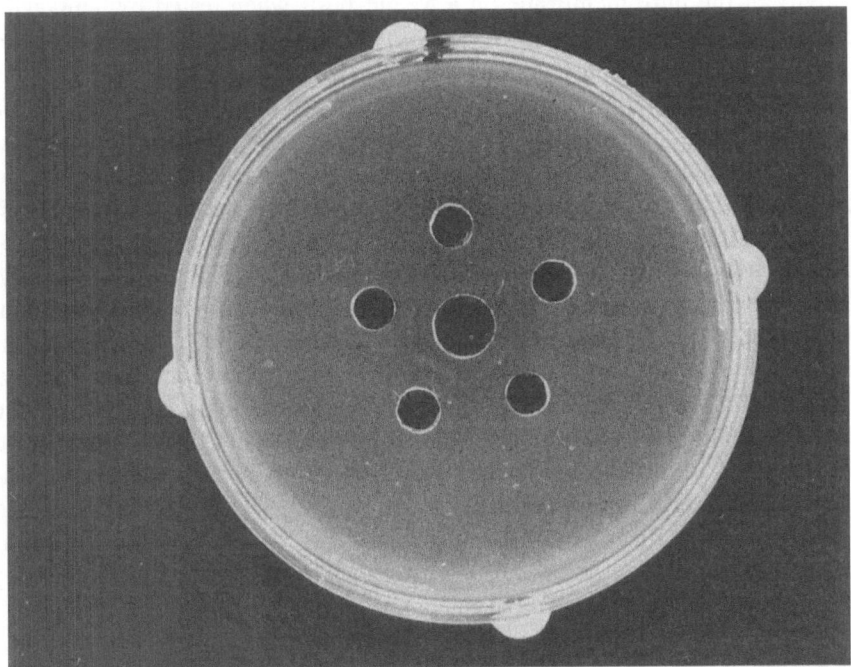

Fig. 6. Immunodiffusion studies of antisera (center well) against (a) human serum upper well—no precipitin band and clockwise; (b) Fraction IIa; (c) antigen carrier component; (d) whole membrane; (e) whole membrane.

in isotonic saline, whole membrane in 0.8 *M* NaCl and the antigen carrier component. The antisera gave no line of precipitation against human sera (well 1), demonstrating that the antisera was not produced against bound human gamma globulin.

Immunoelectrophoresis of the antigen carrier component against the antisera produced a single line of precipitation.

IV. Discussion

In the previous pages a procedure for the isolation of a membrane component having blood group A activity has been described. The isolation procedure involved the initial preparation of pure membrane and then the differential solubilization of the membrane components followed by molecular sieve chromatography and finally isoelectric focusing.

This membrane component comprises approximately 9% of the total membrane protein and has been shown to be a lipoglycoprotein. The molecular weight of the component is greater than 250,000 as it was totally excluded by a P-300 gel column.

This purified component had approximately eight times the antigenic activity of the pure membrane on a weight basis when tested by hemagglutination inhibition against group A cells. This activity is comparable to that of the MN receptor component isolated by SPRINGER (1969). The MN antigen receptor component had 12 to 24 times the activity of the original stroma preparation. However, the original preparation of membrane on which this activity was based could be considered relatively impure compared to the membrane preparation in this series of experiments.

HAKOMORI and STRYCHARZ (1968) and other investigators have isolated glycolipids from erythrocyte membranes having blood group activity. Their approach has been basically to isolate the antigenic determinant rather than the fundamental functional membrane unit or component which has been shown in this study to be a lipoglycoprotein. This lipoglycoprotein has only 0.059 mgm of phospholipid per mgm of protein. The small amount of phospholipid in this active component is comparable to the 1.8 mgm of antigenically active lipid isolated from 500 grams of erythrocyte membrane by HAKOMORI and STRYCHARZ (1968). They separated the active glycolipids into three types, one of which had activity comparable to that of the soluble blood group substances, while the remaining two glycolipids had lesser activity. Thus it is possible that there are at least three different antigenic glycolipids which may account for the presence of residual antigenic activity in fractions I and II of this study. These three different glycolipids may be related to the expression of different strengths of the A blood antigen activity that have been shown to exist, *i.e.,* A_1, A_2, A_3, etc. (SALMON *et al.,* 1965). These glycolipids have been chemically characterized as glycosphingolipids consisting of glucose, galactose, fucose, glucosamine, galactosamine, and sphingomyelin. These carbohydrates were also present in the component described herein.

It is thus possible to propose that the glycosphingolipid (HAKOMORI and STRYCHARZ, 1968) may be bound to the membrane receptor component isolated in this study. With this in mind, it is possible to propose a model for the inclusion of the glycolipid antigen and its protein receptor component within the membrane as a whole. The presence of numerous lipoprotein, glycoprotein, and lipoglycoprotein components within the membrane (DAVIS, 1969) is in keeping with the membrane model proposed by BENSON (1966). According to his model, globular lipoprotein subunits associate to form a two-dimensional membrane with a hydrophobic interior. The stability and effective function of his membrane model require the tertiary structure of the protein to be specific for the association of the various lipids. The structural relationships of his model are consistent with contemporary concepts of membrane function. Figure 7 depicts a concept of the possible arrangement of the antigen carrier protein and the antigen within the overall membrane structure. In Figure 7, four units of the membrane are diagramatically represented. Two are depicted as lipoprotein units and two as glycolipoprotein units. The two glycolipoprotein units are shown to have three residues of carbohydrate attached to

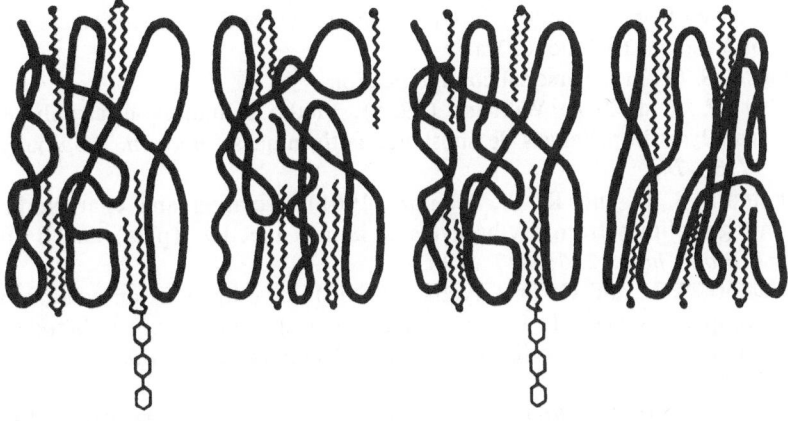

Fig. 7. Proposed model for the inclusion of the Antigen Carrier Component within the overall membrane structure.
———— protein,

∿∿∿ lipid,

◯–◯ carbohydrate

a sphingolipid. The carbohydrate residues are exposed to the outer surface of the membrane. This is supported by the electron microscopic studies of ferritin-labeled anti-A (DOUGLAS, 1971) which show that the antigen is labeled only on one surface of the cell. Enzymatic hydrolysis of the polysaccharide (HAKOMORI and STRYCHARZ, 1968) has shown that the loss of a terminal carbohydrate results in the loss of antigenic activity. This suggests that the carbohydrate is the antigenic determinant and that it may be bound to the membrane by way of the hydrophobic portion of the lipid to which it is attached. DOUGLAS (1971) also demonstrated that the labeling was situated in discrete portions of the membrane, thus supporting the hypothesis of a single component, *i.e.*, the antigen receptor component, specifying the antigen-binding position.

References

Bakerman, S., and G. Wasemiller (1967). Studies on structural units of human erythrocyte membrane. I. Separation, isolation and partial characterization. *Biochemistry, 6*:1100–1113.

Benson, A. A. (1966). On the orientation of lipids in chloroplast and cell membranes. *J. Am. Oil Chem. Soc., 43*:265–270.

Davis, B. J., and L. Ornstein (1959). A new high resolution electrophoresis method. Delivered at Society for the Study of Blood at the New York Academy of Medicine.

Davis, R. F. (1969). Separation and partial characterization of components

derived from human erythrocyte membranes. M.S. Thesis, Medical College
of Virginia, Richmond, Va.

Douglas, S. (1971). Personal communication.

Grabar, R., and C. A. Williams (1955). Méthode immuno-électrophorétique
d'analyse de mélanges de substance antigèniques. *Biochim. Biophys. Acta,*
17:67–74.

Hakomori, S. I., and R. W. Jeanloz (1961). Isolation and characterization
of glycolipids from erythrocytes of human blood A (plus) and B (plus).
J. Biol. Chem., 236:2827–2834.

Hakomori, S. I., and G. D. Strycharz (1968). Investigations on cellular blood
group substances. I. Isolation and chemical composition of blood group
ABH and Le-b isoantigens of sphingoglycolipid nature. *Biochem., 7*:1279–
1286.

Kabat, E. A. (1956). *Blood Group Substances.* New York: Academic Press.

Kosceilak, J. (1963). Blood group A specific glycolipids from human erythrocytes. *Biochim. Biophys. Acta, 78*:313–328.

Koscielak, J., and K. Zakrzweski (1960). Substances from erythrocytes of
blood group A. *Nature, Lond., 187*:516–517.

Limber, G., R. F. Davis, and S. Bakerman (1970). Acrylamide gel electrophoresis studies of human erythrocyte membranes. *Blood, 36*:111–118.

Marcus, D. M. (1969). The ABO and Lewis blood group system. *New Eng.*
J. Med., 280:994–1006.

Ouchterlony, O. (1958). Diffusion in gel methods of immunological analysis.
Prog. Allergy, 5:1–77.

Rosenberg, S. A., and G. Guidotti (1969). Fractionation of the protein components of human erythrocyte membranes. *J. Biol. Chem., 244*:5118–
5124.

Salmon, C., D. Salmon, and J. Reviron (1965). Étude immunologique et
génétique de la variability de phenotype Ax. *Nouv. Rev. Fr. Hematol.,*
5:275–290.

Schneiderman, L. J., and I. G. Junga (1968). Isolation and partial characterization of structural protein derived from human red cell membranes. *Bio-*
chem., 7:2281–2286.

Skipski, V. P., R. F. Peterson, and M. Barclay (1962). Separation of phosphatidyl ethanolamine, phosphatidyl serine and other phospholipids by
thin layer chromatography. *J. Lipid Res., 3*:467–470.

Springer, G. F. (1969). Mammalian erythrocyte receptors: Their nature and
their significance in immunopathology, pp. 47–62. In *Current Problems*
in Immunology, O. Westphal, H. E. Bock and E. Grundmann (eds.),
New York/Heidelberg: Springer-Verlag.

Watkins, W. M. (1966). Blood group substances. *Science, 152*:172–181.

Way, P., and D. J. Hanahan (1964). Characterization and quantification of
red cell lipids in normal man. *J. Lipid Res., 5*:318–328.

Whittemore, N. B., N. C. Trabold, C. F. Reed, and R. I. Weed (1969).
Solubilized glycoprotein from human erythrocyte membranes possessing
blood group A, B, and H activity. *Vox Sang., 17*:289–299.

Yamakawa, T., and S. Suzuki (1952). Chemistry of the lipids of posthemolytic
residue or stroma of erythrocytes. III. Globoside, the sugar-containing
lipid of human blood stroma. *J. Biochem.* (Tokyo), *39*:393–402.

ANTIBODY-, VIRUS- AND ENDOTOXIN- RECEPTORS OF HUMAN ERYTHROCYTE MEMBRANE*

GEORG F. SPRINGER, SHANKAR V. HUPRIKAR and
JAMES C. ADYE

*Department of Immunochemistry Research,
Evanston Hospital and Department of Microbiology,
Northwestern University,
Evanston, Illinois*

The second human blood-group system, the MN system, was discovered by LANDSTEINER and LEVINE (1928a, b). The M and N antigenic specificities are thought to be the expression of products of two allelomorphic genes (WIENER, 1938) and are measured with agglutinins of human, animal, and plant origin (RACE and SANGER, 1968).

The observation that neuraminidases from influenza viruses and from *Vibrio cholerae* inactivate both the M and N as well as the influenza virus receptor activities of human erythrocytes, with the concomitant release of N-acetyl-neuraminic acid (NANA)*, was the first accurate information on the chemical basis of these activities and specificities (SPRINGER and ANSELL, 1958; MÄKELÄ and CANTELL, 1958). A single gene locus appears to determine the specificites of viral receptor *and* isoantigen, suggesting that they share the same molecular basis (SPRINGER and ANSELL, 1958; SPRINGER, 1970). It was then shown that glycoproteins were the carriers of these activities (KLENK and UHLENBRUCK, 1960; STALDER and SPRINGER, 1960; KATHAN et al., 1961; SPRINGER et al., 1966a). No difference in the composition of MM and NN antigens was detected (BARANOWSKI et al., 1959; SPRINGER et al., 1966a; KATHAN and ADAMANY, 1967). Although the myxovirus receptor activity could not be

* Supported by the John A. Hartford Foundation grant SD-340, NIH grants 05681 and 05682 and Chicago Heart Association grant B71-56. The Research Department is maintained by the Susan Rebecca Stone Fund.

* Abbreviations used: NANA, N-acetyl-neuraminic acid; Gal, D-galactose; Gal-NAc, N-acetyl-D-galactosamine; Glc, D-glucose; GlcNAc, N-acetyl-D-glucosamine; Fuc, L-fucose; Lac, lactose; o-NO₂-phenyl β-Gal, orthonitrophenyl β-D-galactopyranoside.

removed from the blood-group MN antigens and could not be destroyed without simultaneous destruction of M and N activities, it was possible to destroy blood-group MN specificity without affecting the myxovirus activity (SPRINGER *et al.,* 1966a; SPRINGER and HUPRIKAR, 1972a).

A reagent thought to detect blood-group N specificity is the agglutinin isolated from the leguminous plant *Vicia graminea* (LEVINE *et al.,* 1955). However, we have shown that the N specificity detected with this reagent is not dependent on sialic acid in contrast to that determined with human and commercial rabbit anti-N sera. Rather, removal of sialic acid under mild conditions increases activity with the *Vicia* reagent; also, isolated blood-group MM antigen generally possesses some *Vicia* specificity (NAGAI and SPRINGER, 1962; STALDER and SPRINGER, 1962; HOTTA and SPRINGER, 1963; LISOWSKA, 1963; SPRINGER and HOTTA, 1963; SPRINGER *et al.,* 1966a). *Vicia* specificity is destroyed by an *Escherichia coli* β-galactosidase, which indicates that the β-D-galactopyranosyl grouping is the immunodominant complementary structure to the combining sites of the *Vicia* reagent (SPRINGER *et al.,* 1966a).

The observations with the *Vicia* reagent are supported by our finding that desialization of the MN antigens also increases their cross-reactivity with horse anti-Pneumococcus type XIV antiserum (SPRINGER *et al.,* 1966a; SPRINGER *et al.,* 1971c, d), which reacts with terminal β-D-galactopyranosyl structures (HEIDELBERGER, 1955).

Thus the NN antigen contains more reactive, terminal β-linked Gal than does the MM antigen. We therefore exposed O,MM and O,NN antigens to pure *E. coli* β-galactosidase. We found that the β-galactosidase decreased human N specificity up to 93%, as determined with several human antisera; Gal was released concomitantly. Our results are summarized in Table 1. They show that more than 20 times as much Gal is released from NN antigen as from MM antigen. Based on a molecular weight of 30,000 for the basic unit of the NN antigen, not more than 2 mole of Gal are affected by the β-galactosidase per mole of antigen, yet this leads, on the average, to 81.5% inactivation of the NN antigen; the MM activity is entirely unaffected by β-galactosidase. Alpha-galactosidase and α- and β-hexosaminidase were without effect on the MN specificities.

Because of the requirement of terminal β-D-galactopyranosyl structures for blood-group N specificity in addition to α-linked NANA, we carried out inhibition studies with substances which possess terminal Gal and in some instances α-NANA as well. It can be seen from Table 2 that approximately one-half of the human anti-N sera tested were inhibited by Gal as well as by oligosaccharides which possessed terminal β-D-galactopyranosyl groups. Only 1 of the 5 rabbit anti-N sera was inhibited by Gal and methyl β-D-galactopyranoside. None of the *Vicia* extracts was inhibited. Ganglioside I (Kuhn & Wiegandt nomenclature) inhibited 4 of 10 human anti-N sera tested and was active with 2 of the 3 rabbit anti-N sera but not with the *Vicia* reagent. More anti-N sera and also the *Vicia*

Table 1. Effect of *E. coli* β-Galactosidase on Human Blood-Group M and N Antigens*

| Antigen and conditions | Blood-group activity | | Influenza virus inhibitory activity | | Gal released moles/mole antigen‡ (Mw 30,000) |
	human antisera†	rabbit antisera	A/PR8	B/Md	
		Percent inactivation of			
NN plus active enzyme	81.5	77.5	Nil	Nil	1.4
NN plus inactivated enzyme	Nil	Nil	Nil	Nil	0.05
MM plus active enzyme	Nil	Nil	Nil	Nil	0.06

* Hemagglutination inhibition assay average of four different experiments.

† Average of duplicate or triplicate measurement with each of four different sera per experiment.

‡ Determined by galactose dehydrogenase as well as chromatography.

reagent were inhibited by asialoganglioside (see also SPINGER *et al.,* 1972c).

The activity range of the different substances on a weight basis did not exceed fourfold among the inhibitable antireagents. Complete inhibition of agglutination was given by approximately 1 mg/ml ganglioside or asialoganglioside and by about 50 mg/ml of any of the active sugars. The

Table 2. Inhibition of Antihuman Blood-Group N Reagents by Compounds with β-D-Galactopyranosyl Termini*

| Substance | Anti-N reagent | | |
	Human	Rabbit	*Vicia graminea*
Gal	5/10†	1/5	
Lac	6/10	none	
Methyl β-Gal	4/10	1/5	
o-NO₂-phenyl β-Gal	4/10	none	none
β-Gal-(1→4)-GlcNAc	4/10	none	
Lacto-N-Tetraose	3/10	none	
Ganglioside I	4/10	2/3‡	
"Asialoganglioside"§	7/10	4/5	all

* Complete inhibition of four hemagglutinating doses; for quantitative data and inactive compounds see text.

† Proportion of sera inhibited.

‡ Insufficient material for tests on all sera.

§ Still contains 1.6% sialic acid.

gangliosides were most active on a molar basis; approximately 0.5 μmole/ml gave complete inhibition of 4 hemagglutinating doses of anti-N reagent; 35 μmoles/ml lacto-N-tetraose and about 100 μmoles/ml of Gal were required for this effect. Inactive at concentrations up to 100 mg/ml were gangliosides II to IV as well as Tay-Sachs ganglioside, methyl α-D-galactopyranoside, melibiose, and stachyose; the latter three contain terminal α-Gal. Also inactive were NANA, GlcNAc, GalNAc, α-D-galacto-pyranosyl-(1 → 3)-D-galactopyranose and colominic acid. Curiously, 4 of the 5 human anti-N sera which were inhibited by Gal were also inhibited by 50 mg/ml Fuc, but none of the rabbit anti-N sera nor the *Vicia* reagent were inhibited (see also SPRINGER *et al.,* 1972c).

The ability of certain gangliosides to inhibit some of the anti-N reagents prompted us to investigate if there was a reciprocal relationship in that anti-ganglioside sera may react with blood-group N or M specific structures.

We found that the rabbit anti-ganglioside serum reactive with ganglio-side I and asialoganglioside (RAPPORT *et al.,* 1970) agglutinated both human O,MM and O,NN erythrocytes nearly as well as commercial anti-M and anti-N sera. The preimmunization serum sample did not agglutinate any human blood-group O erythrocytes. NN erythrocytes were agglutinated to a higher titer and were more efficient absorbents of the anti-erythrocyte agglutinins than MM red cells. The erythrocyte-agglutinating ability of the anti-ganglioside serum was inhibited by isolated, highly purified MM and NN antigens in quantities as small as those needed in the homologous systems. The NN antigen was somewhat more active than the MM antigen.

Several years ago we were able to transform MM into N antigen (NAGAI and SPRINGER, 1962) and we have since isolated N-specific haptens from MM antigen (HOTTA and SPRINGER, 1965). The majority of human anti-N sera, all rabbit anti-N sera, and the *Vicia* reagent were inhibited by not only NN antigens but also by isolated blood-group MM antigens. The degree of inhibition by MM antigen was on the average 12%, as measured with human anti-N sera and for rabbit anti-N sera it amounted to about 30% of that of the NN antigens. On the other hand, none of the human and rabbit anti-M sera tested were inhibited by NN antigen.

Mild acid hydrolysis of the MM antigens at pH 2.0 and 56°C revealed, besides the expected decrease of M activity, concomitant with a NANA content decrease, a surprising increase or even a *de novo* appearance of N activity. Table 3 shows that the release of NANA resulted in the expected decrease of M activity of the MM antigens. The inhibitory activity of the partially hydrolyzed MM antigens decreased faster and more extensively when measured with anti-M of human origin than when determined with rabbit sera (Table 3).

The right-hand section of Table 3 shows the remarkable increase, on mild hydrolysis, of the specific inhibitory effect of the MM antigens versus

Table 3. Effect of Hydrolysis at pH 2.0, 56°C on the Serological Specificity and Sialic Acid Content of the Blood-Group MM Antigen†

MM antigen		Reagent																	
		Anti-M (Fold titer decrease)†		Anti-N (Fold titer increase†)															
		Human	Rabbit	Human											Rabbit				
Hydrolysis (in hours)	NANA (content in percentage)	SM-662	David.	Arm.	6505	Mul.	McN.	Good.	61-317	SN-32	NH-16AH	SN-6	Ehrl.	David.	Wien.	Hyl.	Ortho 83 & 85	Behr.	*Vicia graminea*
0	13.48																		
1	11.58	3	2	>6	>2	>2	4	3	>2	Nil	>6	Nil	Nil	3	3	1.5	Nil	3	2
2	6.74	15	2	>7	>4	>2	2	6	>2	Nil	>4	3	>2	2	Nil	4	4	6	4
4	4.34	80	3	>10	>2	>6	2	2	Nil	>2	>5	2	>2.5	2	Nil	3	2	7	5
8	1.98	95	5	>8	>2	Nil	2	Nil	Nil	>2	>6	2	>2	Nil	Nil	3	Nil	3	4
12	1.14	>150	7	>2	Nil	Nil	2	-2‡	Nil	Nil	>3	Nil	Nil	Nil	-2	3	Nil	Nil	4

* Average of duplicate tests on three different experiments.
† Compared to untreated antigen.
‡ Activity decreased 2-fold.

the various anti-N reagents which increased up to >10-fold. One rabbit anti-N serum, Wien., exhibited only a slight increase in reactivity toward partially desialized MM antigens.

The maximal increase of reactivity with partially desialized MM antigens occurred regularly between 1 and 8 hours. During this time the N activity of the partially hydrolyzed MM antigens as measured with all human sera rose to 25 to 100% (the average was very nearly 50%) of that of the NN antigen preparations. The MM antigens thus became serologically indistinguishable from MN or even NN antigens, depending on which anti-M sera were used (see above). Upon prolonged hydrolysis, the N activity of the MM antigens declined, except versus anti-N serum McN. After 12 hours hydrolysis, it had even fallen below the level of that of the intact MM antigen with 1 human and 1 rabbit antiserum.

Both MM and NN antigens were influenced in a similar way by mild acid hydrolysis as regards their inhibitory activity toward the *V. graminea* reagent (Table 3). None of the O,MM or O,NN antigens acquired ABH(O) or $Rh_o(D)$ activity during any of the hydrolyses described.

The occurrence of N-specific structure on the MM antigen and the large increase of N-determinant groups upon "peeling off" of masking terminal residues—NANA, since no other sugar and no free acetyl were released (SPRINGER *et al.*, 1972c)—favors the view that the M and N antigenic specificities do not result, indirectly via enzymes, from the action of allelomorphic genes as classical theory postulates (LANDSTEINER and LEVINE, 1928b; WIENER, 1938; HIRSCH *et al.*, 1957); rather, our findings indicate that the product of the N blood-group gene is the immediate precursor of the product of the M blood-group gene and that the allele to the M gene is an amorph. The structure reacting with the *Vicia* reagent seems to lie in the biosynthetic pathway of the NM macromolecules and appears before either the N or the M determinants. Figure 1 depicts schematically in its upper half the classical genetic scheme of the relation of MM to NN antigens; to this we have added the myxovirus inhibitory property (If) of the MN antigens (SPRINGER and ANSELL, 1958; SPRINGER *et al.*, 1971d). The new pathway proposed on the basis of our experimental observations is shown in the lower half of Figure 1.

The myxovirus receptor properties of these particular structures appear concurrently with the blood-group N specificity. Compared to the ABH(O) substances, only a relatively small number of M or N determinant groups are distributed over the surface of the glycoproteins carrying these specificities (SPRINGER *et al.*, 1971d; SPRINGER and HUPRIKAR, 1972a).

WIENER *et al.* (1971) came to conclusions similar to ours, insofar as the relation of the M and N antigenic specificities to one another is concerned, from their observation that 11 gorillas were of blood-group N or MN, but not M, a finding presumably not in accord with the Hardy-Weinberg rule.

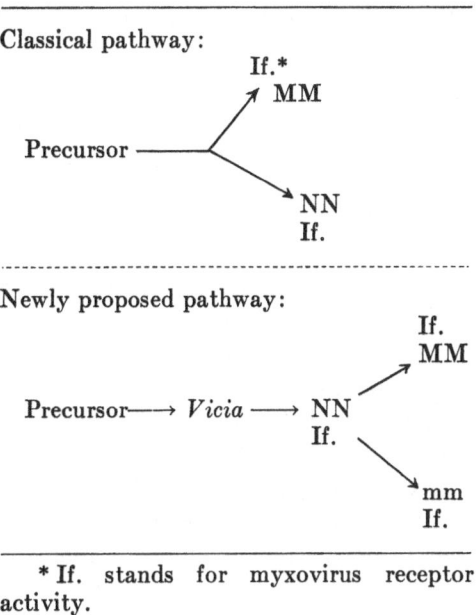

Classical pathway:

Newly proposed pathway:

* If. stands for myxovirus receptor activity.

Fig. 1. Genetic pathways leading to the human blood-group M and N specificities.

M activity also occurs on animal erythrocytes (CALLAHAN and SPRINGER, 1966; SPRINGER, 1969). Table 4 shows the situation for a sheep erythrocyte membrane antigen, where so far it has been inseparable from the antigen which interacts with antibodies occurring in patients suffering from infectious mononucleosis (PAUL and BUNNELL, 1932). The antigen which we have isolated is a glycoprotein (CALLAHAN and SPRINGER, 1966;

Table 4. *In vitro* Biological Activities* of Purified Infectious Mononucleosis Receptor from Sheep Erythrocytes

Hemagglutination inhibition					Precipitation†	
Human anti						Horse anti pneumoc. XIV
I.M.	Serum sick.	Forssman	M‡	*Vicia graminea*	Anti- I.M.	
0.01	0.6	1	0.01	3.5§	2.5	5

* Mg/ml, which inhibit or precipitate corresponding antibody.
† Capillary test.
‡ Active with three of four human anti-M sera, but not with rabbit anti-M sera.
§ After mild degradation: 0.15.

SPRINGER *et al.*, 1972b). It provokes in man the antibodies typical for infectious mononucleosis.

The blood-group MN and infectious mononucleosis antigens inhibit agglutination of a chicken cell receptor by its homologous antiserum (SPRINGER *et al.*, 1971b). This chicken cell receptor has been implicated as the one to which avian leukosis-sarcoma virus attaches in order to infect (CRITTENDEN *et al.*, 1970).

Finally, we have isolated and obtained in homogeneous form a fraction from human erythrocyte ghosts which prevents the attachment of unheated as well as heated lipopolysaccharides (smooth as well as rough) of all Gram-negative bacteria tested to red cells (SPRINGER *et al.*, 1966b; SPRINGER *et al.*, 1970; SPRINGER, 1971a). This material has no significant inhibitory effect either toward the Vi antigen of Gram-negative bacteria or toward the group and common antigens of the Gram-positive bacteria investigated. We therefore named this fraction "lipopolysaccharide receptor" (SPRINGER *et al.*, 1970). This appears to be the first time any substance has been found that specifically inhibits endotoxin attachment without producing other effects.

The specificity of the LPS-receptor is shown in Table 5, where its effect

Table 5. The Lipopolysaccharide Receptor as Inhibitor of Antigen Fixation to Human Erythrocytes

Antigen fixed	Smallest average antigen amount affording optimal titer A μg/ml	Smallest average receptor amount inhibiting coating by >95% B μg/ml	B/A
Gram-negative bacteria			
Isolated O antigens (14)*	3	27	9
Isolated Vi antigens (2)	0.6	700	1170
Gram-positive bacteria			
Streptococcus py., isolated group antigens, stearoyl deriv. (2)	4	1750	440

* Figures in parentheses = numbers of different strains tested.

on the fixation of highly purified bacterial antigens is listed. Column 2 shows the minimal amount of LPS antigen needed to achieve optimal coating of erythrocytes, and column 3 gives the smallest quantity of receptor that completely inhibits erythrocyte coating by the antigens. The receptor was virtually ineffective in preventing the attachment to red cells of the

Vi antigens of Gram-negative bacteria and of the antigens of the Gram-positive bacteria investigated, including the Rantz antigens of *Staphylococcus aureus* and *Bacillus subtilis,* which are not listed in Table 5.

The relationship between the inhibitory unit of receptor and the coating unit of lipopolysaccharide became more clearly recognizable if ratios of these units were established on a weight basis (see Table 5, column 4, B/A). The ratio of inhibiting receptor units to antigen-coating units measured with homologous serum averaged 9 when LPS of Gram-negative bacteria was employed, but was 400 and 130 times higher for the streptococcal and Vi antigens.

We have established on a quantitative basis by means of radioactive

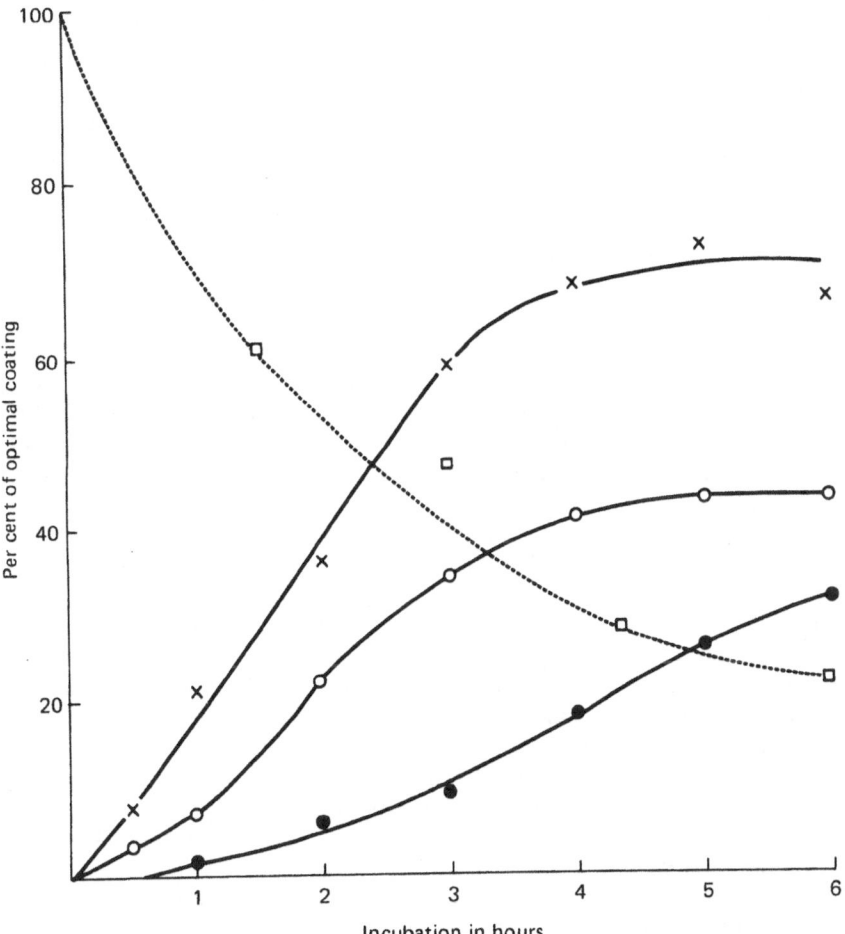

Fig. 2. Equilibration of *E. coli* O₈₆ Lipopolysaccharide between red cells and Receptor. Transfer of L.P.S. from red cells to Receptor, □. Transfer of L.P.S. from Receptor to red cells, Receptor/L.P.S. ratio 1.13, ×; 1.6, ○; 3.0, ●.

tracers that the receptor interacts with lipopolysaccharides and not with erythrocytes and that it forms complexes with and blocks those groupings of lipopolysaccharides which attach to red cells. The effect of the receptor is physical and not enzymatic. The interaction of the receptor with the lipopolysaccharides is reversible, and the receptor removes lipopolysaccharides fixed to erythrocytes and thus "detoxifies" the cells. An equilibrium of lipopolysaccharide distribution between cells and receptor is established when receptor-lipopolysaccharide complexes are incubated with red cells (Figure 2). The receptor is labile toward heat and deviation of the hydrogen-ion concentration from neutrality; its activity is destroyed by proteases and aldehydes (SPRINGER *et al.*, 1970). Sialic acid is not involved in its specificity (SPRINGER, 1971a). We also have obtained model compounds some of them of small molecular size, which exhibit similar action.

This isolation of a cell-bound receptor substance which interacts specifically with the endotoxins of Gram-negative bacteria may have clinical implications in addition to its basic interest.

The examples of cell surface-environment interactions presented in this article demonstrate that true insight into the significance of cell surface components as receptors and mediators of host-environment interactions may only be expected by chemical and physical analyses guided by consideration of biological function.

References

Baranowski, T., E. Lisowska, A. Morawiecki, E. Romanowska, and K. Strozécka (1959). Studies on blood group antigen M and N. III. Chemical composition of purified antigens. *Arch. Immunol. Terapii Dosw.,* 7:15–27.

Callahan, H., and G. F. Springer (1966). Infectious mononucleosis receptor from sheep erythrocytes. *Fed. Proc.,* 25:435.

Crittenden, L. B., W. E. Briles, and H. A. Stone (1970). Susceptibility of an avian leucosis-sarcoma virus: close association with an erythrocyte isoantigen. *Science, 169*:1324–1325.

Heidelberger, M. (1955). Immunological specificities involving multiple units of galactose. *J. Am. Chem. Soc., 77*:4308–4311.

Hirsch, W., P. Moores, R. Sanger, and R. R. Race (1957). Notes on some reactions of human anti-M and anti-N sera. *Brit. J. Haematol., 3*:134–142.

Hotta, K., and G. F. Springer (1963). Blood-group N specificity and sialic acid. *9th Eur. Congr. Haematol. Sang., 9*:183–187.

—— (1965). Isolation and partial characterization of specific haptens from blood-group M and N substances. *Proc. 10the Int. Congr. Soc. Blood Transf.,* 505–509.

Kathan, R. H., and A. Adamany (1967). Comparison of human MM, NN, and MN blood group antigens. *J. Biol. Chem., 242*:1716–1722.

Kathan, R. H., R. J. Winzler, and C. A. Johnson (1961). Preparation of an inhibitor of viral hemagglutination from human erythrocytes. *J. Exp. Med., 113*:37–45.

Klenk, E., and G. Uhlenbruck (1960). Über neuralminsäurehaltige Mucoide aus Menschenerythrocytenstroma, ein Beitrag zur Chemie der Agglutinogene. *Z. Physiol. Chem., 319*:151–160.

Landsteiner, K., and P. Levine (1928a). On individual differences in human blood. *J. Exp. Med., 47*:757–775.

Landsteiner, K., and P. Levine (1928b). On the inheritance of agglutinogens of human blood demonstrable by immune agglutinins. *J. Exp. Med., 48*:731–749.

Levine, P., F. Ottensooser, M. J. Celano, and W. Pollitzer (1955). On reactions of plant anti-N with red cells of chimpanzees and other animals. *Am. J. Phys. Anthropol., 13*:29–36.

Lisowska, E. (1963). Reaction of erythrocyte mucoproteins with anti-N phytoagglutinins from *Vicia graminea* seeds. *Nature, Lond., 198*:865–866.

Mäkelä, O., and K. Cantell (1958). Destruction of M and N blood group receptors of human red cells by some influenza viruses. *Ann. Med. Exp. Biol. Fenn., 36*:366–374.

Nagai, Y., and G. F. Springer (1962). Partial hydrolysis of isolated blood group M antigen. *Fed. Proc., 21*:67d.

Paul, J. R., and W. W. Bunnell (1932). The presence of heterophile antibodies in infectious mononucleosis. *Am. J. Med. Sci., 183*:90–104.

Race, R. R., and R. Sanger (1968). *Blood Groups in Man,* p. 121, 5th edition. Philadelphia: Davis Co.

Rapport, M. M., L. Graf, and R. Ledeen (1970). Reactivity of antiganglioside sera with asialoganglioside. *Fed. Proc., 29*:1834.

Springer, G. F. (1969). Mammalian erythrocyte receptors: their nature and their significance in immunology, pp. 47–62. In *Current Problems in Immunology,* I. O. Westphal, H. E. Book and E. Grundmann (eds.), Berlin/Heidelberg/New York: Springer-Verlag.

—— (1970). Role of human cell surface structures in interactions between man and microbes. *Naturwissenschaften, 57*:162–171.

—— (1971a). 3rd Annual Meeting of German Society for Immunology, in press.

Springer, G. F., and N. J. Ansell (1958). Inactivation of human erythrocyte agglutinogens M and N by influenza viruses and receptor destroying enzyme. *Proc. Nat. Acad. Sci. U.S.A., 44*:182–189.

Springer, G. F., W. E. Briles, and H. Tegtmeyer (1971b). Manuscript in preparation.

Springer, G. F., and K. Hotta (1963). Blood-group N specificity and sialic acid. *Fed. Proc., 22*:2261.

Springer, G. F., and S. V. Huprikar (1972a). On the biochemical and genetic basis of the human blood-group MN specificities. *Haematologia 6*: 81–92.

Springer, G. F., S. V. Huprikar, and E. Neter (1970). Specific inhibition of endotoxin coating of red cells by a human erythrocyte membrane component. *Infect. and Immunity, 1*:98–108.

Springer, G. F., S. V. Huprikar, and H. Tegtmeyer (1971c). Biochemical-genetics of human blood-group MN specificities and their relation to infectious mononucleosis- and oncogenic virus-receptors. *Z. Immunitatsforsch., 142*:99–102.

—— (1971d). The biochemical and genetic basis of human blood-group MN specificities and their relation to infectious mononucleosis- and oncogenic virus-receptors. In *Glycoproteins of Blood Cells and Plasma*, vol. IV, G. A. Jamieson and T. J. Greenwalt (eds.), Philadelphia: J. B. Lippincott, Chapter III, pp. 35–49.

Springer, G. F., Y. Nagai, and H. Tegtmeyer (1966a). Isolation and properties of human blood-group NN and meconium -Vg antigens. *Biochem.*, 5:3254–3272.

Springer, G. F., M. Seifert, and J. C. Adye, 1972b. Immunogenicity of isolated infectious mononucleosis (IM) antigens in man. Fed. Proc. *31*:3242.

Springer, G. F. H. Tegtmeyer, and S. V. Huprikar (1972c). Anti N Reagents in Elucidation of the Genetical Basis of Human Blood-Group MN Specificities. Vox Sanguinis *22*:325–343.

Springer, G. F., E. T. Wang, J. M. Nichols, and J. M. Shear (1966b). Relations between lipopolysaccharide structures and those of human cells. *Ann. N.Y. Acad. Sci., 133(2)*:566–579.

Stalder, K., and G. F. Springer (1960). M and N agglutinin inhibition by human kidney and erythrocyte extracts. *Fed. Proc., 19*:707.

—— (1962). Serological characterization of isolated receptors of the MN erythrocyte system. *Proc. 8th Eur. Congr. Haematol.*, 489.

Wiener, A. S. (1938). The agglutinogens M and N in anthropoid apes. *J. Immunol., 34*:11–18.

Wiener, A. S., J. Moor-Jankowski, and E. B. Gordon (1971). Blood groups of gorillas. *Z. Kriminalistik Forensische Wissenschaften*, in press.

EFFECT OF PERIODATE OXIDATION AND BOROHYDRIDE REDUCTION ON THE BIOLOGY OF THE MAJOR GLYCOPROTEIN OF HUMAN ERYTHROCYTES

RICHARD J. WINZLER

Professor of Chemistry, Florida State University, Tallahassee, Florida

Sialic acids occur as terminal nonreducing sugars in the oligosaccharide chains of the glycoproteins and glycolipids associated with cell membranes. Because of its carboxyl group, sialic acid provides these glycoproteins and glycolipids with a negative charge at physiological pH. Where detailed studies have been carried out, as in erythrocyte membranes, it appears that the carboxyl groups of sialic acid provide most of the ionogenic groups in the cell membrane.

Sialic acids also occur in many nonmembrane glycoproteins. In some instances enzymatic removal of sialic acid modifies the biological activity of glycoproteins. Thus the biological activities of the glycoprotein hormones FSH (GOTTSCHALK and FAZEKAS DE ST. GROTH, 1960), human chorionic gonadotrophin (GOT and BOURRILLON, 1961), and erythropoietin (RAMBACH *et al.*, 1958) are lost after treatment with neuraminidase. The capacity of the intrinsic factor, a sialo-glycoprotein isolated from gastric mucosa, to bind vitamin B_{12}, is decreased or abolished by treatment with neuraminidase (WOLFF and NABET, 1962; FAILLARD *et al.*, 1962). In all instances studied, one or more of the physical properties of glycoproteins is altered by enzymatic removal of N-acetyl-neuraminic acid.

We have been interested in the question of whether the biological significance of the sialic acids in glycoproteins and cell membranes is due primarily to the presence of its negative charge, and whether the intact sialyl group is required for the biological activities of glycoproteins.

Our approach to this problem has been to modify the sialyl groups of biologically active glycoproteins by oxidation with very low concentrations of periodate followed by reduction of the resulting aldehydes to the

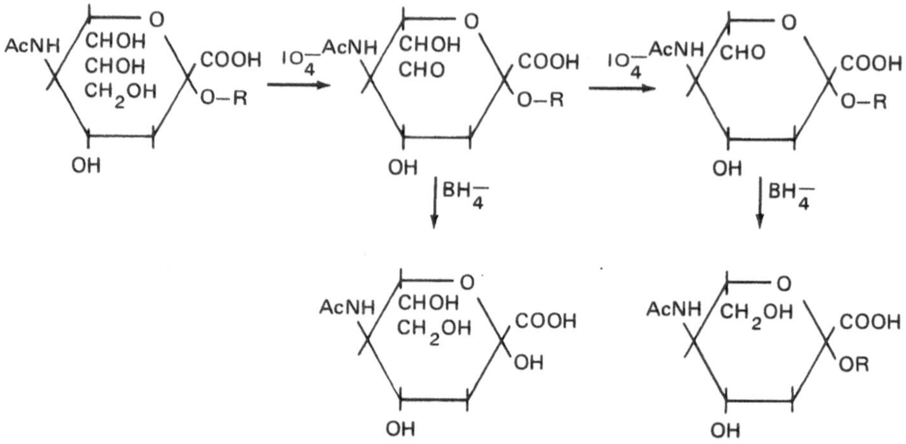

8 carbon analogue   7 carbon analogue

Fig. 1. Reactions in the modification of the sialyl groups of glycoproteins by periodate oxidation and borohydride reduction.

corresponding alcohols. This process produces glycoproteins in which the sialic acid is shortened by one or by two carbon atoms. This procedure has been employed by KUHN and GAUHE (1965), by McLEAN *et al.* (1971), by SUTTAJIT and WINZLER (1971), by SUTTAJIT *et al.* (1971a, b), and by VAN LENTEN and ASHWELL (1971) for the modification of glycoproteins. The reaction is shown in Figure 1.

The work I shall describe was carried out with the glycoprotein extracted from human erythrocyte stroma with 45% phenol at 65°C as previously described (KATHAN and WINZLER, 1963). In this procedure the stromal glycoprotein goes into the aqueous phase from which it can be isolated by alcohol precipitation and gel filtration.

Modification of the glycoprotein by periodate oxidation and borohydride reduction was carried out as shown in Table 1.

The extent to which the sialyl groups were modified by the procedure

Table 1. Modification of Erythrocyte Glycoprotein

1. Dissolve 2 mg of glycoprotein (containing 3.5 μmoles of NANA) in 2 ml of pH 5 cold acetate buffer.
2. Add 0 to 10 μmoles of $NaIO_4$.
3. Incubate for 2 hours in the dark at 4°C.
4. Add 0.1 ml of 10% glycerol to destroy excess periodate.
5. Dialyze against several volumes of water in cold overnight.
6. Bring dialysate to pH 9 to 10 with $NaHCO_3$.
7. Add 2 mg of solid $NaBH_4$; mix and incubate at 20°C for 2 hours.
8. Dialyze and lyophilize.

was determined by gas liquid chromatography of the trimethyl silyl derivatives modified from the procedure of CRAVEN and GERCKE (1968) as previously described (SUTTAJIT and WINZLER, 1971). Gas liquid chromatographs of erythrocyte membrane glycoprotein modified, using a molar ratio of periodate to sialic acid of 0, 1, and 3, is shown in Figure 2.

These data show that N-acetylneuraminic acid is converted to the 8-carbon and 7-carbon analogues by this procedure. No other detectable changes in the neutral sugars, amino sugars, or amino acids could be detected in these samples. Similarly, no changes in electrophoretic mobility, gel electrophoresis or sedimentation constant of the erythrocyte glycoprotein could be detected.

Three biological properties of the modified erythrocyte glycoprotein were studied. These were (1) the capacity of neuraminidases to liberate sialic acid analogues; (2) the capacity to inhibit hemagglutination by influenza virus; and (3) the capacity to combine with anti-M and anti-N antibodies.

In addition, the capacity of the purified 8- and 7-carbon analogues to be cleaved to pyruvate and acetamido sugar by purified neuraminic acid aldolase has been studied. Finally, the effect of similar modification on the biological activity of ovine and human follicle-stimulating hormone, sialoglycoproteins whose activities depend on the presence of sialic acid, have been investigated.

The liberation of sialic acid and its analogues from modified glycoproteins by neuraminidase from *Vibrio chloera* was followed by gas-liquid chromatography. Figure 3 compares the release of sialic acid and its 8- and 7-carbon analogues from the erythrocyte glycoprotein oxidized with an IO_4/NANA ratio of 1, corresponding to the preparation shown in Figure 2. The 8-carbon analogue is split from the glycoprotein at an initial rate about 35% of, and the 7-carbon analogue at a rate about 10% of that of NANA. Very similar results were obtained with neuraminidase from *Clostridium perfringens* and from PR-8 influenza virus. The differences were even more evident with low-molecular-weight substrates. Disaccharides of NANA or its analogues linked $1 \rightarrow 6$ to acetyl galactosaminitol were prepared by modifying ovine submaxillary mucin with periodate oxidation and borohydride reduction, and isolating the reduced disaccharides containing NANA or its 8-carbon or 7-carbon analogues from the supernatant by ion-exchange chromatography after treatment with alkaline borohydride (SUTTAJIT and WINZLER, 1971). Using these disaccharides as substrates for *V. cholera* neuraminidase, it was found that NANA-8 was cleaved at a rate about 30% of that of NANA, while the NANA-7 was scarcely cleaved at all. Similar results were obtained with neuraminidases from *C. perfringens* and from PR-8 influenza virus. It is evident from these results that the complete polyhydroxyl side chain is required for sialic acid to react maximally with neuraminidase.

It was of interest to determine whether similar modification of the

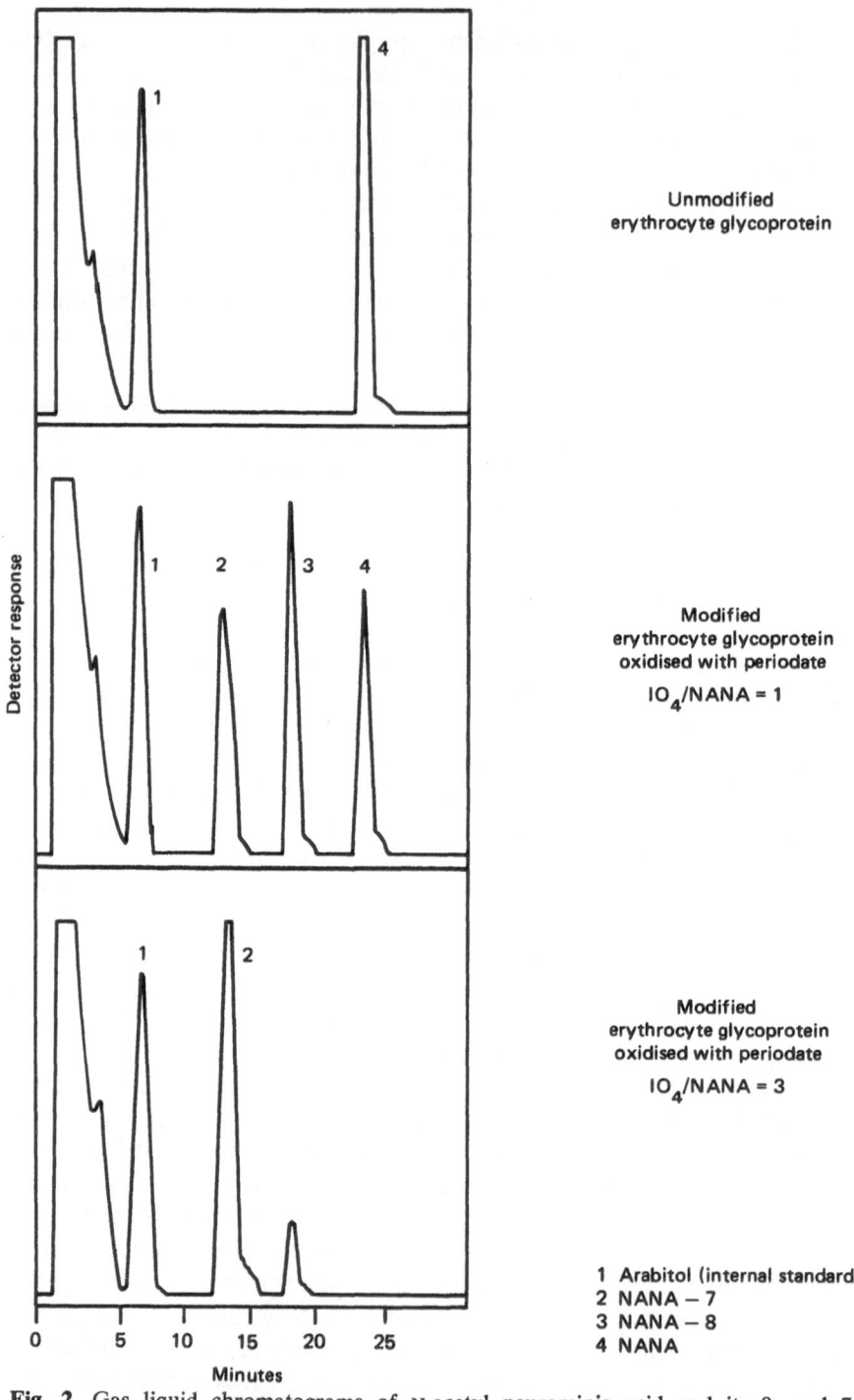

Fig. 2. Gas liquid chromatograms of N-acetyl neuraminic acid and its 8- and 7-carbon analogues in native erythrocyte glycoproteins and that modified by periodate oxidation and borohydride reduction as described in Table 1. 1 = arabitol internal standard; 2 = 7 carbon analogue of NANA; 3 = 8 carbon analogue of NANA; 4 = NANA.

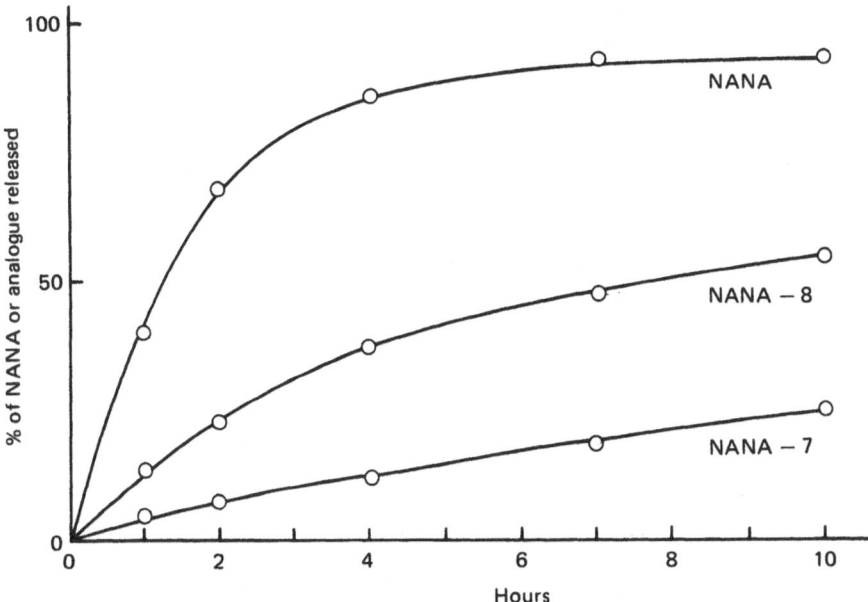

Fig. 3. Release of sialic acid and its 8- and 7-carbon analogues from modified erythrocyte glycoprotein by neuraminidase from *Vibrio cholera*. The erythrocyte glycoprotein was that shown in Figure 2 modified by oxidation with a ratio of periodate to NANA of 1, and reduced with borohydride as described in Table 1.

erythrocyte glycoprotein has any influence on its capacity to combine with the hemagglutinin of influenza virus. This was tested by modifying the glycoprotein at various ratios of periodate to NANA, and determining the minimal amount required to inhibit hemagglutination by one hemagglutination unit of indicator PR-8 influenza virus. Results from such an experiment (shown in Table 2) demonstrate that the capacity of the glycoprotein to combine with influenza virus is markedly reduced but not completely abolished by the modification of its sialyl groups.

Dr. Calderon Howe has kindly tested the native and modified erythrocyte glycoproteins for their capacity to inhibit agglutination of chicken erythrocytes by eight strains of myxovirus. It was found that the inhibitory activity was destroyed by periodate oxidation and borohydride reduction when tested with five strains (FM1, PR301, PR8, A/Jap/57, and GL). It seems quite likely, therefore, that the polyhydroxyl side chain of sialic acid is required for most effective binding of glycoproteins to myxovirus hemagglutinins.

It has been demonstrated by many investigators that the erythrocyte glycoprotein carries the serological M and N specificity of the erythrocytes. We therefore turned to the question of whether modification of the sialyl groups affected the capacity of the erythrocyte glycoprotein to react specifically with M and N antibodies. In this investigation, one hemagglutinat-

Table 2. Effect of Periodate Oxidation and Borohydride Reduction of the Erythrocyte Glycoprotein on the Content of NANA and its Analogues and on its Inhibition of Hemagglutination by Indicator PR-8 Influenza Virus

IO_4/NANA	HAI activity μgm/HA	Sialic acid and its analogues μmoles/mg		
		NANA	NANA-8	NANA-7
Native	0.2	.64	—	—
0	0.2	.64	—	—
0.2	0.3	.71	trace	trace
0.6	0.6	.56	.19	trace
0.8	0.9	.46	.10	.01
1.25	1.3	.10	.31	.28
3.0	1.3	trace	.11	.41
4.0	1.3	0	trace	.43

ing unit of rabbit antibodies directed against the MM or NN antigens on human erythrocytes was mixed with serially diluted native or modified MM or NN erythrocyte glycoprotein. The corresponding MM or NN erythrocytes were then added, and hemagglutination noted.

Results (shown in Table 3) suggest that the capacity of the erythrocyte

Table 3. Effect of Modification of Erythrocyte Glycoprotein on MM and NN Activity

Sample IO_4/NANA	NN antibodies NN glycoproteins NN erythrocytes μgm/HA	MM antibodies MM glycoproteins MM erythrocytes μgm/HA
native	0.25	0.50
0	0.25	0.50
0.25	0.25	1.0
0.5	0.5	2.0
1.0	1.0	4.0
2.0	1.0	16.0
3.0	1.0	32.0

glycoprotein to combine with specific NN and MM antibodies is reduced by modification of the sialyl groups. The sensitivity of the MM antigen, however, is much greater than that of the NN antigen. While these results cannot be considered as conclusive, because of the complexity of the antigens and the lack of knowledge of the chemical basis of the M and N specificity, they do suggest that the intact sialyl group, including its polyhydroxyl side chain, is required for the maximal serological activity of the MN antigens.

Other related studies on the effect of modification of the polyhydroxyl side chain of sialic acid on its matabolic relationships have been carried out. One of these studies, described by SUTTAJIT *et al.* (1971b), involved the study of the capacity of N-acetylneuraminic acid and its crystalline 8-carbon and 7-carbon analogues to be cleaved to pyruvate and acetamido sugars by purified neuraminic acid aldolase from *C. perfringens*. Their results are summarized in Table 4.

Table 4. Action of Neuraminic Acid
Aldolase from *Clostridium
Perfringens* in NANA, NANA-8,
and NANA-7*

Substrate	Relative rate of pyruvate formation
NANA	100
NANA-8	20
NANA-7	0

* SUTTAJIT *et al.*, (1971).

Here again, the capacity of sialic acid to react with an enzyme involved in its metabolism is markedly reduced when the polyhydroxyl side chain is shortened by one carbon and is abolished when it is shortened by two carbons.

On the other hand, SUTTAJIT *et al.* (1971a) have found that modification of the polyhydroxyl side chain of sialic acid of ovine and human follicle-stimulating hormone had relatively little influence on the *in vivo* activity of the hormone, although it is well known that removal of sialic acid completely abolished its *in vivo* activity. Ashwell and his collaborators (personal communication) have made a similar observation with human chorionic gonadrophic hormone.

The exquisite sensitivity of sialic acid to periodate oxidation, along with the external location of the sialic acid in the cell surfaces, should provide a means of selectively labeling the sialic acid-containing glycoproteins in cell membranes. VAN LENTEN and ASHWELL (1971) have already applied this procedure to the labeling of ceruloplasmin and orosomucoid and to a study of their biological half-life. Such labeling would involve a very brief treatment of cells with low concentrations of periodate, followed by reduction with tritium-labeled borohydride to produce cell surface components containing the 8-carbon analogue of sialic acid labeled with tritium.

This procedure should permit studies on isolation and characterization of surface antigens from small numbers of cells.

In summary, it would appear that, in some systems at least, modification of the polyhydroxyl side chain of sialic acid may modify the biological activity and function of glycoproteins in cell membranes.

References

Craven, D. A., and C. W. Gehrke (1968). Quantitative determination of N-acetylneuraminic acid by gas-liquid chromatography. *J. Chromatog.*, *37*:414–421.

Faillard, H., W. Pribilla, and H. E. Posth (1962). Die Einwirkung von Neuraminidase und Papain auf "intrinsic factor"—aktive Mucoid des Mugens von Mensch und Schwein. *Z. Physiol. Chem.*, *327*:100–108.

Got, R., and R. Bourrillon (1961). Enzyme release of sialic acid and inactivation of human menopausal gonadotropin. *Nature, Lond.*, *189*:234–235.

Gottschalk, A., and S. Fazekas De St. Groth (1960). Relation between the indicator profile prosthetic groups of mucoproteins inhibitory for influenza virus hemagglutinin. *J. Gen. Microbiol.*, *22*:690–697.

Kathan, R. H., and R. J. Winzler (1963). Structure studies on the myxovirus hemagglutination inhibitor of human erythrocytes. *J. Biol. Chem.*, *238*:21–25.

Kuhn, R., and A. Gauhe (1965). Bestimmung der Bindungsstelle von Sialinsäureresten in Oligosacchariden mit Hilfe von Perjodat. *Chem. Ber.*, *98*:395–413.

McLean, R. L., M. Suttajit, J. Beidler, and R. J. Winzler (1971). N-acetylneuraminic acid analogues. I. Preparation of the 8-carbon and 7-carbon compounds. *J. Biol. Chem.*, *246*:803–809.

Rambach, W. A., R. A. Shaw, J. A. D. Cooper, and H. Alt (1958). Acid hydrolysis of erythropoietin. *Proc. Soc. Exp. Biol. Med.*, *99*:482–483.

Suttajit, M., and R. J. Winzler (1971). Effect of modification of N-acetylneuraminic acid on the binding of glycoproteins to influenza virus and on susceptibility to cleavage by neuraminidase. *J. Biol. Chem.*, *246*:3398–3404.

Suttajit, M., L. E. Reichert, and R. J. Winzler (1971a). Effect of modification of N-acetylneuraminic acid on the biological activity of human and ovine follicle-stimulating hormone. *J. Biol. Chem.*, *246*:3405–3408.

Suttajit, M., C. Urban, and R. L. McLean (1971b). N-acetylneuraminic acid analogues. II. The action of N-acetylneuraminic acid aldolase on 8-carbon and 7-carbon analogues. *J. Biol. Chem.*, *246*:810–814.

Van Lenten, L., and G. Ashwell (1971). Studies on the chemical and enzymatic modification of glycoproteins: A general method for the tritiation of sialic acid-containing glycoproteins. *J. Biol. Chem.*, *246*:1889–1894.

Wolff, R., and P. Nabet (1962). Action de la neuraminidase sur l'activité— facteur intrinsique de l'extrait purifié de muqueuse gastrique chez l'homme. *Bull. Soc. Chim. Biol.*, *44*:1009–1012.

EFFECT OF CETYLTRIMETHYLAMMONIUM BROMIDE ON HUMAN ERYTHROCYTE MEMBRANES†

A. M. Abdelnoor*, M. Higgins and A. Nowotny

*Department of Microbiology and Immunology,
Temple University School of Medicine,
Philadelphia, Pennsylvania*

I. Introduction

It is generally agreed that membranes have a mosaic-like structure on the surface and are formed by a number of layers. Weinstein (1969) studied the morphology of the red blood cell membrane, using freeze-cleaved preparations. He assumed that the cleavage plane passes along the true outer surface of the membrane and that there is an extraneous coat which covers a layer with granulated structure. Branton (1966) and Pinto da Silva and Branton (1970), on the other hand, assume that freeze etching splits the membrane, exposing its inner matrix. The inner matrix contains particles which appear as granules on the surface of this layer. The granules protrude to the outer surface of the erythrocyte.

Pinto da Silva et al. (1971) have shown that the A antigenic sites of human erythrocytes are associated with the granules occurring on the membranes. Whether only these granules are the antibody-reactive sites or whether there are some hidden receptors in the inner layers of the membrane remains to be investigated. It was found previously that red blood cells (RBC) treated with certain agents such as enzymes, viruses, or bacterial filtrates will expose hitherto masked receptor sites (Burnet et al., 1946; Iseki and Furukawa, 1959; Morton and Pickles, 1947). It has also been reported by others that a decrease in or complete abolishment of existing agglutinability of the RBC will result from such treatment (Watkins and Morgan, 1954; Schiff and Akune, 1931; Springer and Ansell, 1958). Several of these findings have been reviewed earlier (Springer, 1963).

† This work has been supported by PHS grant No. 1 FO2 AI 46928
* PHS postdoctoral fellow.

It has been shown in the authors' laboratories that cetyltrimethylammonium bromide (CTB), a cationic detergent, released blood group active components from RBC stroma. The activity of the CTB-released product was significantly greater than that of the intact stroma, determined and compared on a weight basis, by measuring the capacity of the preparations to inhibit hemagglutination (NOWOTNY and O'NEILL, 1964; RADVANY, 1971). These results suggested that there are hidden iso-antibody reactive sites in the RBC membrane. In the present study, intact erythrocytes were treated with concentrations of CTB that would not cause lysis. Hemagglutination of the treated and untreated cells was measured and compared. Furthermore, electron micrographs of freeze-cleaved preparations of CTB-treated and untreated erythrocytes were made.

II. Materials and Methods

A. Blood

Group A and B blood was drawn from donors in oxalated vacutainers and used on the same day.

B. Anti-A and Anti-B Human Sera

These antisera were obtained from Mr. Joseph Smolens, Director of the Philadelphia Blood Bank.* The titers of the sera were 360.

C. Treatment of Group A and B RBC with CTB

Amounts of 0.0025%, 0.01%, 0.025%, 0.05% and 0.25% CTB solutions in saline were prepared. Both group A and B erythrocytes were treated in the following manner: Cells were separated from plasma and washed three times with saline. To each 0.5 ml of packed-cell sediment, 1.25 ml of the above CTB solutions were added. They were mixed gently, allowed to stand at 25°C for 15 minutes, and then centrifuged. The supernates were dialyzed against distilled water for five days and were then tested for their ability to inhibit agglutination of untreated RBC by the corrsponding antisera. The treated erythrocytes were washed and centrifuged three times with saline, and 2% suspensions in saline were prepared. The ability of these CTB-treated cells to agglutinate with the corresponding antibody and also with influenza virus PR8/E37 strain was tested. Treated and untreated erythrocytes used for freeze cleaving were suspended in 20% glycerol.

D. Hemagglutination Tests

Starting with a 1:10 dilution of antiserum or of an influenza virus suspension, twofold dilutions were carried out in 0.1 ml volumes of saline. 0.1 ml of the RBC suspension to be tested was added to each dilution, mixed, and allowed to stand at room temperature for 1 hour. The degree

---

* The authors gratefully acknowledge the supply of these reagents.

of agglutination was read following an arbitrary 10° scale and expressed as hemagglutination units (NOWOTNY, 1969).

E. Hemagglutination Inhibition

An 0.1 ml amount of supernate obtained from CTB-treated RBC suspensions was mixed with 0.1 ml of a 1:10 dilution of antiserum. 0.1 ml of saline mixed with 0.1 ml of a 1:10 dilution of antiserum served as a control. The mixture was allowed to stand at 25°C for 30 minutes. Twofold dilutions were carried out using saline as a diluent. 0.1 ml of a 2% untreated RBC suspension was added to each dilution, mixed, and allowed to stand at 25°C for 1 hour. The degree of agglutination was read following an arbitrary 10° scale. To express the hemagglutination inhibition activity of the preparations, the dry weight contents of the supernates as well as of other suspensions were determined, and the amount of substance causing a 50% decrease in agglutination (HI_{50}) was calculated. (For details of the procedure, see NOWOTNY, 1969; RADVANY, 1971.)

F. Freeze-etched Preparations of Normal and CTB-treated RBC

Before freezing, glycerol was added to all preparations to achieve a final concentration of 20%. Material was frozen and then freeze-fractured in a Balzers freeze etch unit BA3600M, according to instruction manual No. A11-3992e. The freeze-etching time was 60 seconds. Replicas were examined in a Siemens Elmiskop 1A at instrumental magnifications ranging between 7000 and 30,000 ×.

III. Results

A. Hemagglutination of CTB-treated RBC by Antiserum and Influenza Virus

There was an increase in agglutination of both A and B CTB-treated RBC by anti-A or anti-B sera. This increase was greater in the case of A cells (see Figure 1). CTB treatment of RBC had no effect on agglutination of the cells by influenza virus. To check the specificity of this reaction, group A erythrocytes were treated with CTB and exposed to B antiserum. Similarly, B RBC were incubated with A antiserum. None of these showed any signs of agglutination.

B. Hemagglutination Inhibition

The supernates obtained from both A and B CTB-treated erythrocytes inhibited hemagglutination by the corresponding antisera. Table 1 gives the results expressed in HI_{50}'s. It can be seen from these results that the supernate from B cells appears to be considerably more efficient in inhibiting agglutination than supernate from A cells.

In order to rule out the possibility that enhanced hemagglutination in-

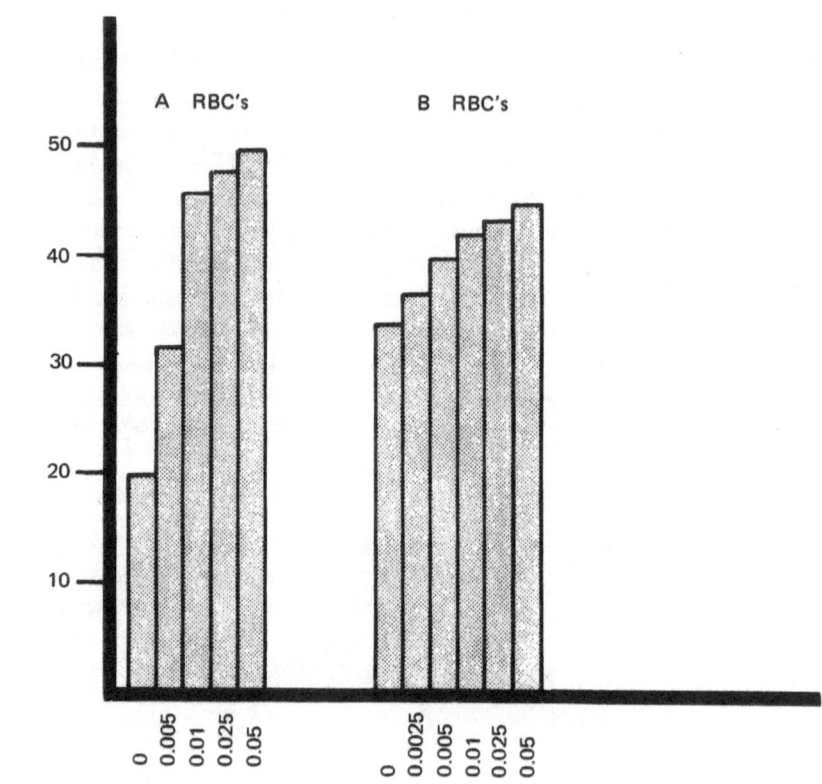

Fig. 1. Hemagglutination of CTB-treated red blood cells. Group A and B erythrocytes were treated with different concentrations of CTB. Hemagglutination of these treated cells by antiserum was tested.

X-axis: Erythrocytes used after treatment with different concentrations of CTB. Y-axis: Degree of hemagglutination obtained, expressed in hemagglutination units (H.U.).

Table 1. HI₅₀ of CTB Supernates from A and B Red Blood Cells*

	0.0025 % CTB	0.005 % CTB	0.01 % CTB	0.025 % CTB	0.05 % CTB
A supernate	>2000	>2000	316	ND†	ND†
B supernate	251	159	100	159	ND†

* HI₅₀ values are expressed as µg dry weights.
† Not done.

hibitory capacity of detergent solubilized RBC membrane preparations is due to the presence of non-dialyzable, bound detergent, the following experiment was carried out. The supernates were incubated not with their corresponding antisera, but A supernate was mixed to anti B serum and vice versa. The hemagglutination was not inhibited at all.

C. Electron Microscopy of Freeze-cleaved Preparations of Normal and CTB-treated Erythrocytes

Electron micrographs of freeze-etched preparations of normal RBC showed two types of membrane layers. One contains numerous granules, while the other shows relatively fewer of these granules (Figure 2a). The average size of these granules ranged between 85 and 138 Å. Similar results were obtained with erythrocytes treated with 0.01% CTB. Cells treated with 0.25% CTB appeared to have larger granules in addition to very fine particles dispersed between them. The average size of the larger granules ranged from 194 to 242 Å. In addition, extensive multi-layered vesicle formation could be observed in the 0.25% CTB-treated cells (see Figure 2b).

IV. Discussion

Several possible mechanisms can be considered to explain the enhanced agglutinability of the CTB-treated RBC by their corresponding antibodies. POLLACK and RECKEL (1970) have shown that changes in the zeta potential of erythrocytes will cause changes in their hemagglutinating ability. It is possible that CTB caused a change in the zeta potential of the erythrocyte surfaces. However, if these changes occurred, it did not seem to influence the agglutinability of the erythrocytes by influenza viruses. The only visible difference between the treated and untreated cell surface was the average diameter of the granules, seen on electron micrographs. Furthermore, there was fragmentation and vesicle formation of the membranes which obviously added to the increase in the specific surface of the substance. These may have contributed to the enhanced reactivity with the corresponding antibodies.

NICOLSON et al. (1971) compared the distribution of $RH_o(D)$ antigenic sites on the membrane and the closeness of the receptor sites. These were measured by ferritin-labeled antibodies. If the number and size of the granules were directly proportional to the antibody receptor sites, one could visualize a greater degree of clumping of erythrocytes caused by the same number of antibodies in the treated cell system compared to the untreated one.

CTB washing removes not only inert materials from the cell surfaces, but seems to wash away antibody-reactive receptors as well. This has been shown by the hemagglutination-inhibiting activity of the supernates. One should also mention here that the activity of these supernates expressed

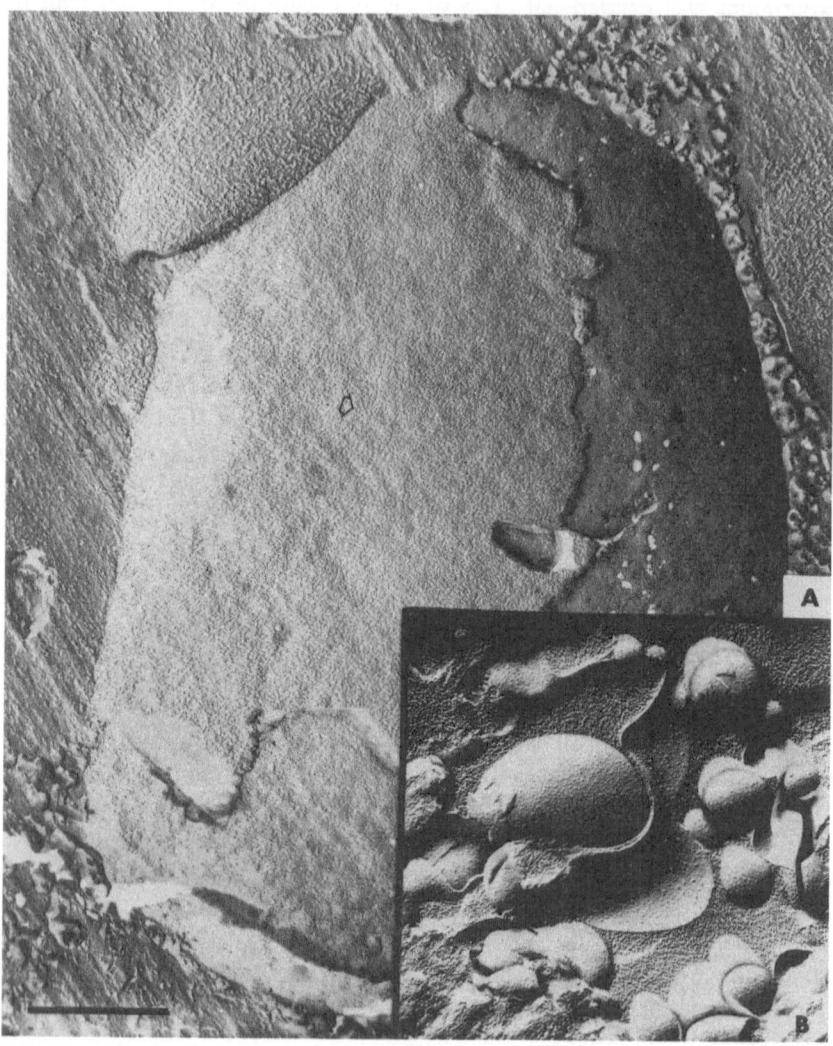

Fig. 2. Freeze fractures of normal and CTB-treated blood cells. A is a convex fracture of a normal cell showing a membrane surface with few granules (dark arrow) and another surface with numerous granules (light arrow). B is a preparation that has been treated with 0.25% of CTB. Note the presence of small multilayered vesicles. The bar in A equals 1 μ and also applies to B.

as HI_{50} was approximately 50 times lower than the activity of partially purified blood group substances isolated from erythrocyte membranes by different procedures (RADVANY 1971).

Whether or not the granules on the exposed surfaces are rich in antibody receptor sites cannot be determined from these experiments. The appearance of larger and better-exposed granules after CTB treatment, on the other hand, indicates that erosion and removal of some membrane constitutents follows the treatment with dilute detergents. One can visualize the exposure of more resistant structural elements of the erythrocyte membrane to be more apparent after this erosive treatment.

FRIEDENREICH (1928) described the existence of T antigens, common to many erythrocytes belonging to different blood groups. CTB treatment apparently did not uncover the T antigenic sites, since the group specificity of the cells was fully maintained.

Earlier studies on some physico-chemical properties of the A and B cells as well as their sensitivity to proteolytic enzymes revealed differences between group A and B RBC (NOWOTNY, 1954a, b). These differences could not be explained solely by the presence of A and B blood group substances in the chemical make-up of the erythrocyte surfaces. The hemagglutination assay described here revealed another difference between A and B cells. The increase in agglutination of group A cells treated with CTB was always greater than that obtained with group B cells. While group B cells start from a relatively high level of agglutinability, the group A cells have a much lower activity, but due to CTB treatment they rapidly approach the activity levels of the B cells. The fact that the released substances present in the CTB washings from A cells were less active than washings obtained from B cells indicates that this treatment removed more inert material from group A. It seems that in group A erythrocytes more receptors are hidden by some inert material than is the case in group B.

V. Summary

Group A and B erythrocytes were treated with low concentrations of cetyltrimethylammonium bromide (CTB). The hemagglutinability of the treated cells by the corresponding anti-A or anti-B sera was determined. The group activity of the CTB-eluted substance was measured by using the hemagglutination inhibition procedure. Freeze-cleaved electron microscopy was carried out to investigate changes on the cell surfaces. The results indicate that mild washing with the above detergent uncovers hidden antibody receptor sites in the membrane.

References

Branton, D. (1966). Fracture faces of frozen membranes. *Proc. Nat. Acad. Sci. U.S.A.*, *55*:1048–1056.

Burnet, F. M., J. F. McCrea, and J. D. Stone (1946). Modification of human red cells by virus action. I. The receptor gradient for virus action in human red cells. *Brit. J. Exp. Path.*, *27*:228–236.

Friedenreich, V. (1928). Investigations into the Thomsen hemagglutination phenomena. *Acta Path. Microbiol. Scand.*, *5*:59–101.

Iseki, S., and K. Furukawa. (1959). On blood group specific decomposing enzymes derived from bacteria. *Proc. Jap. Acad.*, *53*:620–625.

Morton, J. A., and M. M. Pickles (1951). The proteolytic enzyme test for detecting incomplete antibodies. *J. Clin. Path.*, *4*:189–199.

Nicolson, G. L., S. P. Masouredis, and S. J. Singer (1971). Quantitative two-dimensional ultrastructural distribution of $Rh_o(D)$ antigenic sites on human erythrocyte membranes. *Proc. Nat. Acad. Sic. U.S.A.*, 1416–1420.

Nowotny, A. (1955a). Studies on the chemical structure of A, B and O blood group antigens. I. Physico chemical examination of the A and B group red cell stroma. *Acta Physiol. Acad. Sci. Hung.*, *7*:31–42.

—— (1955b). Studies on the chemical structure of A, B and O blood group antigens. II. Effect of proteolytic enzymes on the blood group antigens of the erythrocyte membrane. *Acta Physiol. Acad. Sci. Hung.*, *8*:25–31.

—— (1969). *Basic Exercises in Immunochemistry*. New York: Springer-Verlag.

Nowotny, A., and G. J. O'Neill (1964). Immunochemical investigations of A and B blood group antigens localized in the erythrocyte membrane, K;21. Lecture, Xth Congr. Int. Soc. Hematol., Stockholm.

Pinto da Silva, P., and D. Branton (1970). Membrane splitting in freeze-etching. Covalently bound ferritin as a membrane marker. *J. Cell Biol.*, *45*:598–605.

Pinto da Silva, P., S. D. Douglas, and D. Branton (1971). Localization of A antigenic sites on human erythrocyte ghosts. *Nature, Lond.*, *232*:194–196.

Pollack, W., and R. P. Reckel (1970). The zeta potential and hemagglutination with Rh antibodies. *Int. Arch. Allergy Appl. Immunol.*, *38*:482–496.

Radvany, R. M. (1971). Isolation of blood group A and B antigens from human erythrocyte membranes. Ph.D. Thesis, Temple University, Philadelphia, Pennsylvania.

Schiff, F., and M. Akune (1931). Blutgruppen und Physiologie. *Muench. Med. Wichschr.*, *78*:657–660.

Springer, G. F. (1963). Enzymatic and non-enzymatic alterations of erythrocyte surface antigens. *Bact. Rev.*, *27*:191–227.

Springer, G. F., and N. J. Ansell (1958). Inactivation of human erythrocyte agglutinogens M and N by influenza viruses and receptor destroying enzyme. *Proc. Nat. Acad. Sci. U.S.A.*, *44*:182–189.

Watkins, W. M., and T. J. Morgan (1954). Inactivation of the H receptors on human erythrocytes by an enzyme obtained from *Trichomonas foetus*. *Brit. J. Exp. Path.*, *35*:181–190.

Weinstein, R. S. (1969). Electron microscopy of surfaces of red cell membranes, pp. 36–76. In *Red Cell Membrane Structure and Function*, G. A. Jamieson and T. J. Greenwalt (eds.), Philadelphia/Toronto: J. B. Lippincott Co.

ISOLATION OF ERYTHROCYTE MEMBRANE ANTIGENS*

A. NOWOTNY

Department of Microbiology and Immunology,
Temple University School of Medicine,
Philadelphia, Pennsylvania

I. Introduction

The membrane of human erythrocytes contains 40 to 50% proteins, 35 to 45% lipids, and approximately 10 to 14% carbohydrates. The arrangement of the different components in the membrane has been and still is the subject of active investigation as well as of controversy. Several theories have been offered which used the results of ultrastructural studies as well as chemical analyses and suggested structure models which would perform the multifaceted biological functions of membranes (SJÖSTRAND, 1963; LUCY, 1964; WHITTAM, 1964; KAVANAU, 1966; CHAPMAN, 1968; ZAHLER, 1968, 1969). Both these theoretical considerations and ultrastructural studies (HILLIER and HOFFMAN, 1953; GLAESER et al., 1966; BRANTON, 1967; BRANTON and PARK, 1967; PINTO DA SILVA et al., 1971) concluded that the great majority of biological membranes, including the red blood cells (RBC), have a mosaic-like structure on the surface and are formed by a number of layers.

Since the first reported observations of hemagglutination in the nineteenth century, the existence of several biologically active groups, antigens and other receptor sites were described on erythrocyte membranes. Table 1 gives an incomplete account of the antigens according to which human erythrocytes are grouped. Receptors for viruses are also present, and a number of different chemicals and natural products are known to adhere in a nonspecific manner to erythrocyte membranes. In this summary, we shall attempt to review the most efficient procedures for the isolation of the antigens and other receptors of the RBC.

---

* Chairman's summary of the session on "Erythrocyte Antigens."

Table 1. Some of the Antigens Found on the Human Red Blood
Cell Membrane

System	Antigens detected by specific antiserum	Chemical nature
Heterophile	Forssman	Fatty acids, Sphingosine, Carbohydrates
Species specific	I	Protein?
A_1, A_2, O	A_1, B, H	Glycoproteins, Glycolipids
M, N, S, d	M, N, S, s, U, Mg, M, T_m, 16 others	Glycoproteins
P	P_1, P^k, Luke	Unknown
Rh	D, C, c, C^w, C^x, E, e, 13 others	Glycolipid?
Lutheran	Lu^a, Lu^b	Glycoprotein?
Kell	K, k, Kp^a, Kp^b, Js^a, Js^b	Unknown
Lewis	Le^a, Le^b	Unknown
Duffy	Fy^a, Fy^b	Unknown
Kidd	Jk^a, Jk^b	Unknown
Diego	Di^a, Di^b	Unknown
Yt	Yt^a, Yt^b	Unknown
Xg	Xg^a	Unknown
Dombrock	Do^a	Unknown

II. Extraction Procedures and Findings

A. Solubilization and Isolation of Blood Group Antigens and Other Receptor Sites

Different ethanol concentrations were first used for the isolation of AB blood group substances from the RBC (LANDSTEINER et al., 1925), and are still used to obtain crude extracts suitable for further purification (KOSCIELAK and ZAKRZEWSKI, 1960; HAKOMORI and JEANLOZ, 1964; KORN, 1966; ZVILICHOVSKY et al., 1971). Other organic solvents, such as alcohol-ether or alcohol-chloroform mixtures, were also used (KLENK and LAUENSTEIN, 1951; NOWOTNY and BACKHAUSZ, 1957; RADIN, 1957; YAMAKAWA and IRIE, 1960; YAMAKAWA et al., 1960). Klenk and coworkers (KLENK and LAUENSTEIN, 1951) and simultaneously but independently Yamakawa and associates (YAMAKAWA and SUZUKI, 1951) were the first to isolate blood group active sphingolipids from human erythrocytes, using combinations of different organic solvents. All these methods gave a low yield of active fractions. Yamakawa's laboratory used 50 liters of human blood to obtain a few milligrams of active sphingolipid; others reported similarly low yields of active material (HAKOMORI and STRYCHARZ, 1968).

In spite of the low yields, the isolated substances were sufficient for structural analysis using microanalytical procedures. Their chemical structure has been extensively studied and greatly clarified (KLENK and HEUER, 1960; HANDA and YAMAKAWA, 1964). In addition to the aforementioned, the thorough studies of Hakomori and associates must be referred to here. HAKOMORI (1970) showed that a trihexoside ceramide structure, NAcGlu $1 \overset{\beta}{\to} 3$ Gal $1 \to 4$ Glu-ceramide, is present in A, B, Le^a blood group RBC, as well as in parenchimatous organs and in glandular tissues. Furthermore, the same trihexoside ceramide was found in adenocarcinomatous tumor tissue (HAKOMORI and JEANLOZ, 1964; HAKOMORI et al., 1967). The addition of one or more carbohydrates to this "core" gives group specificities to the cells. In tumors, the incompleteness of the carbohydrate chain was found by HAKOMORI and MURAKAMI (1968), and the authors assumed that this may also play a role in the lack of contact inhibition phenomenon.

In our laboratory, hemoglobin-free group A and B erythrocyte stroma have been prepared by washing the hemolytic residue either with 0.001 N acetic acid or with CO_2-saturated distilled water. If such preparations were exposed to an 0.01 N NaOH or LiOH concentration before lyophilization, partial solubilization could be achieved. When A or B activity of such solubilized substances was determined by hemagglutination inhibition, an enhanced activity could be measured (NOWOTNY, 1955).

A large variety of organic solvents such as ethanol, dimethyl sulfoxide, formamide, pyridine, formic acid, urea, phenol, and trichloroacetic acid in varying concentrations and in different combinations was also used in more recent experiments. Detergents, especially cetyltrimethylammonium bromide (CTB), were introduced by us in 1964 (NOWOTNY and O'NEILL, 1964) as the most effective solubilizing agents for RBC (see Table 2). The use of CTB not only released blood group active components of the erythrocyte, but seemed to enhance significantly the activity of dissolved stroma, most probably by exposing antibody-reactive sites hitherto hidden in the membrane structure. For further details, we refer to the paper of ABDELNOOR et al. in this volume (p. 153), as well as to another publication (RADVANY, 1971).

It seems important to emphasize here that the initial condition of the red blood cells before the extraction starts is quite critical. Some procedures, such as our solubilization with CTB, are much less effective if the starting material—the isolated membrane of the RBC—has been lyophilized before the extraction. While in this case approximately 40 to 50% of the total dry weight will go into "solution", over 90% of the stroma's dry weight can be solubilized if the stroma has not been lyophilized. Similarly, the butanol-water extraction procedure will not give significant quantities of water-soluble proteins if the isolated membranes have been lyophilized. The same could be observed earlier by attempting to dissolve the erythrocyte membranes in diluted alkali, as mentioned above. Lyophilization of cells alters their immunogenic properties also. Lyophilization appears, therefore, to have a profound effect on the conformation of membrane structures.

Disc electrophoresis of CTB-extracted RBC gave several distinguishable bands, but a larger number of more distinct bands could be obtained by the butanol-water extraction process developed for RBC protein solubilization by MADDY (1964). This procedure was used to obtain a protein fraction with A, M, and N activity from group A cells by POULIK and

Table 2. Extraction of Blood Group Antigen Containing RBC Membrane Components, Using Organic Solvents (Nowotny and O'Neill, 1964)

Extraction with*	% Total Extractable Material		% of Total Starting Activity Extracted†	
	Group A	Group B	Group A	Group B
45% phenol at 68°C	3.3	2.4	5.0	1.1
5% trichloroacetic acid at 4°C	10.4	11.3	10.0	15.2
0.5% cetyltrimethyl-ammonium bromide	91.0	79.1	450.0	506.0
Formamide at 150°C	100.0	100.0	31.4	34.6
N,N-dimethyl formamide	22.2	22.8	9.0	12.0
Ethanol‡	20.4	21.8	23.0	44.5
Dimethylsulfoxide	23.0	34.0	40.0	61.5
Pyridinium formate	17.2	27.2	49.5	68.5
Pyridine	20.6	22.2	23.4	34.3
Urea, 40%	36.6	45.2	45	42

* Extraction procedure of the wet, hemoglobin-free RBC membrane sediment varied. It was carried out in most cases by using glass tissue grinders, at room temperature, unless otherwise noted. Separation of soluble material was achieved by centrifugation at 10,000 × g for 60 min. Further purification involved precipitation with ethanol or with acetone and dialysis.

† The activity of the washed, hemoglobin-free RBC membrane was determined on dry weight basis by semi-quantitative hemagglutination inhibition method (Nowotny and Backhausz, 1957) and was taken as 100%.

‡ 20, 40, 60, 80 and 100% ethanol were used. Activity of the extracts was measured individually. The last columns give the sum of all percentages recovered.

LAUF (1969) and also by WHITTEMORE et al. (1969). No B cells were subjected to the same treatment. In our laboratories, Abdelnoor used the same procedure to extract A and B cells, and found that the water phase contains highly active A or B substances among a great number of other solubilized RBC components.

As reported at this conference, Davis and Bakerman used EDTA and 2-mercaptoethanol to solubilize group A RBC membrane. Fractionation involved gel-filtration chromatography and isoelectric focusing. The fraction, homogeneous in the above systems as well as in immunoelectrophoresis, showed high A activity and is characterized as lipoglycoprotein. Full details of their procedure are published in this volume (p. 121).

Other blood group antigens or receptors of RBC were also isolated. MN antigens were obtained from erythrocytes by the phenol-water method (SPRINGER et al., 1966), which procedure was found by us to be of limited use for the extraction of A or B antigens.

The chemical nature of the complicated Rh system has been less successfully investigated. CARTER (1949) reported the isolation of an Rh hapten lipid from erythrocytes. MOSKOWITZ and CALVIN (1952) isolated a lipoprotein called "elinin" with both Rh and A activities.

Other authors could not confirm these findings. WOLF and SPRINGER (1964) solubilized the RBC with alkali and fractionated it on a Sephadex column. MILGROM and LOZA (1969) solubilized Rh antigen of human RBC by tryptic digestion, and the preparation obtained reacted with Rh antisera in gel diffusion. GREEN (1968) obtained an active Rh extract from RBC by organic solvents and purified it on column chromatography. The active material was identified as phosphatidylcholine with some unsaturated fatty acids. A low-molecular-weight peptide, consisting of 13 amino acids, was isolated by WEICKER (1968). This preparation inhibited the hemagglutination of papain-treated Rh^+ RBC by anti-D serum. NICOLSON et al. (1971) described a method for quantitative ultrastructural studies of D-antigen distribution on RBC surface using isotope and ferritin-labeled antibodies. It was found that this antigen forms small clusters.

The Forssman antigen is present in sheep erythrocytes, among others, and in human A RBC. Its isolation and chemical analysis has been reported by several authors (PAPIRMEISTER and MALLETTE, 1955; RAPP and BORSOS, 1966). MAKITA et al. (1966) isolated Forssman hapten from equine organs and characterized it chemically and immunologically. ANDO and YAMAKAWA (1970) published the structure of the oligosaccharide portion of the Forssman hapten from equine organs and also characterized it chemically and immunologically.

Isolation of virus receptors from RBC was carried out by KLENK and LEMPFRID (1957), GOTTSCHALK (1960), KATHAN and WINZLER (1963), and SPRINGER et al. (1967).

A part of Winzler's lecture at this conference (p. 145) deals with the viral receptors of erythrocytes. These studies involve the effect of periodate oxidation of sialic acid on the reactivity of the membrane under the influence of a virus. Similarly, eight strains of myxovirus receptor sites were also investigated. It has been found that shortening of the carbon skeleton of the sialic acid by one or two carbon atoms is sufficient to abolish the reactivity of the membrane with the above viruses.

The paper given by Springer and coworkers at this conference (p. 133) reviews their studies on the isolation and chemical characterization of antibody, virus, and endotoxin receptor sites of human erythrocytes. An earlier extensive summary of RBC receptors was compiled by Springer in 1963.

B. Occurrence and Loss of Blood Group Antigens

Several RBC antigens or receptors occur in other mammalian cells (IVANYI, 1966; HAKOMORI et al., 1967), and they can also be detected on some bacteria (SPRINGER et al., 1966; PARDOE et al., 1968; SPRINGER and FLETCHER, 1969). The fact that some blood group active receptors disappear from human cells during development of malignancy has been used as a diagnostic tool to detect malignant tumors. The findings of Dr. Levine refer to this observation.

Dr. Levine who, with Dr. Landsteiner, pioneered the foundations of modern immunohematology many years ago, also participated at our conference. The remarks he made during the panel discussion are published here in their entirety.

"The loss of A and B antigens in fucose-containing glycoproteins from human malignant tissue has been shown by HAKOMORI and JEANLOZ (1970), who demonstrated that some of their 7 subfractions contained Le^a and Le^b substances not present on the red cells. This represents either a block in synthesis or a redifferentiation. Loss of antigens A and B was also demonstrated by the technique of specific red cell adhesion, even in old paraffin sections of malignant tissues, while adjacent normal tissues retained their normal complement of A and B antigens (DAVIDSOHN and NI, 1970). Tests for Lewis antigens by this technique are still to be carried out. It is uncertain whether or not these observations relate also to alcohol-soluble blood substances.

The complete absence of Rh antigens in Rh$_{null}$ is associated with an obvious defect in the red cell membrane, as indicated by a compensated hemolytic anemia (shortened red cell survival, increased fragility, reactions with both anti-I and anti-i, mild spherocytosis and high reticulocyte count), referred to as Rh$_{null}$ disease. However, there were no associated abnormalities in any of the 20 intraerythrocytic enzymes tested. In the parallel case of Bombay blood (failure of AB gene expression by the action of an independent suppressor gene in the homozygous state), of which there are many more examples than Rh$_{null}$, the membrane appears to be normal, as indicated by the absence of abnormal hematologic signs and symptoms. This suggests that the receptor site for Rh is strategically located on the red cell membrane so that its antigenic structure—probably a glycolipoprotein—is essential for maintaining the integrity of the membrane. That Rh$_{null}$ blood exerts a widespread effect on the membrane is indicated by its aberrant reactions on the SsU determinants; also, reactions for M and Fy^b appear to be increased (STURGEON, 1970).

A defect in the red cell membrane was observed in two very rare bloods (English and Finnish) called En (envelope) because they lack the normal antigen En^a. The individuals were recognized by their production of anti-En^a as a result of pregnancies or transfusions. Their red cells exhibit greatly diminished activity with anti-M, anti-N, and *Vicia graminea,* while their Rh activity is increased when tested in saline suspension with incomplete (7S) antibodies. Apparently their circulating red cells behave as though they have been pretreated with enzymes, and yet these individuals appear to be physically normal, and one of them is 70 years of age. Hematologic studies are still to be carried out. EnEn red cells have considerably diminishing sialic acid and slower electrophoretic mobility. (Heterozygous cells (En^{a}En) show intermediate values.) The only Rh$_{null}$ blood tested showed normal values."

Dr. Levine said that these observations are in harmony with Springer's findings (HUPRIKAR and SPRINGER, 1970) that (1) the active terminal group for M activity is sialic acid, and (2) β-galactose is required for N but especially for *V. graminea* activity. Springer found that antigen M contains a very small quantity of β-galactose and therefore a small amount of antigen N, thus providing a firm biochemical basis for hitherto unexplained serologic observations. The initial observation that M antigen contained some N was made in 1928 when it was found that anti-N sera rapidly lost their activity on absorption with M blood (LANDSTEINER and LEVINE, 1928).

"The antigen M^k determined by an operator (?) gene which fails to produce M, N, S, and s and, to a lesser degree the antigen M^g determined by a gene which does not make M or N, resemble En since they have lower-than-normal values of both sialic acid and electrophoretic mobility. In common with En, their Rh antigens also react in saline suspensions with 7S Rh antibodies, but not quite as strongly as En cells."

Dr. Levine concluded his comments by saying that it was hoped that future investigations would be extended to include also studies of membranes of the rare red cells discussed above.

C. Isolation of Other Immunogenic
Components of RBC Membranes

Erythrocyte membrane components, mostly proteinaceous in nature, which could elicit the production of antibodies, were studied by a few authors only. SANTOS-BUCH and GUZMAN-RIVERO (1968) isolated two proteins from sheep erythrocytes. One had an MW over 200,000 and initiated the output of hemolytic IgM immunoglobulins in two days after injection into rabbits. Production of IgG could be detected only 24 days after primary immunization. MARCHESI and STEERS (1968) could solubilize a portion of the RBC proteins by dialyzing membranes against ATP and 2-mercaptoethanol. The protein obtained, called "spectrin", gave a single band in gel diffusion. ADACHI and FURUSAWA (1968) described a protein antigen of RBC which is not accessible on the cell surface. FURTHMAYR and

TIMPL (1970) used water adjusted to pH 9.5, pyridine, ATP and 2-mercaptoethanol or urea at pH 11 to obtain protein immunogens, which are not located on the outside of the cells.

The work of BRON et al., from the Institut de Biochimie, Université de Lausanne, Lausanne, Switzerland, and the Child Research Center of Michigan, Wayne State University, Detroit, Michigan, as reported at this conference, showed that several protein antigens can be detected by immunoelectrophoresis, using a rabbit antiserum to membrane proteins of human red cell stroma (POULIK and BRON, 1969). This antiserum reacted also with human RBC membrane proteins obtained by several different solubilization methods. Strong cross-reactivity was found with lipid-free stromal proteins from eight mammalian species, suggesting that they may have similar chemical structures. Repeated absorptions of this antiserum with human plasma, globin, pure blood group substances, sheep and human red cells did not affect the immunoelectrophoretic pattern, thus indicating that most of the antigens detected were not readily available on the surface of the cell and may therefore be related to structural function.

Bron and coworkers also reported that one of these antigens was demonstrated in normal human urine by immunoelectrophoresis with the same antiserum (BRON and POULIK, 1971). Its presence on the surface of the red cell of any major blood group specificity and among the precipitable antigens of stromal proteins was established by absorption of antiserum with human RBC as well as by direct immunofluorescence or agglutination of red cells with anti-human urine rabbit serum. Complete identity was shown between the precipitin lines obtained with both anti-urine and anti-stroma using the same urinary fraction as antigen. These data were confirmed by the microcomplement fixation test (LEVINE, 1967), whereby it was possible to detect strong antigenic activity, both in normal human urine using anti-stromal protein serum and in lipidfree stromal proteins using anti-urine serum.

The origin and occurrence of these antigens on other cell surfaces is under investigation. Isolation and characterization of red cell membrane proteins using antibodies as markers is in progress.

In connection with the above important findings of Bron and associates, one should mention here that the existence of a major "structural protein" component of the RBC as well as some other membranes was assumed by MAZIA and RUBY (1968), the existence of which was questioned by ROSENBERG and GUIDOTTI in 1968.

D. Methods for the Isolation of Various RBC Membrane Components with Unknown Immunological Role

There are a number of other procedures described for the isolation of RBC proteins. One of them was developed by BLUMENFELD (1968) who used 33% pyridine. The mechanism of sodium deoxycholate solubilization of RBC was studied by PHILIPPOT (1971). SCHNEIDERMAN and JUNGA (1968) applied Triton-X-100. LENARD (1970) carried out acrylamide gel electrophoresis of entire membranes in the presence of Na dodecyl sulfate. Fourteen different protein fractions were identified and their molecular weights were estimated. ROSENBERG and GUIDOTTI (1968) used EDTA and different salts combined with lipid solvents to obtain protein fractions.

Guidotti reported at this conference that two proteinaceous components are exposed on the RBC surface. One of them is a glycoprotein containing sialic acid and the other is also a glycoprotein with 11% carbohydrates. The latter has an unusually high content of nonpolar amino acids which may be embedded in the nonpolar region of the multilayered membrane. GWYNNE and TANFORD (1970) isolated a similarly long single polypeptide chain from human erythrocytes with an approximate MW of 200,000. FAIRBANKS *et al.* (1971) isolated a major polypeptide from human RBC and studied its electrophoretic properties. MITCHELL and HANAHAN (1966) used hypertonic NaCl solutions to solubilize some proteins from RBC. HOOGEVEEN *et al.* (1970) used distilled water. Glucose binding components of human erythrocytes were successfully isolated by BOBINSKI and STEIN (1966) and also by BONSALL and HUNT (1966). Although none of the above publications reported data regarding the antigenicity of the preparations, they provide new procedures for immunochemists to fractionate membrane components.

III. Conclusion

Considering what was said in the introduction about the mosaic-like arrangement of the membrane components arrayed into a multilayered structure and held together by noncovalent forces, one would assume that it is relatively easy to dissociate them and obtain these substances in pure form without introducing a great deal of denaturation. This is true for the different phospholipids, but isolation of proteins, glycoproteins, glycolipids, and lipoproteins proved to be quite difficult. This becomes understandable if one considers that the interaction between these components is responsible for the creation of the very flexible and remarkably enduring erythrocyte wall. Similarly, the same interactions stabilize certain conformations of the building substances to provide biologically functioning active sites. Drastic outside interference with these interactions will lead not only to dissolution of the architecture but also to conformational changes in some of the active sites. What are needed are mild but efficient methods which would liberate all components without altering their structure and thus their biological activities.

Immunogenicity may be very dependent on the conformation of tertiary and quaternary structure. If so, it must also depend on interaction of membrane constituents. This makes isolation of active and structurally unaltered components quite difficult, if not impossible, unless one either isolates the immunogenic structures together with those which interact and thus activate them, or attempts to form active mixtures *in vitro*. That this may be the case can be substantiated by the following examples.

HAMASATO (1950) and later KOSCIELAK (1963) found that blood group A and B activity of RBC extracts could be enhanced if hydrophobic lipids from the same cell were mixed with the extracts. Another example for this may be the Rh system, where butanol extraction inactivates Rh antigen D, but addition of certain lipid extract fractions to the inactivated preparations restores some of their Rh activity (GREEN, 1968).

We observed that isolated and lyophilized sheep erythrocyte membranes are poor hemolytic or agglutinating antibody producers in comparison to fresh intact cells. The same is true for tumor antigens, as found by us recently, where lyophilized cells or cell membranes at a certain dose given rise to immune enhancement, while injection of subthreshold live inocula provoke immune resistance against lethal tumor challenge (unpublished data). One of the possible explanations is that simple dehydration by freeze drying may alter the conformation of some immunologically active sites which cannot be restored by rehydration *in vivo*.

On the other hand, outside interference with the interactions between the membrane constituents may also result in the uncovering of hitherto hidden receptor sites or even in the formation of immunogenic sites with new characteristics. A few publications which assume the existence of such covered components in RBC have already been cited in this review. We think that our finding of the enhanced antibody-binding capacity of RBC after detergent treatment is one of the most plausible examples of this possibility. One should bear in mind another possible effect of dissolution and fractionation of membranes, which is the removal of strongly competitive antigens and thus the enhancement of immunogenicity of weaker, hitherto masked or suppressed immunogens. Examples of this are also available, such as the suppression of "common antigens" in Gram-negative bacteria by lipopolysaccharides (see NETER's paper in this volume, p. 14) unless they are separated, or the lack of immune response to weak histocompatibility antigens in the presence of strong ones. One wonders if the relative weakness of some tumor-specific transplantation antigens as immunogens *in vivo* is not due to similar circumstances, and could be enhanced by proper outside, *in vitro* interference with the interaction of the tumor cell membrane constituents.

All these possibilities raise new views of mammalian cellular antigens. One is the necessity to investigate immunogenicity both *in situ* on unaltered cell surfaces and by using isolated components. The other, and the most important, is the realization of the enormous difficulties to be expected in attempting to isolate membrane constituents without irreversibly altering their active sites.

References

Adachi, H., and M. Furusawa (1968). Immunological analysis of the structural molecules of erythrocyte membrane in mice. I. Analysis of the aqueous phase molecules obtained by butanol fractionation of erythrocyte membrane. *Exp. Cell Res., 50*:490–496.

Ando, S., and T. Yamakawa (1970). On the oligosaccharide of Forssman-active sheep red cell glycolipid. *Chem. Phys. Lipids, 5*:91–95.

Blumenfeld, O. (1968). The proteins of erythrocyte membrane obtained by solubilization with aqueous pyridine solution. *Biochem. Biophys. Res. Commun., 30*:200–205.

Bobinski, H., and W. D. Stein (1966). Isolation of a glucose-binding component from human erythrocyte membranes. *Nature, Lond., 211*:1366–1368.

Bonsall, R. W., and S. Hunt (1966). Solubilization of a glucose-binding component of the red cell membrane. *Nature, Lond., 211*:1368–1370.

Branton, D. (1967). Fracture faces of frozen myelin. *Exp. Cell Res., 45*:703–707.

Branton, D., and R. B. Park (1967). Subunits in chloroplast lamellae. *J. Ultrastruct. Res., 19*:283–303.

Bron, C., and M. D. Poulik (1971). Urinary glycoprotein antigenically related to human red cell membrane. *Immunochemistry, 8*:447–449.

Carter, B. B. (1949). Rh hapten: Its preparation, assay and nature. *J. Immunol., 61*:79–88.

Chapman, D. (1968). Physical studies of biological membranes and their constituents, pp. 6–18. In *Membrane Models and the Formation of Biological Membranes*, L. Bales and B. A. Pethida (eds.), Amsterdam: North Holland Publ. Co.

Davidsohn, I., and L. Y. Ni (1970). Immunocytology of cancer. *Acta Cytol., 14*:276–282.

Fairbanks, G., T. L. Steck, and D. F. H. Wallach (1971). Electrophoretic analysis of the major polypeptides of the human erythrocyte membrane. *Biochem., 10*:2606–2617.

Furthmayr, H., and R. Timpl (1970). Immunochemical studies on structural proteins of red cell membrane. *Eur. J. Biochem., 15*:301–310.

Glaeser, R. M., T. Hayes, H. Mel, and C. Tobias (1966). Membrane structure of OsO₄-fixed erythrocytes viewed "face on" by electron microscope techniques. *Exp. Cell Res., 42*:467–477.

Gottschalk, A. (1960). *The Chemistry and Biology of Sialic Acid and Related Substances.* Cambridge: The University Press.

Green, F. A. (1968). Phospholipid requirement for Rh antigenic activity. *J. Biol. Chem., 243*:5519–5521.

Gwynne, J. T., and C. Tanford (1970). A polypeptide chain of very high molecular weight from red blood cell membranes. *J. Biol. Chem., 245*:3269–3271.

Hakomori, S.-I. (1970). Glycosphingolipids having blood-group ABH and Lewis specificities. *Chem. Phys. Lipids, 5*:96–115.

Hakomori, S.-I., and R. W. Jeanloz (1964). Isolation of a glycolipid containing fucose, galactose, glucose and glucosamine from human cancerous tissue. *J. Biol. Chem., 239*:3606–3607.

—— (1970). Glycolipids as membrane antigens, pp. 149–161. In *Blood and Tissue Antigens*, D. Aminoff (ed.), New York: Academic Press.

Hakomori, S.-I., J. Koscielak, K. J. Bloch, and R. W. Jeanloz (1967). Immunogenic relationship between blood group substances and a fucose-containing glycolipid of human adenocarcinoma. *J. Immunol., 98*:31–38.

Hakomori, S.-I., and W. T. Murakami (1968). Glycolipids of hamster fibroblasts and derived malignant-transformed cell lines. *Proc. Nat. Acad. Sci. U.S.A., 59*:254–261.

Hakomori, S.-I., and G. D. Strycharz (1968). Investigations on cellular blood-group substances. I. Isolation and chemical composition of blood-group ABH and Le^b isoantigens of sphingoglycolipid nature. *Biochem., 7*:1279–1286.

Hamasato, I. (1950). The blood group substance in red bloods. *Tohoku J. Exp. Med., 52*:17–27, 29–33 and 35–36.

Handa, S., and T. Yamakawa (1964). Chemistry of posthemolytic residue on stroma of erythrocytes. XII. Chemical structure and chromatographic behaviour of hematosides obtained from equine and dog erythrocytes. *Jap. J. Exp. Med., 34*:293–304.

Hillier, J., and J. F. Hoffman (1953). On the ultrastructure of the plasma membrane as determined by the electron microscope. *J. Cell. Comp. Physiol., 42*:203–247.

Hoogeveen, Th. J., R. Juliano, J. Coleman, and A. Rothstein (1970). Water-soluble proteins of human red cell membrane. *J. Membrane Biol., 3*:156–172.

Huprikar, S. V., and G. F. Springer (1970). Structural aspects of human blood-group M and N specificity, pp. 327–335. In *Blood and Tissue Antigens,* D. Aminoff (ed.), New York: Academic Press.

Ivanyi, P. (1966). Blood groups and transplantation antigens. *Ann. Inst. Pasteur, 110*:144–154.

Kathan, R. H., and R. J. Winzler (1963). Structure studies on the myxovirus hemagglutination inhibitor of human erythrocytes. *J. Biol. Chem., 238*:21–25.

Kavanau, L. J. (1966). Membrane structure and function. *Fed. Proc., 25*:1096–1107.

Klenk, E., and K. Heuer (1960). Gangliosides of dog erythrocytes. *Dtsch. Z. Verdan. Stoffwechselkr., 20*:180–183.

Klenk, E., and K. Lauenstein (1951). Über die zukerhaltigen Lipoide der Formbestandteile menschlichen Blutes. *Hoppe-Seyler's Z. Physiol. Chem., 288*:220–228.

Klenk, E., and H. Lempfrid (1957). Über die Natur der Zellreceptoren für das Influenzavirus. *Hoppe-Seyler's Z. Physiol. Chem., 307*:278–281.

Korn, E. D. (1966). Structure of biological membranes. *Science, 153*:1491–1498.

Koscielak, J. (1963). Blood group A specific glycolipids of human erythrocytes. *Biochim. Biophys. Acta, 78*:313–328.

Koscielak, J., and K. Zakrzewski (1960). Substances from erythrocytes of blood group A. *Nature, Lond., 187*:516–517.

Landsteiner, K., and P. Levine (1928). On individual differences in human blood. *J. Exp. Med., 47*:757–775.

Landsteiner, K., J. van der Scheer, and D. H. Witt (1925). Group specific flocculation reactions with alcohol extracts of human blood. *Proc. Soc. Exp. Biol. Med., 22*:289–291.

Lenard, J. (1970). Protein and glycolipid components of human erythrocyte membranes. *Biochem., 9*:1129–1132.

Levine, L. (1967). Micro-complement fixation, pp. 707–719. In *Handbook of Experimental Immunology*, N. B. Weir (ed.), Philadelphia: F. A. Davis.

Lucy, J. A. (1964). Globular lipid micelles and cell membranes. *J. Theor. Biol., 7*:360–373.

Maddy, A. H. (1964). The solubilization of the protein of the ox-erythrocyte ghost. *Biochim. Biophys. Acta, 88*:448–449.

Makita, A., C. Suzuki, and Z. Yoshizawa (1966). Chemical and immunological characterization of the Forssman hapten isolated from equine organs. *J. Biochem.* (Tokyo), *60*:502–513.

Marchesi, V. T., and E. Steers, Jr. (1968). Selective solubilization of a protein component of the red cell membrane. *Science, 159*:203–204.

Mazia, D., and A. Ruby (1968). Dissolutions of erythrocyte membranes in water and comparison of the membrane protein with other structural proteins. *Proc. Nat. Acad. Sci. U.S.A., 61*:1005–1012.

Milgrom, F., and U. Loza (1969). Immunodiffusion tests with Rh antigens and antibodies. *Vox Sang., 16*:470–477.

Mitchell, C. D., and D. J. Hanahan (1966). Solubilization of certain proteins from the human erythrocyte stroma. *Biochem., 5*:51–57.

Moskowitz, M., and M. Calvin (1952). On the components and structure of the human red cell membrane. *Exp. Cell Res., 3*:33–46.

Nicolson, G. L., S. P. Masouredis, and S. J. Singer (1971). Quantitative two-dimensional ultrastructural distribution of Rh₀(D) antigenic sites on human erythrocyte membranes. *Proc. Nat. Acad. Sci. U.S.A., 68*:1416–1420.

Nowotny, A. (1955). Studies on the chemical structure of A, B and O blood group antigens. I. Physicochemical examination of the A and B group red cell stroma. *Acta Physiol. Hung., 7*:31–42 (1955).

Nowotny, A., and E. Backhausz (1957). Studies of the chemical structure A, B and blood group antigens. III. The role of the erythrocyte membrane lipoids in group specificity. *Acta Physiol. Hung., 12*:53–64.

Nowotny, A., and G. J. O'Neill (1964). Immunochemical investigations of A and B group antigens localized in the erythrocyte membrane. Xth Congr. Int. Soc. Haematol. 1963. *Scand. J. Haematol. Suppl. 1–10.*

Papirmeister, B., and M. F. Mallette (1955). The isolation and some properties of the Forssman hapten from sheep erythrocytes. *Arch. Biochem. Biophys., 57*:94–105.

Pardoe, G. I., G. W. G. Bird, and G. Uhlenbruck (1968). Structural and serological studies of the lipopolysaccharide of *Proteus vulgaris* OX 19. *Z. Immunitätsf., 136*:488–495.

Philippot, J. (1971). Study of human red blood cell membrane using sodium deoxycholate. I. Mechanism of solubilization. *Biochim. Biophys. Acta, 225*:201–213.

Pinto da Silva, P., S. D. Douglas, and D. Branton (1971). Localization of A antigenic sites on human erythrocyte ghosts. *Nature, Lond., 232*:194–196.

Poulik, M. D., and C. Bron (1969). Immunology of red cell membrane proteins, pp. 131–153. In *Red Cell Membrane Structure and Function*, G. A. Jamieson and T. J. Greenwalt (eds.), Philadelphia/Toronto: J. B. Lippincott Co.

Poulik, M. D., and P. K. Lauf (1969). Some physicochemical and serological properties of isolated protein components of red cell membranes. *Clin. Exp. Immunol.*, *4*:165–175.

Radin, N. S. (1957). Glycolipid chromatography. *Fed. Proc.*, *16*:825–826.

Radvany, R. M. (1971). Isolation of blood group A and B antigens from human erythrocyte membranes. Ph.D. Thesis, Temple University, Philadelphia, Pennsylvania.

Rapp, H. J., and T. Borsos (1966). Forssman antigen and antibody. *J. Immunol.*, *96*:913–919.

Rosenberg, S. A., and G. Guidotti (1968). The protein of human erythrocyte membranes. I. Preparation, solubilization and partial characterization. *J. Biol. Chem.*, *243*:1985–1992.

Santos-Buch, C. A., and J. R. Guzman-Rivero (1968). Isolation of a protein antigen following extraction of lipids from sheep red blood cell stroma. I. Hemolytic IgM primary response and hemolytic IgG secondary response in rabbits. *J. Immunol.*, *100*:475–484.

Schneiderman, L. J., and I. G. Junga (1968). Isolation and partial characterization of structural protein derived from human red cell membranes. *Biochem.*, *7*:2281–2286.

Sjöstrand, F. G. (1963). A new ultrastructural element of the membranes in mitrochondria and of some cytoplasmic membranes. *J. Ultrastruct. Res.*, *9*:340–361.

Springer, G. F. (1963). Enzymatic and nonenzymatic alterations of erythrocyte surface antigens. *Bact. Rev.*, *27*:191–227.

Springer, G. F., and M. A. Fletcher (1969). Cell surface receptors: Structural aspects and importance in immunopathology. Reprint from Organstransplantation Immunologie und Klinik, Schatter-Verlag, Stuttgart, pp. 35–45.

Springer, G. F., M. A. Fletcher, and O. Pavlovskis (1967). Homogeneous erythrocyte glycoproteins with blood-group-, virus-, and endotoxin-receptor activities, pp. 109–122. In *Protides of the Biological Fluids*, H. Peeters (ed.), Amsterdam: Elsevier Co.

Springer, G. F., E. T. Wang, J. H. Nicholas, and J. M. Shear (1966). Relations between bacterial lipopolysaccharide structures and those of human cells. *Ann. N.Y. Acad. Sci.*, *133*:566–579.

Sturgeon, P. (1970). Hematological observations on the anemia associated with blood type R_{null}. *Blood*, *36*:310–320.

Weicker, H. (1968). Isolierung eines Kleinmolekularen Peptides mit Rh-Blutgruppeneigenschaften als der Erythrocyten-Membran des Menschen. *Klin. Wochenschr.*, *46*:824.

Whittam, R. (1964). *Transport and Diffusion in Red Blood Cells*. London: Edward Arnold, p. 34.

Whittemore, N. B., N. C. Trabald, C. F. Reed, and R. I. Weed (1969). Solubilized glycoprotein from human erythrocyte membranes possessing blood group A, B and H activities. *Vox Sang.*, *17*:289–299.

Wolf, I., and G. F. Springer (1964). Attempts to characterize the Rh antigen. *Fed. Proc.*, *23*:296.

Yamakawa, T., and R. Irie (1960). On the mucolipid nature of ABO group substance of erythrocytes. *J. Biochem.*, *48*:919–920.

Yamakawa, T., R. Irie, and M. Iwanga (1960). The chemistry of lipid of posthemolytic residue or stroma of erythrocytes. IX. Silicic acid chromatography of mammalian stroma glycolipids. *J. Biochem.*, *48*:490–507.

Yamakawa, T., and S. Suzuki (1951). The chemistry of the lipids of posthemolytic residue or stroma of erythrocytes. III. Globoside, the sugar-containing lipid of human blood stroma. *J. Biochem.*, *39*:393–402.

Zahler, P. (1968). Blood group antigens in relation to chemical and structural properties of the red cell membrane. *Vox Sang.*, *15*:81–101.

Zahler, P. (1969). The structure of the erythrocyte membrane. *Experientia*, *25*:449–456.

Zvilichovsky, B., P. M. Gallop, and O. O. Blumenfeld (1971). Isolation of a glycoprotein-glycolipid fraction from human erythrocyte membranes. *Biochem. Biophys. Res. Commun.*, *44*:1234–1243.

Part 4

White Blood Cell Antigens

ISOLATION AND PROPERTIES OF HL-A ANTIGENS AND ANTIGENS OF NEOPLASTIC MOUSE PLASMA CELLS*

DAVID PRESSMAN

Department of Biochemistry Research, Roswell Park Memorial Institute, Buffalo, New York

In connection with the subject of antigens of white cells, I would like to report on two pertinent lines of effort in progress at Roswell Park. These are concerned with (1) the isolation and properties of HL-A antigens from hematopoietic cell lines and (2) the antigens of neoplastic mouse plasma cells and the physiological effects of antibodies to these antigens, preventing the formation of hemolytic plaques by spleen cells from mice injected with sheep red blood cells.

These projects were carried out with my colleagues, Drs. Y. Yagi, N. Tanigaki, T. Watanabe, and Y. Miyakawa.

I. HL-A Antigens

Established lines of human hematopoietic cells appear to be ideal sources of HL-A antigens. Lines possessing various phenotypes of HL-A antigens are available, and a supply of uniform cells, relatively unlimited in amounts as compared with those from normal organs, are available from culture as required (MOORE and MINOWADA, 1969). In addition, the yield of the antigen from these cells is apparently high (MANN et al., 1969), and the antigenic activity of the final product seems to be stable during storage for long periods (REISFELD et al., 1970).

Several laboratories including ours have reported previously on the isolation of some HL-A antigens from these cultured cells (MANN et al., 1969; REISFELD et al., 1970; MIYAKAWA et al., 1971) as well as from normal organs (DAVIES et al., 1968; SANDERSON, 1968; BOYLE, 1969;

* Supported in part by the John A. Hartford Foundation, U.S. Atomic Energy Commission Contract AT-(30-1)-2561, and National Institute of Allergy and Infectious Diseases Grant AI-8899.

KAHAN and REISFELD, 1969; COLOMBANI *et al.*, 1970). The work summarized here was undertaken to characterize the molecular size and electrophoretic mobility of papain-solubilized HL-A antigens from cultured cells and to determine whether HL-A antigens of the same specificity which are derived from different cell lines are similar with respect to these properties. Such information would be useful in establishing efficient, general procedures for the isolation of various HL-A antigens and might indicate whether structural changes in the HL-A antigens occur in a long-term culture.

Cultured cells from hematopoietic cell lines which are of HL-A phenotypes suitable for the present purpose were chosen. These cells lines, RPMI 8235, 1788, 5287, 8057, and 4265, were originally established by Dr. G. E. Moore of this Institute (MOORE and MINOWADA, 1969).

The isolation procedure used gave a high yield of relatively pure HL-A antigens. HL-A antigens in particulate form were obtained from cells disrupted by shaking in buffer. These particulates were digested with papain in the presence of cysteine, and the soluble papain fragments carrying antigenic activity were purified by a succession of steps starting with hypotonic dialysis, followed by ion-exchange chromatography on DEAE-Sephadex, ultrafiltration, gel filtration, and ending with column electrophoresis.

Several preparations of soluble HL-A antigens, each carrying one of the representative antigenic activities—HL-A1, HL-A2, or HL-A7—were obtained from three different cell lines. These preparations had a narrow molecular size distribution and a narrow electrophoretic mobility in the α_2-globulin region. In these preparations, the overall recovery of HL-A antigenic activity was 30 to 37% for each isolated specificity, the recovery of OD_{280} units was 0.08 to 1.14%, the OD_{280}/OD_{260} ratio was approximately 1.7, and the increase in the specific antigenic activity was approximately 300 times, as compared with the papain digests.

Of the five alloantigenic specificities followed during the isolation procedures, the four identifiable HL-A specificities (HL-A1, HL-A2, HL-A7, and HL-A8) were each found to be eluted exclusively within a single peak upon gel filtration, indicating a high degree of homogeneity of molecular size. In addition, all four HL-A activities were eluted at identical positions corresponding to the same molecular size of 48,000 Daltons. Thus, HL-A1 and HL-A2 activities, under control of the first sublocus, were carried by fragments having the same molecular size as the fragments carrying HL-A7 and HL-A8 activities, which are under control of the second sublocus. The unidentified alloantigenic activity in some of our cell lines and found on fragments of definitely larger size (74,000 Daltons) than those carrying the identifiable HL-A specificities, may or may not be an HL-A activity. The molecular weight observed for our preparations of fragments carrying identifiable HL-A activities is similar to that reported by SANDERSON (1968) for papain fragments with HL-A2 (BT5) activity.

However, this is larger than the molecular weight of the fragments reported
here by PAPERMASTER *et al.* (1969), which were isolated by use of 3 *M*
KCl. As has been pointed out at this meeting, the action of HCl is more
complicated than the breaking of noncovalent bonds and depends on autolytic action by cellular enzymes.

Our results do not confirm the existence of differences in molecular
size among HL-A active fragments controlled by different subloci as reported by MANN *et al.* (1969). They found that papain-solubilized HL-A
antigens from cultured cells, spleen cells, and peripheral leukocytes could
be separated by gel filtration. In the pattern of elution, the alloantigens
of the "LA" series (HL-A1, HL-A2, HL-A3, and HL-A9 controlled by
the first sublocus) were regularly found in peaks 1 (excluded) and 3, and
were separated from the alloantigens of the "4" series (HL-A5 and HL-A7
controlled by the second sublocus), which were found in peak 2. Peaks
2 and 3 were in the elution volumes corresponding to molecular weights
of 50,000 to 60,000. We do not know the reasons for the discrepancy
between these results and our observation that there is a single molecular
size common to the identifiable HL-A activities we examined.

Our observation that soluble HL-A2 antigen fragments from cells in
culture have a lower electrophoretic mobility than HL-A1, HL-A7, and
HL-A8 antigen fragments is in accord with the observations of others who
isolated these activities from tissues or peripheral leukocytes. Thus DAVIES
et al. (1968) and SANDERSON (1968) showed that some soluble HL-A
antigens with different specificities prepared either by autolysis or papain-
digestion from spleen can be separated from one another by ion-exchange
chromatography. COLOMBANI *et al.* (1970) extended these observations
and eluted HL-A-soluble antigenic fragments from DEAE-Sephadex
columns. The sequence of elution was HL-A2, HL-A5, HL-A9, and
HL-A1. Although these data were obtained with soluble HL-A antigens
prepared from spleen cells by autolysis, the sequence of elution seems to
be comparable to our findings, *i.e.,* papain fragments carrying HL-A2
specificity from hematopoietic cell lines have an electrophoretic mobility
that is lower than those of fragments carrying the other specificities. Generally, it would seem, HL-A-active fragments obtained from hematopoietic
cell lines have properties very similar to those obtained from normal
tissues.

We have also shown that HL-A-active fragments have consistent
molecular properties for each specificity, even if derived from different cell
lines. Since some of the hematopoietic cell lines used in the present study
have been maintained for more than five years, these findings may also
be indicative of the stability of expression of HL-A antigens in an *in vitro*
environment, as is seen by the fact that human hematopoietic cell lines
retain all the HL-A phenotypic expression of the original donors
(PAPERMASTER *et al.,* 1969).

II. Surface Antigens of Mouse Neoplastic Plasma Cells

Antigenic differences between mouse myeloma cells and normal mouse lymph node cells were investigated by comparing the reaction patterns of rabbit antisera prepared against cellular components of myelomas of BALB/c mice and of antisera prepared against lymph node cells. The results of this study with six transplantable myeloma cells lines indicated that surface antigens are present on mouse myeloma cells which are not detected on cells of thymus, liver, and kidney, and are present in insignificant amounts on the surfaces of spleen cells and lymph node cells. On the other hand, lymph node cells apparently contain surface antigens which are not detected on myeloma cells. Thus rabbit antibodies specifically cytotoxic to myeloma cells of any of the lines tested, yet not cytotoxic to lymphocytes from lymph node or spleen, have been prepared from antisera against myeloma cells.

The nature of these surface antigens on myeloma cells is not clear. Five alloantigens have been reported on the surface of lymphocytes and thymocytes. They are: theta (θ) (REIF and ALLEN, 1964), TL (BOYSE and OLD, 1969), Ly-A, Ly-B (BOYSE et al., 1968), and H-2 (SNELL and STIMPFLING, 1966). In addition, a "mouse-specific lymphocyte antigen (MSLA)" has been characterized by the use of rabbit anti-mouse thymocyte serum (SHIGENO et al., 1968). On the other hand, myeloma cells have been reported that lack all of these surface antigens except H-2 but possess PC-1 alloantigen, which is undetected on lymphocytes or thymocytes (TAKAHASHI et al., 1970). Although the antigen responsible for the cytotoxic antibodies in our rabbit anti-myeloma sera was also found on myeloma cells and not on thymocytes, PC-1 alloantigen and the antigen described here are not the same, because PC-1 antigen was found also in liver and kidney, while our antigen was not.

Large amounts of spleen and lymph node homogenates completely absorbed cytotoxic antibodies from anti-myeloma sera. Spleen and lymph node cells from immunized mice (even after only one injection with an unrelated antigen) were more efficient than those from normal mice. This probably indicates that plasma cells in spleen and lymph nodes possess, in common with myeloma cells, the antigen responsible for the formation of myeloma-cytotoxic antibodies in anti-myeloma sera. Indeed, exposure of immune spleen cells to anti-myeloma antisera in the presence of rabbit complement causes complete suppression of plaque formation, even though little or no change is seen in the viability of spleen cells. Similar effects on plaque formation were reported with alloantiserum (presumably anti-PC-1 antiserum) produced in DBA/2 mice against BALB/c myeloma cells (TAKAHASHI et al., 1970).

The antiserum against lymph node cells contained antibodies cytotoxic to lymph node and spleen cells, but not to myeloma cells. The antigen responsible for these cytotoxic antibodies was found in spleen, lymph node, and thymus, and in much smaller amounts in myeloma, liver, and kidney. Thus, this antigen seems to differ from the MSLA antigen described by SHIGENO *et al.* (1968), which was not detected in liver and kidney. Its relation to other lymphocyte alloantigens is not clear from the present study.

YAMADA *et al.* (1969) demonstrated that isologous antibodies cytotoxic to transplantable myeloma cells were formed in mineral oil-injected mice which were prevented from developing plasma cell tumors by treatment with glycoprotein pituitary hormones (GPH). The cytotoxicity pattern of individual sera against various cell lines of transplantable plasma cell tumors differed one from the other. Their results seem to suggest individual antigenic differences among the transplantable tumor cell lines. Although our data on the cytotoxicity pattern of antiserum against MOPC-104E myeloma also suggest the possibility of the presence of individual-specific antigens on the surface of 104E myeloma cells, the difference seems to be quantitative rather than qualitative, because large amounts of other myeloma cells can remove the cytotoxicity of anti-104E serum against 104E cells.

Anti-myeloma sera contained antibodies reactive toward mouse and sheep erythrocytes and antibodies against mouse serum proteins. However, the cytotoxic activity of anti-myeloma sera is not due to these antibodies since no reduction in titer was observed after treatment of these antibodies with mouse serum and with mouse and sheep erythrocytes.

The antigen responsible for cytotoxic antibodies in anti-myeloma sera is probably not associated with virus-like particles reported to be present in some transplantable myeloma cell lines (PARSONS *et al.*, 1961a, b; COHN, 1967; WATSON *et al.*, 1970). Although similar particles were found in cytoplasm of myeloma cells of all the lines used here, no such particles were detected in normal mouse spleen, as reported by PARSONS *et al.*, (1961b). KAJIMA and POLLARD (1968) reported the presence of virus-like particles in the thymus of conventional mouse strains. Complete removal of cytotoxic antibodies in antimyeloma sera by normal spleen and the total ineffectiveness of thymus would exclude the possibility that the responsible antigen is of viral origin.

We have also found that formation of hemolytic plaques by spleen cells from mice immunized with sheep erythrocytes is strongly suppressed following treatment of the cells with rabbit anti-myeloma sera in the presence of complement, although the apparent viability of the spleen cells in general remained unchanged. Both direct and indirect plaque-forming cells (PFC) were affected in both the primary and secondary immune response. The antigens which react with PFC-suppressive antibody were found on myeloma cells and normal plasma cells. Few, if any, of these antigens were detected on lymphocytes, thymocytes, or on liver or kidney cells. The presence of these plasma cell antigens is not restricted to certain strains of mice. Suppression of PFC was seen in all of 9 inbred strains of mice tested.

References

Boyle, W. (1969). Soluble HL-A iso-antigen preparations. *Transpl. Proc.,* *1*:491–493.

Boyse, E. A., M. Miyazawa, T. Aoki, and L. J. Old (1968). Ly-A and Ly-B: two systems of lymphocyte isoantigens in the mouse. *Proc. R. Soc. Ser. B, 170*:175–193.

Boyse, E. A., and L. J. Old (1969). Some aspects of normal and abnormal cell surface genetics. *Ann. Rev. Genet., 3*:269–290.

Cohn, M. (1967). Natural history of the myeloma. Cold Spring Harbor Symp. Quant. Biol., *32*:211–221.

Colombani, J., M. Colombani, D. C. Viza, O. Degani-Bernard, J. Dausset, and D. A. L. Davies (1970). Separation of HL-A transplantation antigen specificities. *Transplantation, 9*:228–239.

Davies, D. A. L., J. Colombani, D. C. Viza, and J. Dausset (1968). Human HL-A transplantation antigens: separation of molecules carrying different immunological specificities determined by a single genotype. *Biochem. Biophys. Res. Commun., 33*:88–93.

Kahan, B. D., and R. A. Reisfeld (1969). Advances in the chemistry of transplantation antigens. *Transpl. Proc., 1*:483–488.

Kajima, M., and M. Pollard (1968). Wide distribution of leukaemia virus in strains of laboratory mice. *Nature, Lond., 218*:188–189.

Mann, D. L., G. N. Rogentine, J. L. Fahey, and J. Nathenson (1969). Human lymphocyte membrane (HL-A) alloantigens: isolation, purification and properties. *J. Immunol., 103*:282–292.

Miyakawa, Y., N. Tanigaki, Y. Yagi, and D. Pressman (1971). Human transplantation antigens: isolation and radioimmunoassay. *Proc. Soc. Exp. Biol. Med., 136*:899–902.

Moore, G. E., and J. Minowada (1969). Human hematopoietic cell lines: A progress report, pp. 100–114. In *Hemic Cells In Vitro*. New York: Williams and Wilkins.

Papermaster, V. M., B. W. Papermaster, and G. E. Moore (1969). Histocompatibility antigens of human lymphocytes in long-term culture. *Fed. Proc., 28*:379.

Parsons, D. F., M. A. Bender, E. B. Darden, Jr., G. T. Pratt, and D. L. Lindsley (1961a). Electron microscopy of plasma-cell tumors of the mouse. II. Tissue cultures of the X5563 tumor. *J. Biophys. Biochem. Cytol., 9*:369–381.

Parsons, D. F., E. B. Darden, Jr., D. L. Lindsley, and G. T. Pratt (1961b). Electron microscopy of plasma-cell tumors of the mouse. I. MPC-1 and X5563 tumors. *J. Biophys. Biochem. Cytol., 9*:353–368.

Reif, A. E., and J. M. V. Allen (1964). The AKR thymic antigen and its distribution in leukemias and nervous tissues. *J. Exp. Med., 120*:413–433.

Reisfeld, R. A., M. Pellegrino, B. W. Papermaster, and B. D. Kahan, (1970). HL-A antigens from a continuous lymphoid cell line derived from a normal donor. I. Solubilization and serologic characterization. *J. Immunol., 104*:560–565.

Sanderson, A. R. (1968). HL-A substances from human spleens. *Nature, Lond., 220*:192–195.

Shigeno, N., U. Hammerling, C. Arpels, E. A. Boyse, and L. J. Old (1968). Preparation of lymphocyte-specific antibody from anti-lymphocyte serum. *Lancet, ii*:320–323.

Snell, G. D., and J. H. Stimpfling (1966). Genetics of tissue transplantation, pp. 457–491. In *Biology of the Laboratory Mouse*. New York: McGraw-Hill.

Takahashi, T., L. J. Old, and E. A. Boyse (1970). Surface alloantigens of plasma cells. *J. Exp. Med., 131*:1325–1341.

Watson, J., P. Ralph, S. Sarkar, and M. Cohn (1970). Leukemia viruses associated with mouse myeloma cells. *Proc. Nat. Acad. Sci. U.S.A., 66*:344–351.

Yamada, H., A. Yamada, and V. P. Hollander (1969). Role of cellular and humoral factors in the destruction of nascent plasma cell tumors. *Cancer Res., 29*:1420–1427.

CHARACTERIZATION AND ISOLATION OF HL-A ANTIGENS FROM CONTINUOUS CULTURED HUMAN LYMPHOCYTE CELL LINES: A REPORT OF CURRENT PROGRESS

B. W. Papermaster,* V. M. Papermaster,†
R. A. Reisfeld,‡ M. A. Pellegrino,* S. Ferrone,‡
B. D. Kahan,§ P. I. Terasaki,‖ M. Takasugi‖
and E. A. Albert‖

I. Introduction and Review

As the genetics of the HL-A antigenic system became systematized through the efforts of many investigators (Curtoni et al., 1967) our interest developed in the task of isolation and characterization of HL-A antigens for determining the chemical nature of the gene products of the HL-A chromosomal region (Reisfeld et al., 1970a, b). Various human tissues were apparent sources of antigen (e.g., peripheral blood lymphocytes, spleen, liver, kidney, and other organs obtainable at surgical operations), but these sources were limited because of the small amount of material available from a single donor. Moreover, the extensive genetic polymorphism within the HL-A system precluded the likelihood of obtaining material from two different individuals with identical antigenic patterns. Therefore, we became interested in isolating HL-A antigens from cultured human lymphocytic cell lines in order to determine whether they were an adequate source of material.

Cultured lymphocytic cell lines have been established from a growing

* Associated Biomedic Systems, Inc., Buffalo, New York.
† Department of Graduate Studies, Roswell Park Division, SUNY, Buffalo, New York.
‡ Department of Experimental Pathology, Scripps Clinic and Research Foundation, La Jolla, California.
§ Department of Surgery, University of Colorado Medical Center, Denver, Colorado.
‖ Department of Surgery, U.C.L.A., Los Angeles, California.

pool of donor sources over the past several years. Establishment of continuous cultured lymphoid cell lines was first reported by IWAKATA and GRACE (1964) and FOLEY et al. (1965). The first successful long-term cultured cell lines were established from the peripheral blood of donors with leukemia, and reports soon followed describing lymphocytic cell lines derived from the peripheral leukocytes of normal donors (MOORE et al., 1967a; GERBER and MONROE, 1968; BRODER et al., 1970; CHOI and BLOOM, 1970).

Reviews of the literature and methodology on isolation of defined histocompatibility antigens from disrupted cells have been published (KAHAN and REISFELD, 1969; REISFELD and KAHAN, 1971). PALM and MANSON (1965) showed that the H-2 antigens are distributed on the plasma membrane of liver and membranes of endoplasmic reticulum and other subcellular organelles, and chemical characterization studies have been carried out on murine H-2 antigens by SHIMADA and NATHENSON (1969). Further information on human and murine antigens can be found in REISFELD and KAHAN (1971), TERASAKI (1970), and GOODMAN (1971).

KAHAN et al. (1968) isolated guinea pig histocompatibility antigens by low frequency sound treatment of spleen cells. Following ultracentrifugation (134,000 $\times$ g for 2 hours), the supernatant was concentrated against Aquacide and purified by gel exclusion chromatography and acrylamide gel electrophoresis. This material accelerated graft rejection in allogeneic recipients, evoked delayed-type hypersensitivity reactions, and stimulated allogeneic lymphocyte transformation *in vitro*. Spleens from five individual donor patients were extracted by the same techniques, and the antigenic extract was purified by semipreparative acrylamide gel electrophoresis. The antigenic principle had the identical HL-A determinants as those found upon the donor's peripheral leukocytes when assayed by both direct and absorption typing. The purified HL-A antigens specifically inhibited 11 cytotoxic alloantisera, 9 of which were operationally monospecific. Each of the isolated antigens contained different mosaics of HL-A antigenic determinants 1, 2, 3, 5, 7, 8, 9 and 12. Moreover, these materials were immunogenic *in vitro* and elicited hypersensitivity reactions only in individuals who were producing cytotoxic antibodies directed against antigenic specificities present on the donor's cells.

The major limitation of this study was an insufficient amount of uniform source material to permit accumulation of antigen for thorough biological, chemical, and clinical characterization. As a consequence, we began our collaborative efforts in 1969, when it became clear that cultured human lymphocytic cell lines derived from the peripheral blood of normal donors provided an excellent source of material with relatively high antigen content (REISFELD et al., 1970a). Studies have also been reported by MANN et al. (1969), who utilized cultured human cell lines derived from donors with lymphoid malignancies.

In our studies the objectives have been 3-fold: (1) to isolate HL-A antigens with different mosaics of antigenic determinants in large enough quantity to produce operationally monospecific alloantisera in animals; (2) to obtain enough antigens to begin biological studies on specific immunosuppression in man and the mouse; and (3) to elucidate the chemical structure of the gene products of the HL-A locus by localizing the determinants of specificity on the antigen molecule, and determining their chemical structure.

Our current progress toward achieving these goals is reported here.

II.   Antigen Source Material

Antigen source material consisted of cultured human lymphocyte cells derived from healthy laboratory personnel. The morphology of these cells is extremely heterogeneous; these cells differ greatly in size, mass, buoyant density, and the state of differentiation, which includes both immunoglobulin production and differentiation of cell organelles. The chromosome morphology has remained within normal limits. In the case of one of the cell lines, 1788, studied over a two-year period, the karyotype has remained normal with the number of chromosomes remaining at 46, without any evidence of translocations or aberrations. Many of the cell lines are considered to be clonal in origin, and either secrete a specific immunoglobulin or have immunoglobulins on the cell surface (FAHEY et al., 1966; TANIGAKI et al., 1966; GLADE and CHESSIN, 1968; BLOOM et al., 1971). Cloning studies indicate that the cultured cell lines derived from healthy donors can be cloned with less efficiency than malignant cell lines (IMUMURA and MOORE, 1968; MAURER et al., 1970; CHOI and BLOOM, 1970). Additional studies indicate that they contain the complement component C'3 and are also a source of macrophage inhibition factor (MIF) (GLADE and HIRSCHHORN, 1970). The procedures for establishing cultured cell lines have been discussed previously by MOORE et al. (1967b) and CHOI and BLOOM (1970). Briefly, these cell lines are started from a sample of blood obtained by venipuncture or from larger amounts of cells harvested by leukophoresis. Once the cell lines are established they exist as freely floating suspension cultures growing singly or in clumps. The growth of large-scale cultures can be carried out in a steady-state culture (BURNS, 1969) by other methods, as described previously (MOORE et al., 1967b).

III.   The HL-A Antigen Profile of Cultured Lymphoid Cell Lines

A number of techniques have been useful for the direct typing of cultured cells (see CURTONI et al., 1967). The two methods which we have used for typing the cell lines indicated in Table 1 have been the fluorochromasia (BODMER et al., 1967; PAPERMASTER et al., 1969; TAKASUGI, 1971) and dye exclusion techniques of TERASAKI and McCLELLAND, as

Table 1. HL-A Typed Cultured Lymphocyte Cell Lines†

Cell line	First segregant series							Second segregant series								
	1	2	3	9	10	11	Te 40	5	7	8	12	13	50	Te 54	57	58
1788*		+			(+)			+						+		
4098*			+													
7249*	+	+								+						
1120		+	+													
6237		+	+													
5028																
7350		+						+			+					
8057					+											(66)
1036					+							+				
7470			+		+			+	+							
7216		+						+								
P3J (HRIK)		+											+		+	+

* Donors peripheral blood is available and types identically except where indicated by ( ).

† Current consensus of collaborating laboratories. These designations may change as new antigenic specificities are found or new techniques are developed.

modified by FERRONE *et al.* (1971). Because of greater sensitivity to complement by cultured cells, typing techniques must be modified. Cultured cell lines may also carry more antigen, or may be more susceptible to the lytic action of antibody and complement, or both. In three cell lines in Table 1, indicated by asterisks, the peripheral blood has been studied extensively as well as the cultured cell lines. In the case of cell lines 1788, the cells have been genotyped, as indicated in Table 2. In addition to direct

Table 2. Genotype Analysis

	1	2	3	9	10	11	5	7	12	13	50	54	57
Paternal		+						+			+		
Maternal	+											+	+
Maternal aunt (I)	+												+
Maternal aunt (II)	+											+	+
Sibling to donor (♂)		+						+				+	
Donor P.B. (♂)		+						+				+	
Cell line 1788		+						+				+	

Haplotypes		a	2	7
On Line 1788		b	—	54

typing, absorption studies have been carried out on a number of the cell lines, and an example of the studies on cell lines 1788 are shown in Table 3. Problems in typing the cultured cell lines revolve around the nonspecific cytotoxic action of complement (TAKASUGI, 1971; FERRONE *et al.,* 1971). Figure 1 illustrates an example in the typing of two cultured cell lines compared with peripheral blood lymphocytes, showing that the cell lines are

Table 3. Quantitative Absorption of Donor Peripheral Lymphocytes and Cultured Cells with Alloantisera

	HL-A2 to 11.03 (1:32)	HL-A7 Cutten (1:8)	HL-A5 D-66 (1:2)
Cultured cells RPMI 1788 (HL-A2+, 7+, 5−)	850*	1100	30,000
Donor cells (HL-A2+, 7+, 5−)	3000	3500	12,000†

* Number of cells required to obtain a 50% reduction in the cytotoxic titer of alloantisera (AD$_{50}$).

† Endpoint of titration not yet determined.

more sensitive to dilute antisera than peripheral lymphocytes (PAPERMASTER *et al.,* 1969; BERNOCO *et al.,* 1969; ROGENTINE and GERBER, 1970). A quantitative basis for the differences could involve the expression of antigen on the cell surface or on the larger cell mass. Cell typing was not considered to be definitive in these studies until the cell line had been typed in all the participating laboratories in this report. Uniform agreement has been reached on the antigens of the cell lines in Table 1 by the participating laboratories.

IV. Assay Systems for HL-A Antigen Activity

The methods of HL-A antigen isolation devised at first by Kahan and Reisfeld have been simplified from the initial method of sonication, ammonium sulfate precipitation, gel chromatography, and semipreparative gel electrophoresis to salt extraction in 3 *M* KCl and preparative acrylamide gel electrophoresis (REISFELD and KAHAN, 1970a; REISFELD *et al.,* 1971).

Soluble HL-A antigens are most commonly detected by a serologic method that measures the combining capacity of the antigen with cytotoxic alloantibody in a complement-dependent test system.

The reaction mixture consists of peripheral lymphocytes or cultured lymphoid cells as target cells, human alloantisera as a source of antibody,

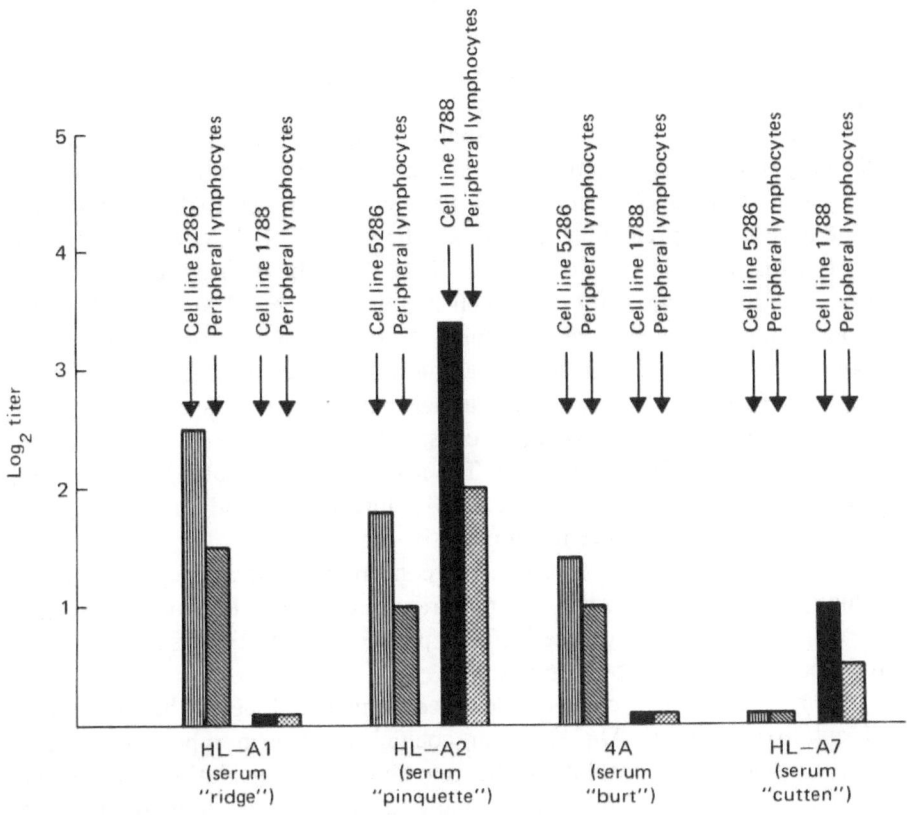

Fig. 1. Differential reactivity of cultured lymphocyte cell lines and peripheral blood lymphocytes to HL-A sera under identical conditions of incubation.

and fresh rabbit or mixtures of rabbit and human sera as the origin of complement (FERRONE *et al.*, 1971). The interaction of antibody and complement produces functional damage on the target cell surface, which can be evaluated in different ways. Cell viability is commonly evaluated by either (1) phase contrast microscopy; (2) exclusion of vital dyes; (3) fluorochromasia; (4) release of intracellular isotope markers, *e.g.*, ^{51}Cr, from labeled cells. The potency of soluble antigen is measured by inhibition of alloantibody-mediated cytotoxicity detected as an increased percentage of viable cells and defined by any one of the three methods described below. In order to permit meaningful comparisons of data among different laboratories, the following serological parameters were adopted in our studies.

(1) The inhibition dose (ID_{50}) represents the dose of antigen re-

quired to halve the cytotoxic power of a specific alloantiserum. The 50%
inhibitory dose can be calculated from a linear transformation of the van
Krogh equation as modified by Reif and described previously (REISFELD
and KAHAN, 1970a). In this assay the amount of antigen inhibiting 50%
of the cytotoxic activity of a reference antiserum is determined.

In the assay itself, 1 μl of alloantiserum is diluted at concentrations
of zero and 2 cytotoxicity units, *i.e.*, zero units being the end point of
95% cell death and 2 units being the next lower two-fold dilution. The
diluted sera are preincubated with duplicate sets of 9 serial dilutions of
antigen dissolved in 1 μl of 0.95% NaCl for 1 hour at 25°C. After this
time 3 μl containing 3000 target cell lymphocytes and 3 μl of rabbit complement are added and an additional 4-hour incubation is performed at
room temperature. The reactions are terminated by the addition of 1 μl
of 36% formalin and are read under an inverted phase contrast microscope for the number of dead cells. Control readings of alloantiserum
preincubated with buffer are used. The sigmoid curve resulting from inhibition of cytotoxicity is transformed to a linear plot, and the 50% point
of inhibitory antigen concentration can be read directly.

(2) The specificity ratio, i.e., the concentration of antigen required to
inhibit a non-specific serum vs. that required for the homologous serum,
measures the potency of the antigen preparation. Ratios as high as 300
have been observed in our studies.

(3) Avidity units can be calculated from the slope of the linear plot
of cytotoxic activity. The slope or coefficient of $1/m$ represents avidity
or the association constant of the reaction. This calculation gives some
idea of the overall avidity of the antibody preparation used to measure
antigen activity.

At the present time no international unit of HL-A antigen activity has
been established. However, it would seem that the 50% inhibitory dose
and specificity ratio offer reasonable bases for such standards, since they
are semiquantitative indications of biologic activity *in vitro* and provide
an indication of both the activity and specificity of the isolated product.

V. Extraction and Solubilization

In all recent studies we have employed the KCl extraction procedure
(REISFELD *et al.*, 1971) because of its simplicity and high rate of repro-
ducibility. We have carefully examined some of the variables of this
extraction procedure, *i.e.*, time, temperature, and molarity of the KCl
solution. Extraction of cells seems to be optimal with 3 *M* KCl during
a period of 16 hours at 4°C. Additional experiments indicated that higher
salt concentration (4 *M*) and high temperature of extraction (40°C)
combined with shorter extraction periods (15 minutes, 1 hour, 4 hours)
did not result in improved antigenic yields. Moreover, extraction with 4 *M*

KCl at 40°C over periods in excess of 4 hours caused progressive loss of antigenic activity. Repeated extraction with 3 *M* KCl over a period of 96 hours did not result in any appreciable increase in antigen yield.

Table 4 depicts average yields of soluble antigen achieved by 3 *M* KCl extraction of cultured cells (RPMI 1788; HL-A2+, 7+5−). It is appar-

Table 4. Solubilization of HL-A Antigens from Cultured Lymphoid Cells (RPMI 1788)

Sample	ID_{50} units*/mg HL-A2	ID_{50} units/10^9 cells HL-A2	ID_{50} units/mg HL-A7	ID_{50} units/10^9 cells HL-A7	Specificity ratio† HL-A2	Percent recovery‡ HL-A2	HL-A7
1	25,000	450,000	9000	162,000	70	37	29
2	33,000	462,000	5500	170,500	75	38	31
3	20,000	620,000	5500	165,000	70	51	30
4	40,000	920,000	11,000	341,000	100	85	64

* The reciprocal of the soluble protein antigen dose which inhibits cytotoxic antisera (TO 11.03, anti-HL-A2) and (Cutten, anti-HL-A7) at zero cytotoxic units to 50% of the untreated level.

† Ratio of the concentration of antigen required to inhibit an indifferent antisera *vs.* that required for a homologous antiserum.

‡ Percent recovery = (AD_{50}/CE_{50}) 100.

ent that with this method of extraction one can obtain good recoveries of HL-A2 and HL-A7 antigenic specificities possessing similar specificity ratios.

VI. Details of the KCl Procedure

Up to 50×10^9 dispersed cultured human lymphoid cells suspended in phosphate-buffered saline containing 3 *M* KCl, pH 7.4 (10 ml solvent per 10^9 cells), were gently agitated for 16 hours at 4°C on an Eberbach shaker and then centrifuged at $163,000 \times g$ for 1 hour (R_{max} 9.9 cm, $235,000 \times g$). During dialysis (24 hours) against three changes of 200 volumes of saline a gelatinous material formed, which comprised 18% of the ultracentrifugal supernate, and contained primarily DNA. This contaminant was removed by centrifugation at $1500 \times g$ for 20 minutes. DNA could not be detected in the $1500 \times g$ supernate either by diphenylamine test of a hot trichloracetic acid extract (REISFELD *et al.*, 1971) or by radioactive label experiments. The $1500 \times g$ supernatant antigen derived from two generations of cultured WIL-2 lymphoid cells which had been uniformly labeled with thymidine 2-^{14}C (New England Nuclear, 50 mCu/m mole, utilizing 0.1 μCu 5×10^5 cells), contained less than 1% of the total labeled DNA. Crude antigen preparations were stable for at least several weeks when stored in 3 *M* KCl at 4°C or up to 1 year in 0.9% NaCl at −20°C.

Hypertonic salt extraction yields considerably greater recoveries of potent HL-A antigens from lymphoid cell lines than do other methods when tested in our laboratories. The active principle in the KCl extract was soluble since (1) it did not sediment upon ultracentrifugation at 164,000 × g; (2) it could be passed through a Millipore filter (0.45 μ); (3) it did not show any ultrastructural components upon electron microscopic analysis; and (4) it could be purified to a well-defined protein by preparative acrylamide gel electrophoresis.

VII. Electrophoretic Purification

Preparative acrylamide electrophoresis provides a powerful tool for the purification of HL-A antigens solubilized by extraction with 3 *M* KCl (REISFELD and KAHAN, 1970b). Moreover, the incorporation of radioactively labeled amino acids into cultured lymphocytic cell lines makes it feasible to work with relatively small quantities of cells. Furthermore, the use of the radioactively labeled cells, coupled with Kjeldahl nitrogen analyses, makes it feasible to determine minute amounts of material accurately.

In one experiment, cultured cells (WIL-2; HL-A 2 + 5 + 7 −) were allowed to reach a concentration 5×10^8/ml while in log phase and were then placed for 15 minutes in Eagle's minimal essential medium containing 1% of its normal amino acid content. Then a mixture of ^{3}H-labeled amino acids was added (2.5 mc) and reacted with the cells for 4 hours at 37°C. The cell suspension was centrifuged at 500 × g and the cells were thoroughly washed with Hanks' balanced salt solution. Approximately 12% of radioactive label added to the cells in culture was incorporated into the cells. The washed cells were placed in 10 ml 3 *M* KCl and extracted for 16 hours at 4°C. The extract was centrifuged at 130,000 × g for 6 hours. The supernate was concentrated to a 2-ml volume after which a small aliquot was removed for radioactive label and nitrogen determination. The solution was then dialyzed against "upper gel buffer" (0.05 Tris-phosphate, pH 6.9, containing 5% (w/v) sucrose) and applied to a Buchler-Polyprep 100 column (100 mm² gel surface). The column consisted of 60 ml lower gel (5% acrylamide) and 25 ml upper gel. Electrophoresis was carred out at 0°C at a constant current of 30 mA. Eluate fractions were collected at 10-minute intervals at a flow rate of 0.8 ml/minute. The elution buffer was 0.15 *M* Tris-HCl, pH 7.8, containing 10% (w/v) sucrose.

Typical elution profiles as visualized by analytical acrylamide gel electrophoresis (pH 9.4; 7.5% acrylamide) have been published elsewhere (REISFELD and KAHAN, 1970b; REISFELD et al., 1971). Antigenic activity is found in early eluate fractions migrating with the fastest anodic band (Rf 0.78 − 0.80).

The yield of the electrophoretically purified fraction was approximately 2% of the material initially applied with a 43% recovery of the total ID_{50} units.

VIII. Chemical Characterization

Electrophoretically purified HL-A antigens derived from human spleens were found to be 94% monodisperse and to have a molecular weight of 31,000 when analyzed by Yphantis sedimentation equilibrium techniques. Thorough characterization by discontinuous polyacrylamide gel electrophoresis at pH 9.4, both in the presence and absence of 8.4 M urea and at varying pore sizes, showed the antigenic principle to be a single electrophoretic moiety with an Rf of 0.78. No carbohydrate or lipid ($>1\%$) could be detected within the limitations of the methods employed, *viz.*, Folch partition thin-layer chromatography on silica gel G was performed following lipid extraction with chloroform:methanol (2:1 v/v) while staining selectively for lipids, glycolipids, and cholesterol esters. Hexose and pentose content was determined by the cysteine-sulfuric acid method (REISFELD *et al.*, 1971). Amino acid analyses of electrophoretically purified HL-A antigens isolated from two different cell lines (RPMI 1788; HL-A2+, 7+, and RPMI 4098; HL-A3+), respectively, showed reproducible and significant differences in five amino acids, aspartic acid serine, proline, alanine, and tyrosine (REISFELD and KAHAN, 1971).

XI. Discussion

These findings support the hypothesis that genetically segregating allospecificities of HL-A antigens are related to protein structure.

Although during the past 15 years the chemical nature of H antigens has been attributed to virtually every common biochemical entity, recent progress strongly suggests that these antigenic determinants are predominantly polypeptides (GOODMAN, 1971). A large body of evidence suggests that polypeptides are absolutely essential for the expression of histocompatibility antigen activity (REISFELD and KAHAN, 1971). Recent reports indicate that most of the carbohydrate found on papain-solubilized H-2 antigens could be removed by carbohydrase treatment without seriously affecting antigenic activity (GOODMAN, 1971). Thus it seems that carbohydrate has functions other than that of contributing to the antigenicity of these molecules. In additional, it has been suggested by CEPPELLINI *et al.* (1967) that the great number of HL-A haplotypes are rather similar to immunoglobulin allotypes where many mutants are distributed along a limited number of linked structural genes corresponding to many antigenic determinants of a few polypeptide chains. This is in contrast to the ABO system in which a more limited number of antigenic variants are controlled by a few enzymes acting on a common substrate.

The KCl extraction method offers certain advantages such as (1) simplicity; (2) greater overall yields of alloantigens; and (3) lack of detectable lipids in the initial extract.

Although sonication has previously been shown to be a powerful tool for solubilizing transplantation antigens, its successful application requires considerable skill as well as complex equipment. On the other hand, the great simplicity of the KCl procedure has already permitted several investigators to successfully test our method in the short period of time since we first introduced it (ETHEREDGE and NAJARIAN, 1971; MELTZER and RAPP, 1971). Although quantitative recovery studies have not yet been performed which employ aliquots of a single batch of cells treated with different extraction methods, comparable data from a number of our experiments suggest from two- to threefold greater total yields of antigen with KCl (REISFELD and KAHAN, 1971).

One of the principal obstacles to the efficient purification of the HL-A antigenic principle is the presence of lipids and lipoproteins. These materials markedly interfere with preparative acrylamide electrophoresis, resulting in extensive aggregation of applied materials interfering with the function of the acrylamide gel while binding antigen to lipoprotein complexes. While crude sonicates of splenic cells contain relatively modest amounts of lipid, sonicates of cultured cells are rich in these compounds, which effectively decreases the yield of purified antigen obtainable by preparative acrylamide electrophoresis. KCl extracts of cultured cells contain relatively little lipid and thus make it possible to attain larger yields of purified antigens.

KCl probably dissociates hydrogen bonds and salt linkages releasing alloantigen, possibly in a manner analogous to the depolymerization of membrane components by sonication. These relatively mild methods seem preferable to papain digestion, which breaks peptide linkages and is difficult to control, and to extraction with complex salts and detergents, which are difficult to remove from the solubilized product (REISFELD and KAHAN, 1971). The KCl method is undoubtedly applicable to solubilization of other cell membrane activities. Being analogous to the solubilization of tumor-specific antigens by sonication, 3 M KCl has been shown to effectively solubilize tumor-specific antigenic determinants of diethylnitrosamine-induced guinea pig sarcomas (MELTZER and RAPP, 1971).

The physical nature of the chemical expression of the HL-A antigenic determinants on polypeptide chains is still unclear. For example, it is not certain whether the 4 antigenic determinants of a heterozygote controlled by the two major genetic regions of the HL-A locus are present on separate molecules, or on single molecules, and whether a *cis-trans* effect may operate in the expression of determinants in a single molecule. Many of the problems involved in obtaining definitive information on this aspect are related to the low titer of the antisera available for HL-A studies which so far have proven inapplicable to quantitative radioimmuno-assay procedures in our hands. The availability of purified antigen now opens up opportunities for producing high titer isoantisera in humans and has already shown promise for production of heterologous antisera of high titer (FERRONE et al., 1972).

It is quite clear that cultured cell lines offer many advantages not available with human tissue material or inbred animal strains. For example, radioactively labeled precursor amino acids can be fed to cultured cells in log-growth phase, making it feasible to quantitate minute amounts of antigen. Pulse-labeling of synchronous cells affords an approach to characterizing the biosynthetic pathways of HL-A antigen assembly. Such information on regulation of biosynthesis may lead to a mechanism for selectively increasing the concentration of certain HL-A determinants and for promoting their release from the cell membrane into the medium in membrane-precursor form. For this purpose, massive screening programs may eventually yield cultured lymphoid cell lines from members of a single family.

Although most lymphoid cell lines yield material possessing multiple antigenic determinants, variant lines can be isolated from antigenically negative individuals, and possibly by somatic cell genetic manipulation *in vitro* (PAPERMASTER and HERZENBERG, 1966). Selection by lytic viruses (KAPLAN and BEN-PORAT, 1968) or immunoselection with isoantisera may produce variant lines which lack some antigenic determinants, or are monodeterminant, and which could be used for the extraction of single HL-A determinants.

In conclusion, the development of long-term lymphoid cell lines with defined HL-A phenotype and genotype, together with new isolation and purification methods, have led to a clarification of the chemical and physical nature of the gene products of the HL-A locus. The approaches developed have already yielded large-scale production of antigens in purified form, and their availability now affords an opportunity for more detailed chemical and clinical studies. With increased availability of these materials, the chemical basis of the complex genetic polymorphism of the HL-A system in man and its role in transplantation individuality will more quickly be elucidated.

References

Bernoco, D., P. R. Glade, S. Broder, V. Miggiano, K. Hirschhorn, and R. Ceppellini (1969). Stability of HL-A and appearance of other antigens (LIVA) at the surface of lymphoblasts grown *in vitro*. *Acta Haematol.*, 54:795–812.

Bloom, A. D., K. W. Choi, and B. J. Lamb (1971). Immunoglobulin production by human lymphocytoid lines and clones: Absence of gene exclusion. *Science*, 172:382–383.

Bodmer, W., M. Tripp, and J. Bodmer (1967). Application of a fluoro-chromatic cytotoxicity assay to human leukocyte typing, pp. 341–350.

In *Histocompatibility Testing,* E. S. Curtoni, P. L. Mattiuz and R. M. Tosi (eds.), Copenhagen: Munksgaard.

Broder, S. W., P. R. Glade, and K. Hirschhorn (1970). Establishment of long-term lines from small aliquots of normal lymphocytes. *Blood, 35*:539–542.

Burns, A. A. (1969). Unpublished observations.

Ceppellini, R., E. S. Curtoni, P. L. Mattiuz, V. Miggiano, G. Scudeller, and A. Serra (1967). Genetics of leukocyte antigens: A family study of segregation and linkage, pp. 149–187. In *Histocompatibility Testing, 1967,* E. S. Curtoni, P. L. Mattiuz and R. M. Tosi (eds.), Copenhagen: Munksgaard.

Choi, K. W., and A. D. Bloom (1970). Cloning human lymphocytes *in vitro. Nature, Lond., 227*:171–173.

Curtoni, E. S., P. L. Mattiuz, and R. M. Tosi (eds.). (1967). *Histocompatibility Testing,* 1967, Copenhagen: Munksgaard.

Etheredge, E. E., and J. S. Najarian (1971). Solubilizaton of human histocompatibility substances. *Transpl. Proc., 3*:224–226.

Fahey, J. L., I. Feingold, A. S. Rabson, and R. A. Manaker (1966). Immunoglobulin synthesis *in vitro* by established human cell lines. *Science, 152*:1259–1261.

Ferrone, S., P. G. Natali, A. Hunter, P. I. Terasaki, and R. A. Reisfeld (1972). Immunogenicity of soluble HL-A alloantigens. J. Immunol. (in press).

Ferrone, S., M. Pellegrino, and R. A. Reisfeld (1971). A rapid method for direct HL-A typing of cultured lymphoid cells. *J. Immunol., 107*:613–615.

Foley, G. E., H. Lazarus, S. Farber, B. G. Uzman, B. A. Boone, and R. E. McCarthy (1965). Continuous culture of human lymphoblasts from peripheral blood of a child with acute leukemia. *Cancer, 18*:522–529.

Gerber, P., and J. H. Monroe (1968). Studies on leukocytes growing in continuous culture derived from normal human donors. *J. Nat. Cancer Inst., 40*:855–866.

Glade, P. R., and L. N. Chessin (1968). Infectious mononucleosis: Immunoglobulin synthesis by cell lines. *J. Clin. Invest., 47*:2391–2401.

Glade, P. R., and K. Hirschhorn (1970). Products of lymphoid cells in continuous culture. *Am. J. Path., 60*:483–492.

Goodman, H. (ed.). (1971). Biological significances of histocompatibility antigens. Fogarty International Center Proceedings No. 15. Bethesda National Institutes of Health, in press.

Imumura, T., and G. E. Moore (1968). Ability of human hematopoietic cell lines to form colonies in soft agar. *Proc. Soc. Exp. Biol. Med., 128*:1179–1183.

Iwakata, S., and J. T. Grace (1964). Cultivation *in vitro* of myeloblasts from human leukemia. *N.Y. State J. Med., 64*:2279–2282.

Kahan, B. D., and R. A. Reisfeld (1969). Transplantation antigens. *Science, 164*:514–521.

Kahan, B. D., R. A. Reisfeld, M. A. Pellegrino, E. S. Curtoni, P. L. Mattiuz, and R. Ceppellini (1968). Water-soluble human transplanation antigen. *Proc. Nat. Acad. Sci. U.S.A., 61*:879–904.

Kaplan, A. S., and T. Ben-Porat (1968). Metabolism of animal cells infected with nuclear DNA viruses. *Ann. Rev. Microbiol., 22*:427–450.

Mann, D. L., G. N. Rogentine, J. L. Fahey, and S. G. Nathenson (1969). Human lymphocyte membrane (HL-A) alloantigens: Isolation, purification and properties. *J. Immunol., 103*:282–292.

Maurer, B. A., T. Imumura, and S. M. Wilbert (1970). Incidence of EB virus containing cells in primary and secondary clones of several Burkitt lymphoma cell lines. *Cancer Res., 30*:2870–2875.

Meltzer, M. S., and H. J. Rapp (1971). Personal communication.

Moore, G. E., R. E. Gerner, and H. A. Franklin (1967a). Culture of normal human leukocytes. *J. Am. Med. Assoc., 199*:519–524.

Moore, G. E., R. E. Gerner, and J. Minowada (1967b). Studies of normal and neoplastic human hemopoietic cells *in vitro*, pp. 41–63. In *The Proliferation and Spread of Neoplastic Cells*, 21st Annual Symp. Found. Cancer Research, Houston.

Palm, J., and L. A. Manson (1965). Tissue distribution and intracellular sites of some mouse isoantigens, pp. 21–35. In *Isoantigens and Cell Interactions*, J. Palm (ed.), Philadelphia: Wistar Institute Press.

Papermaster, B. W., and L. A. Herzenberg (1966). Isolation and characterization of an isoantigenic variant from a heterozygous mouse lymphoma in cultures. *J. Cell Physiol., 67*:407–420.

Papermaster, V. M., B. W. Papermaster, and G. E. Moore (1969). Histocompatibility antigens of human lymphocytes in long-term culture. *Fed. Proc., 28*:379.

Reisfeld, R. A., and B. D. Kahan (1970a). Biological and chemical characterization of human histocompatibility antigens. *Fed. Proc., 29*:2034–2039.

—— (1970b). Transplantation antigens, pp. 117–200. In *Advances in Immunology*, vol. 12, F. I. Dixon and H. G. Kunkel (eds.), New York: Academic Press.

—— (1971). Extraction and purification of soluble histocompatibility antigens. *Transpl. Rev., 6*:81–112.

Reisfeld, R. A., M. Pellegrino, and B. D. Kahan (1971). Salt extraction of soluble HL-A antigens. *Science, 172*:1134–1136.

Reisfeld, R. A., M. Pellegrino, B. W. Papermaster, and B. D. Kahan (1970a). HL-A antigens from a continuous lymphoid cell line derived from a normal donor. *J. Immunol., 104*:560–565.

—— (1970b). Serologic characterization of soluble HL-A antigens from continuous lymphoid cell lines derived from normal donors, pp. 455–460. In *Histocompatibility Testing*, P. I. Terasaki (ed.), Copenhagen: Munksgaard.

Rogentine, G. N., and P. Gerber (1970). Qualitative and quantitative comparisons of HL-A antigens on different lymphoid cell types from the same individuals, pp. 333–338. In *Histocompatibility Testing*, P. I. Terasaki (ed.), Copenhagen: Munksgaard.

Shimada, A., and S. G. Nathenson (1969). Murine histocompatibility H-2 alloantigens. *Biochem., 8*:4048–4062.

Takasugi, M. (1971). An improved fluorochromatic cytotoxic test. *Transplantation, 12*:148–151.

Tanigaki, N., Y. Yagi, G. E. Moore, and D. Pressman (1966). Immunoglobulin production in human leukemia cell lines. *J. Immunol., 97*:634–649.

Terasaki, P. I. (ed.). (1970). *Histocompatibility Testing*, Copenhagen: Munksgaard.

CURRENT CONCEPTS OF THE HL-A SYSTEM

C. M. ZMIJEWSKI

*Ortho Research Foundation,
Raritan, New Jersey*

The HL-A system is comprised of a series of antigens associated with human leukocytes and controlled by a complex genetic locus. The antigens of this system are present not only on leukocytes but may be found on platelets and various other body tissues (AMOS and PEACOCKE, 1964). As a result of their distribution, the HL-A determinants have been implicated in histocompatibility testing for human organ transplantation, and it is apparent from current evidence that compatibility for these antigens may play some significant role in the outcome of an organ graft (KISSMEYER-NIELSEN *et al.,* 1970).

During the initial phases of the leukocyte antigen study, a large number of antigenic determinants were found by various laboratories. It was only later, after an intensive and cooperative genetic study by the most eminent investigators in the world, at Professor Ceppellini's laboratory in 1967 in Turin, Italy, that a conclusion was reached concerning the inheritance of these determinants. From family studies, it became apparent that all of the antigens being detected were under the control of a single genetic locus. This locus was named HL-A to signify the first or A locus of human leukocyte antigens. Additional work in the ensuing years confirmed the presence of the HL-A antigens on the surface of granulocytes, lymphocytes, platelets, and various other tissues, such as kidney, heart, skin, spleen, liver, and lung.

The various determinants of the HL-A system seem to be under the control of allelic genes comprising two segregant series, as shown in Table 1. Since chromosomes are paired, no individual can have more than two antigens from each series—one having come from the father, and the other from the mother.

Table 2 is taken from DAUSSET (1971) and presents a much more complex picture of the HL-A system. In addition to the simple number terminology of the World Health Organization, various other antigens with so-called local designations can be seen. Antigens, such as ILN*, Bª*, LND, Mapi, and Maki, have become household words among workers in

Table 1. The HL-A Antigens

First Segregant Series	Second Segregant Series
HL-A1	HL-A5
HL-A2	HL-A7
HL-A3	HL-A8
HL-A9	HL-A12
HL-A10	HL-A13
HL-A11	

the leukocyte field. In addition, some other antigens can be seen grouped in brackets. This means that sera obtained from some donors and directed against a particular antigen will cross-react with another antigen, which may be genetically similar. These are what are normally called "broad specificities."

Serologically it is possible to detect these antigens in more than one way. Three major methods are normally employed: (1) agglutination using leukocytes harvested from blood collected in EDTA; (2) agglutination using leukocytes harvested from blood which has been defibrinated; and (3) cytotoxicity. Because of the extent of reactivity of various antisera by these different methods, some people claim that one technique is more sensitive than another. Actually it is more likely that antibodies produced by different individuals, even with the same specificity, react preferentially in a different manner. This is a situation which is completely analogous to that found in red cell serology, where in some isolated cases anti-Fy[a]

Table 2. The Expanded HL-A Antigenic Determinants

First Segregant Series	Second Segregant Series
HL-A1 ⎱ HL-A3 ⎰ HL-A11 (Da21, ILN*) ⎱ DA12	HL-A12 ⎱ HL-A13 ⎰ Te18
HL-A2 ⎱ Ba* ⎰ 8ᵃ	HL-A5 ⎱ R* ⎰ 4ᶜ
HL-A9	LND
HL-A10	Mapi
Da22 ⎱ Da25 ⎰ Li ⎱ Te19 ⎰ Ao28 ⎰ LA-W	HL-A7 ⎱ BB ⎰ FJH ⎰ Da9 ⎰ 6ᵇ
	HL-A8
	Maki
	AA

sera, for example, although reacting preferentially by the antiglobulin technique occasionally will react even when the red cells are suspended in saline diluent.

The physicochemical parameters of leukocyte reactions are not well understood because of the lack of potent monospecific antisera which would be required in order to study them. However, if one considers the ingredients of a typical agglutination reaction or a typical cytotoxicity test, it becomes clear as to why different results may be obtained. For example, in EDTA agglutination, the final reaction mixture consists of granulocytes, EDTA resulting in complement inactivation because of a lack of calcium and magnesium ions, plasma ingredients plus the gelatin which is used as a sedimenting agent. In defibrinated agglutination the final reaction mixture contains, in addition to granulocytes, both calcium and magnesium and therefore, fully active complement, a lack of platelets, serum lacking fibrinogen, and either dextran or PVP. Thus even without experimentation, it is possible to speculate that the dielectric increments of these two sus-

Table 3. Reactions of Five Anti-4ª Sera (EDTA Agglutination)

Cells	Sera				
	8	10	22	MC	DK
1	++++	+++	++++	++	++++
2	+++	+++	++++	+++	++++
3	++++	++++	++++	+++	++++
4	--	--	--	--	--
5	++	+++	++++	++++	+++
6	--	--	--	--	--
7	--	--	--	--	--
8	++++	++++	+++	++	++++
9	++++	++++	+++	++++	++++
10	++++	++++	++++	++	++++
11	++++	++++	++++	+++	+++
12	--	--	--	--	--
13	--	--	--	--	--
14	++	+++	++++	++	++++
15	+++	++++	+++	++++	++
16	++++	+++	++++	++++	++++
17	--	--	--	--	--
18	+++	+++	++++	++++	+++
19	+++	++++	+++	++++	++++
20	++++	++++	++++	++++	+++
21	++++	++++	+++	++++	++++
22	++++	++++	+++	+++	+++
23	++++	++++	+++	++++	++
24	--	--	--	--	--
25	++++	++++	++++	++++	++++

pending media are entirely different, and should complement dependency be required in any way, it is present only in the defibrinated test. The cytotoxicity test, on the other hand, is a complement-dependent reaction and employs only lymphocytes as the target cell. In view of the fact that it is complement-dependent, only antibodies capable of activating complement can be employed.

A resume of the reactions of five sera which give identical reaction with the cells of the test panel is shown in Table 3. It is obvious that there is no difference among these sera. However, in Table 4 it can be

Table 4. Reactions of Anti-4ᵃ Sera by Different Methods

Cells	Serum 8			Serum 10		
	EDTA	Defib	Cytotox	EDTA	Defib	Cytotox
1	+++	++++	++++	++++	+++	──
2	++++	+++	++	+++	++++	+++
3	──	──	──	──	──	──
4	──	──	──	+++	──	──
5	──	──	──	──	──	──
6	+++	++++	++++	++++	──	──
7	──	──	──	++	──	──
8	++++	+++	++	++++	++++	+++
9	──	──	──	++	+++	──
10	──	──	──	──	──	──

seen that when the sera have been tested by three different methods, even though the first serum is identical, the second shows inclusions. In other words, a cell is never positive by either defibrinated agglutination or cytotoxicity unless it is also positive in the EDTA method. These strange findings are difficult to explain. One explanation would be to conclude that the sera contain antibodies which are complex, capable of reacting with many cross-reacting antigenic determinants. Another explanation would be to postulate the existence of complex antigens. In other words, a single antigen would have many determinants. Finally, one could conclude that the sera used for the detection of these antigens are exceedingly complex, containing antibodies against many small portions of the antigenic determinant. This is the so-called "alphabet soup" type of explanation.

As a matter of evolutionary interest, it can be noted that these same antigens, identifiable with human isoantisera, are also present on the white blood cells of chimpanzees and, probably, other subhuman primates (DORF and METZGAR, 1970). This could be a clue to the extreme polymorphism found in this system, and further studies may shed some light on the biological purpose of this polymorphic series of antigens.

Needless to say, the studies of leukocyte antigens are at present in their most infantile and embryonic stage. We have not even considered at the present moment the problems associated with autoneutralization due to the presence of soluble antigens in the serum or plasma (CHARLTON and ZMIJEWSKI, 1970). Secondly, we must consider the level of antibody detection. At present it seems as though the amount of antibody we can detect in our serological tests is far below the actual amount of immunoglobulin present in a particular serum and directed against the particular antigenic determinant being tested. Until these questions can be answered, it is very difficult to make any assumptions regarding the chemical nature or distribution of the HL-A antigens on cellular membranes.

References

Amos, D. B., and N. Peacocke (1964). Leukoagglutination technique, pp. 161–162. In *Histocompatibility Testing*. Copenhagen: Munksgaard.

Charlton, R. K., and C. M. Zmijewski (1970). Soluble HL-A7 antigen: localization in the -lipoprotein fraction of human serum. *Science, 170*:636–637.

Dausset, J. (1971). The genetics of transplantation antigens. *Transpl. Proc., III No. 1*:8.

Dorf, M. E., and R. Metzgar (1970). Serological relationships of human, chimpanzee, and gorilla lymphocyte isoantigens, pp. 287–297. In *Histocompatibility Testing*. Copenhagen: Munksgaard.

Kissmeyer-Nielsen, F., L. Staub-Nielsen, A. Lindholm, L. Sandberg, A. Svejgaard, and E. Thorsby (1970). The HL-A system in relation to human transplantations, pp. 105–135. In *Histocompatibility Testing*. Copenhagen: Munksgaard.

Workshop Data. (1967), pp. 435–449. In *Histocompatibility Testing*. Copenhagen: Munksgaard.

INTERACTION OF SYNTHETIC ANTIGENS WITH THE UNIQUE RIBONUCLEOPROTEIN OF MACROPHAGE CELLS

A. Arthur Gottlieb, S. R. Waldman and
R. H. Schwartz

*Institute of Microbiology, Rutgers, The State
University, New Brunswick, New Jersey*

We have previously described an unusual ribonucleoprotein (RNP) in adherent peritoneal exudate cells. This complex, which comprises 5% of the bulk RNA, is characterized by a density in cesium sulfate solution of 1.588 g/cc, a protein content of 30%, and a molecular weight of 12,000 daltons (Gottlieb *et al.,* 1967). When adherent peritoneal exudate cells (80% of which appear morphologically to be large mononuclear cells) are exposed to a variety of antigens *in vitro,* fragments of these antigens are found in this RNP complex (Gottlieb, 1969a). Fragments of bacteriophage antigens, when complexed to the RNP complex, manifest increased immunogenicity as compared with free antigen.

Since the macrophage RNP complex is relatively resistant to the action of pancreatic ribonuclease A, preparations of RNA from a variety of sources can be examined for the presence of this complex by treatment of the bulk RNA (pretreated with DNAase to eliminate contaminating DNA) with RNAase-A, followed by banding in cesium sulfate solution in the analytical centrifuge. Using this method, which eliminates artifactual DNA-RNA complexes, we have never detected an RNP complex in rat thymus, thyroid, kidney, or in the murine myeloma tumor. A small amount of RNP exists in spleen and liver, and this is consistent with the presence in these tissues of small numbers of macrophage or macrophage-like cells.

We have found that the RNP complex has a distinct electrophoretic mobility and is homogeneous by gel electrophoresis at pH 7.7 and 3.9 (Gottlieb and Strauss, 1969; Gottlieb, 1971). These systems permit us to evaluate the binding of antigenic fragments to the RNP species in the bulk RNA preparation without the necessity of isolating the RNP species. For these studies, the use of linear random synthetic antigens of the type $Glu^{60}Ala^{30}Tyr^{10}$ and $Glu^{52}Lys^{33}Tyr^{15}$ permits us to obtain antigens of

very high specific activity (3–30 $\mu c/\mu g$) by the use of the iodination procedure of GREENWOOD et al. (1963). Since these polymers are essentially
neutral at pH 3.9, the migration of ^{125}I label representing antigen on these
gel systems must represent antigen associated with mobile species in this
system, i.e., RNA's. Since RNP moves ahead of sRNA at pH 7.7 but behind sRNA at pH 3.9, it is possible to have a high degree of confidence
in the determination of the amount of antigen specifically complexed to
the RNP species. In our hands, at the dosages of antigen employed, we
rarely see labeled antigen associated with species of RNA other than the
RNP species. It is perhaps worthwhile to point out that the simplistic determination of amount of label per μg of bulk RNA may lead to serious errors
in the determination of the amount of iodinated copolymer attached to
the RNP species. This is due to the fact that the bulk RNA preparations
contain labeled polymer, which may be attached to species of RNA other
the RNP, as well as free labeled polymer. ROELANTS and GOODMAN
(1969) have clearly shown that such bulk RNA preparations may contain
complexes of RNA and negatively charged antigens which are not related
to the RNP species and that these artifactual complexes may be disrupted
by heating in versene solution. It is critical to point out that the gel electrophoresis systems employed in the studies presented in this report enable
us to specifically look at the binding of labeled copolymers to the RNP
species alone. Under these conditions, we have found that this binding is
unaffected by heating the bulk RNA preparation to 70°C for 3 minutes
in 20 m M disodium EDTA.

By the use of these techniques, it is possible to evaluate the binding
of optical enantiomorphs of the synthetic polymers to RNP. As shown in
Table 1, the binding of the L-forms of both polymers is considerably greater
than that of the analogous D-form. This observation, by itself, is sufficient
evidence to conclude that the binding of the undegraded forms of these
polymers to RNP cannot be a simple electrostatic interaction, as claimed
by ROELANTS and GOODMAN (1969), since such interaction should not
discriminate between optical forms of these polymers. Although there is
a parallelism in this case between the binding of these optical forms and
their immunogenicity, this relationship has not been found to be generally
true for all antigens studied thus far.

A second observation in this regard concerns the effect of simultaneous
exposure of peritoneal exudate cells to labeled copolymer of the L-form
and unlabeled copolymer of the D-form. A most unexpected result is
shown in Table 2. While addition of unlabeled L-copolymer results in a
reduction of binding of the L-form as expected, the D-copolymer is nearly
as effective in this regard. Since the D-form is not bound to the RNP complex, this effect cannot reflect competition for binding at the level of the
RNP complex itself, but must be due to a suppressive effect of the
D-copolymer, at a point before linkage of the L-fragment to the RNP com-
plex. The effect is seen with both D-Glu^{60}Ala^{30}Tyr^{10} and D-Glu^{52}Lys^{33}Tyr^{15}

against the homologous L-copolymer. D-Glu⁶⁰Ala³⁰Tyr¹⁰ exerts a similar effect on the binding of L-Glu⁵²Lys³³Tyr¹⁵. It is unlikely that this represents a toxic effect of D-copolymer on the catabolism of the L-copolymers, since the conversion of the L-copolymers to acid-soluble fragments by these cells was, if anything, increased by exposure to the D-copolymers. Moreover,

Table 1. Comparative Binding of L and D Stereoisomers of the Synthetic Copolymers Glu⁶⁰Ala³⁰Tyr¹⁰ and Glu⁵²Lys³³Tyr¹⁵

	Relative specific activity*
Experiment I	
4 μg L-GAT	90
10 μg L-GAT	281
25 μg L-GAT	278
6 μg D-GAT	52
10 μg D-GAT	50
25 μg D-GAT	51
Experiment II	
20 μg L-GLT	114
20 μg D-GLT	10
30 μg L-GLT	136
30 μg D-GLT	11
Experiment III	
20 μg L-GLT	153
20 μg D-GLT	9
50 μg L-GLT	273
50 μg D-GLT	8

$$* \text{ Relative specific activity} = \frac{\text{cpm}/\mu\text{g RNP}}{\begin{array}{c}\text{specific activity of copolymer} \times \\ \text{mean residue weight of polymer}\end{array}}.$$

no reduction in viability, as judged by Trypan Blue staining, was observed under these conditions.

We are currently attempting to determine what characteristics of the D-copolymer are required for this effect. Similar competition experments (Table 3) carried out with an analogous copolymer of GAT in which the negative charges of the glutamic acid residues have been neutralized by amidification with hydroxy-propanolamine [Poly N⁵-(3-hydroxy propyl)-L-Glutamine⁶⁰Ala³⁰Tyr¹⁰] (hp GₙAT), and with polyvinylpyrollidone (an undegradable molecule of similar molecular weight), demonstrate that neither of these compounds exhibits the same effect as does the D-copolymer. It may be that both nondegradability and net negative charge are required for this effect. The evidence present here suggests that there is an enzyme in the adherent peritoneal exudate population which is required

Table 2. Competition of Synthetic Copolymers for Binding to Macrophage RNP

	Percent of ^{125}I polymer bound to RNP	sup TCA/ppt TCA
Experiment I		
15 μg [125]I L-GAT	100	0.171
15 μg [125]I L-GAT + 250 μg cold D-GAT	40	0.227
15 μg [125]I L-GAT + 500 μg cold D-GAT	22	0.346
15 μg [125]I L-GAT + 100 μg cold L-GAT	59	0.186
15 μg [125]I L-GAT + 250 μg cold L-GAT	7	0.170
15 μg [125]I L-GAT + 500 μg cold L-GAT	0	0.163
Experiment II		
20 μg [125]I L-GLT	100	0.123
20 μg [125]I L-GLT + 500 μg cold D-GLT	66	0.151
20 μg [125]I L-GLT + 1000 μg cold D-GLT	45	0.139
20 μg [125]I L-GLT + 500 μg cold L-GLT	36	0.120
20 μg [125]I L-GLT + 1000 μg cold L-GLT	6	0.141
Experiment III		
25 μg [125]I L-GLT	100	0.146
25 μg [125]I L-GLT + 200 μg D-GAT	87	0.170
25 μg [125]I L-GLT + 400 μg D-GAT	41	0.137

The effect of cold L- or D- polymers on the binding of standard doses of [125]I L-GAT is displayed. Sup TCA/ppt TCA refers to the ratio of TCA-soluble to TCA-precipitable ^{125}I label in the cells.

for linkage of fragments of antigen to the RNP complex and that this enzyme is blocked by the D-copolymers.

These observations suggest, but do not prove, that there is a role for these antigen-RNP complexes in the immune response *in vivo*. We currently view these cells as representative of a mechanism which is complementary to the primary interaction of antigen with antigen-sensitive cells. This mechanism may be important in the control of antigen dosage and the balance between immunity and tolerance. Since it is clear that the antigenic fragments in this complex represent pieces of the original antigen, these antigen-RNP complexes are not likely to play a role in the primary interaction of native antigens with lymphocytes, since such fragments would clearly not be native. This does not exclude the possibility that

Table 3. Selective Suppression of Binding of ^{125}I L-GAT to RNP by D-GAT

	Relative specific activity	sup TCA/ppt TCA
Experiment I		
18 μg ^{125}I L-GAT	212	0.201
18 μg ^{125}I L-GAT	224	0.191
18 μg ^{125}I L-GAT + 200 μg D-GAT	127	0.275
18 μg ^{125}I L-GAT + 600 μg D-GAT	109	0.325
18 μg ^{125}I L-GAT + 200 μg PVP	198	0.204
18 μg ^{125}I L-GAT + 600 μg PVP	226	0.177
18 μg ^{125}I L-GAT + 260 μg hpG$_n$AT	251	0.212
18 μg ^{125}I L-GAT + 600 μg hpG$_n$AT	196	0.227
Experiment II		
21 μg ^{125}I L-GAT	272	
21 μg ^{125}I L-GAT	330	
21 μg ^{125}I L-GAT + 200 μg D-GAT	144	
21 μg ^{125}I L-GAT + 500 μg D-GAT	109	
21 μg ^{125}I L-GAT + 200 μg PVP	397	
21 μg ^{125}I L-GAT + 500 μg PVP	282	
21 μg ^{125}I L-GAT + 260 μg hpG$_n$AT	269	
21 μg ^{125}I L-GAT + 520 μg hpG$_n$AT	331	

The effect of D-GAT, polyvinylpyrollidone (PVP) and a neutral analogue of L-GAT (hp GnAT) on the binding of standard doses of ^{125}I L-GAT is shown. Sup TCA/ppt TCA refers to the ratio of TCA-soluble to TCA-precipitable ^{125}I label in the cells. Relative specific activity is calculated as described in Table 1.

local regions of tertiary structure as large as 3600 daltons can be preserved during "antigen processing." Such fragments of the T2 bacteriophage linked to RNP have indeed been shown to be immunogenic (GOTTLIEB, 1969), and may play a role in modulating the immune response.

As a final thought, it is perhaps worth re-emphasizing the point that the binding of antigens to RNA may result in gross electrostatic aggregations of antigen with varied species of RNA (ROELANTS and GOODMAN type). These complexes can be dissociated by heating in versene solution or by extraction with hot phenol. The binding of antigens to RNP cannot be dissociated by these methods and represent a specific set of interactions of antigens with the unique ribonucleoprotein moiety of macrophages.

References

Gottlieb, A. A. (1969a). Studies on the binding of soluble antigens to a unique ribonucleoprotein fraction of macrophage cells. *Biochem.,* 8:2111–2116.

—— (1969b). Macrophage ribonucleoprotein: Nature of the antigenic fragment. *Science, 165*:592–594.

—— (1971). The capture of antigens by macrophage ribonucleoprotein. In *Biological Effects of Polynucleotides,* R. Beers and W. Braun (eds.), New York: Springer-Verlag, pp. 293–301.

Gottlieb, A. A., V. Glisin, and P. Doty (1967). Studies on macrophage RNA involved in antibody production. *Proc. Nat. Acad. Sci. U.S.A., 57*:1849–1856.

Gottlieb, A. A., and D. S. Strauss (1969). Physical studies on the light density ribonucleoprotein complex of macrophage cells. *J. Biol. Chem., 244*:3324–3329.

Greenwood, F. C., W. M. Hunter, and J. S. Glover (1963). The preparation of ^{131}I-labeled human growth hormone of high specific activity. *Biochem. J., 84*:114.

Roelants, G. E., and J. W. Goodman (1969). The chemical nature of macrophage in RNA-antigen complexes and their relevance to immune induction. *J. Exp. Med., 130*:557.

THE TARGET (OR MODE OF ACTION) OF ANTILYMPHOCYTIC ANTIBODIES*

KEITH JAMES

*Department of Surgery, University of Edinburgh
Medical School, Edinburgh, EH8 9AG, Scotland*

I. Introduction

The advantages of understanding the target or mode of action of antilymphocytic antibodies are widely appreciated. In the first instance this knowledge should permit a more effective clinical application of these products and in addition it should greatly improve our understanding of immunological processes.

Prior to absorption, almost all antilymphocytic sera cross-react with cells other than the lymphoid cells. This cross-reactivity is attributable, in part, to the common antigen determinants found on lymphoid and other hemopoietic cells and also to the contamination of the inoculum used with cells other than lymphocytes. While the cross-reactivity creates problems in connection with the production of antilymphocytic antibodies for clinical use and may be of interest in relationship to other papers presented at this symposium, I propose to restrict the present communication to the effect of antilymphocytic antibodies on cells or factors directly involved in immunological processes.

II. The Cellular Basis of the Immune Response

Before considering the target of action of antilymphocytic antibodies, it is necessary to consider, albeit briefly, certain basic aspects of the immune responses, some of which are illustrated in Figure 1.

The lymphoid cells involved in immunological processes comprise two

* The author wishes to acknowledge both the generous financial assistance he has received from the Wellcome Trust throughout his investigations and the advice and encouragement of his colleagues in the Department of Surgery. He is also greatly indebted to Professor I. Roitt and to the Editor of The Lancet for allowing him to include Figure 1 in this paper.

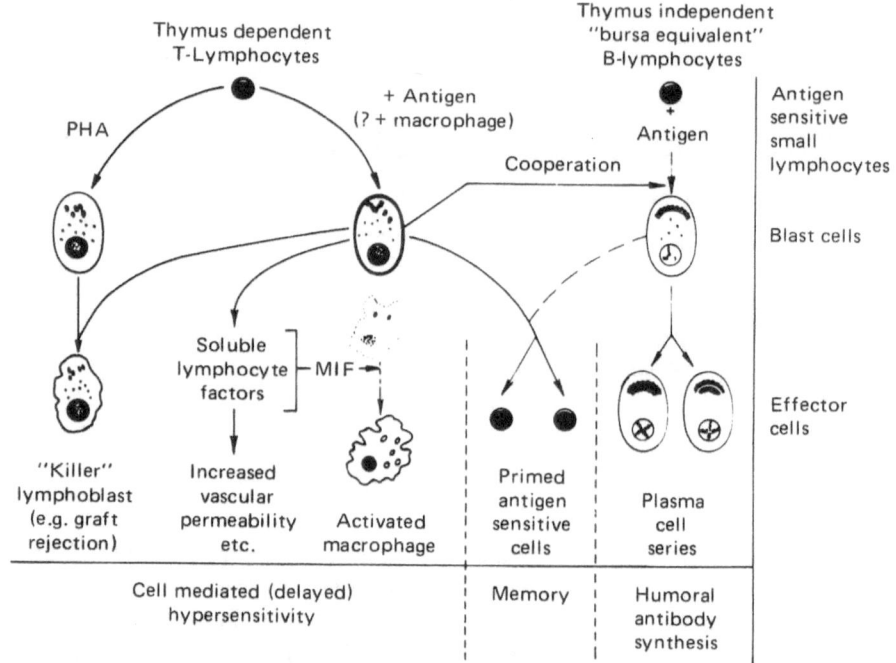

Fig. 1. The role of thymus-dependent and independent lymphocytes in immunological processes. (This figure is a slightly modified version of that published by ROITT *et al.*, 1969.)

distinct populations which originate from stem cells arising in the bone marrow. These are known as the thymus-dependent (T) or thymus-independent or bursa-equivalent (B) lymphocytes.

The T lymphocytes acquire their capacity to elicit immune responses as the result of some complex process which requires the presence of the thymus. These cells reside in the so-called thymus-dependent areas of lymphoid tissues, namely the paracortical area of lymph nodes and the periarteriolar regions of the spleen, and form what is known as the recirculating lymphocyte pool. Furthermore, it is these cells which engage in cell-mediated immune responses. The antigen-sensitive T lymphocytes may be triggered either by coming into contact with antigen at the periphery, *e.g.*, in a graft bed, or by antigen carried to draining lymphoid organs. Following this stimulation the cell differentiate into blast forms and multiplication precedes the appearance of sensitized lymphocytes and activated macrophages which are able to react with and destroy the sensitizing antigen.

The cells of the B population, on the other hand, are not dependent upon the thymus for the development of their immunopotential. They are also distinguishable from the T cell population in that they lack the so-called θ isoantigen and are located in anatomically different parts of the lymphoid organs, namely the medulla of lymph nodes and the red pulp of spleen. The antigen-sensitive B cells may be stimulated (triggered) directly by antigens, as in the case of so-called thymus-independent antigens, or in the presence of high concentrations of antigen. Alternatively, stimulation may be a less direct process involving the cooperation of T lymphocytes and macrophages. Following stimulation, the B cells differentiate and divide to give rise to antibody-producing or plasma cells.

Both cell-mediated and humoral immune processes give rise to lymphoid cells known as memory or primed cells. These are generally derived from the thymus-dependent population but on occasions the memory may reside in the thymus-independent population. They do not differ fundamentally from the originally antigen-sensitive (reactive) cells but there are more of them and they may have been selected on the basis of their avidity and specificity for immunogen. The cells are able to migrate from their site of origin and populate other lymphoid tissue.

Further details on the cellular basis of the immune response may be found in excellent articles by HUMPHREY (1969), ROITT et al. (1969), and TURK (1970), which formed the basis of this summary.

III. The Mode of Action of Antilymphocytic Antibody

A. The Target Cell—The T-dependent Small Lymphocyte

Since the development of renewed interest in antilymphocytic antibodies in the mid-sixties, an appreciable number of theories have been advanced to explain their mode of action, and the evidence for and against the various alternatives has been extensively discussed elsewhere (JAMES, 1969; LANCE, 1970a). At the present time the most widely held view is that these reagents exert their immunosuppressive effect by inactivating the so-called antigen-sensitive, thymus-dependent, long-lived small lymphocytes and the evidence for this hypothesis is briefly as follows:

(1) Detailed histological studies indicated that antilymphocytic antibody caused a marked depletion of the thymus-dependent areas of lymphoid organs and had little, if any, effect on the thymus-independent areas (TURK and WILLOUGHBY, 1967; TAUB and LANCE, 1968a). This specific depletion was confirmed by subsequent studies indicating that antilymphocytic antibody treatment resulted in a fall in the number of spleen and lymph node cells carrying the theta isoantigen characteristic of T-dependent lymphocytes (RAFF, 1969; SCHLESINGER and YRON, 1969), and a reduction in the number of antigen-induced mitoses in the thymus-dependent areas of the spleen (LEUCHARS et al., 1968).

(2) Studies on the rate of reappearance of tritiated thymidine-labeled lymphoid cells into the peripheral blood and lymph nodes of antilympho-

cytic antibody-treated mice revealed that the long-lived thymus-dependent population was depleted for an appreciable period of time, while the short-lived population rapidly recovered (DENMAN *et al.,* 1968).

(3) Studies in thymectomized, and thymectomized-irradiated, bone-marrow-reconstituted CBA/H mice, treated with antilymphocytic antibody revealed that the recovery of the immune response to sheep erythrocytes and skin homografts was thymus-dependent, suggesting that the antilymphocytic antibody temporarily inactivated thymus-dependent cells (LEUCHARS *et al.,* 1968).

(4) Furthermore, spleen cells from antilymphocytic antibody-treated CBA mice have to be supplemented with thymocytes before they can restore to normal levels the immune response to sheep erythrocytes of X-irradiated syngeneic recipients. In contrast, the bone marrow cells are ineffective (MARTIN and MILLER, 1968).

(5) Lymphocytes from antilymphocytic antibody-treated animals are often incapable of responding in a variety of immunological processes attributed to thymus-dependent lymphocytes. For example, they fail to evoke graft-versus-host reactions (see reviews: JAMES, 1969; LANCE, 1970) or respond to phytohemagglutin and other mitogens (TURSI *et al.,* 1969).

(6) Finally there is the widespread belief that antilymphocytic antibodies are more effective at suppressing thymus-dependent, cell-mediated immune responses than humoral immunity (LANCE, 1970a; MEDAWAR, 1969). It has been implied that the antilymphocytic antibody may only suppress humoral responses which are dependent upon the cooperation of thymus-dependent helper cells with thymus-independent lymphocytes (MEDAWAR, 1969).

B. Mechanism of Inactivation

A number of mechanisms have been suggested whereby antilymphocytic globulin may interfere with lymphocyte function. These include cytolysis, sterile inactivation, blindfolding, and antigenic competition (for further details see JAMES, 1969; LANCE, 1970a). The failure of a variety of noncytotoxic antilymphocytic antibody derivatives to suppress both cellular and humoral immune responses suggested to us that the immunosuppressive properties of this reagent were due, at least in part, to the cytolytic destruction of lymphocytes, albeit a subpopulation (JAMES, 1970a). However, subsequent studies revealed that alternative mechanisms existed, for antilymphocytic antibody was found to exert an immunosuppressive effect in mice deficient in the fifth component of complement, namely B 10 D2 old line, DBA/2 and AKR/N mice (BARTH and CARROLL, 1970; WEITZEL and ROTHER, 1970). It now appears that nonlysed antibody-coated lymphocytes may be removed by phagocytosis (GREAVES *et al.,* 1969; MARTIN, 1969; TAUB and LANCE, 1968). This process involves the fixation of the first four components of complement (C1, 4, 2, 3), thus explaining why antilymphocytic antibody is effective in C5-deficient mice and why the cytotoxic titer does not always correlate with the immunosuppressive effect (BACH, 1970a, b). It would thus appear that the failure of noncytotoxic antilymphocytic antibody derivatives to suppress immune responses is due to their inability to fix the complement components involved in phagocytosis (C1 — C4) rather than to their inability to fix all the components required for cell lysis (C1 — 9).

In general the interaction of antilymphocytic antibody and target cells occurs in the peripheral blood, the central lymphoid organs being affected only insofar as they replenish the recirculating lymph pool. However, there is evidence that a more direct effect on central lymphoid tissue and bone marrow may occur, especially when large doses of antilymphocytic antibody are used (see later section).

IV. The Effect of Antilymphocytic Antibody on Immunological Processes

A. Cell-Mediated Responses

The effect of antilymphocytic antibody on cell-mediated responses can perhaps be best illustrated by considering its effect on the rejection of first set allografts. It may interfere with both the afferent and efferent arcs of this process. For example, it may inhibit or suppress the afferent arc by:

(1) inactivating the recirculating thymus-dependent, antigen-sensitive small lymphocytes before they come into contact with graft antigens;

(2) preventing cells which have come into contact with the graft from homing to regional lymph nodes and setting in motion the proliferative events which lead to the production of "killer" lymphoblasts;

(3) inhibiting any sensitization which may occur as a result of lymphoid cells in the donor graft migrating to the recipient lymphoid tissues;

(4) it may also suppress the efferent arc of the immune response by inactivating sensitized lymphoid cells (killer lymphoblasts) in transit between the central lymphoid tissue and the graft.

This ability to interfere with both the afferent and efferent arcs of the immune response may also be operative following the application of second set allografts.

B. Humoral Immune Processes

A number of investigators have noted that antilymphocytic antibodies are, in general, less effective at inhibiting this type of immunological response. For example, they frequently fail to suppress primary humoral responses if administered shortly after antigenic challenge (JAMES, 1967; BERENBAUM, 1967; MÖLLER and ZUKOWSKI, 1968) and have little effect on secondary responses (JAMES and ANDERSON, 1967; JAMES and JUBB, 1967; also see reviews by JAMES, 1969, and LANCE, 1970a). Furthermore, as previously mentioned, its effect on primary humoral responses is less consistent than its effect on primary cell-mediated reactions (LANCE, 1970a; MEDAWAR, 1969). These observations have led to the following conclusions:

(1) Antilymphocytic antibody suppresses primary humoral responses by the peripheral inactivation (LEVEY and MEDAWAR, 1967) of thymus-dependent, antigen-sensitive small lymphocytes (MÖLLER and ZUKOWSKI, 1968) which cooperate with thymus-independent cells in the production of antibody (PLAYFAIR, 1971). Once these cells have been "triggered" and/or homed to draining lymph nodes they appear to be refractory to antilymphocytic antibody treatment, presumably because they are relatively inaccessible (KALDEN and JAMES, 1971).

(2) On the same basis, many of the memory cells involved in secondary immune responses would also be relatively inaccessible, thus explaining why antilymphocytic antibody has little effect on these processes (JAMES, 1970b).

(3) The inconsistent effect of antilymphocytic antibody on humoral responses may be due to its inability to suppress so-called thymus-independent immune responses, *i.e.,* those not requiring the cooperation of thymic helper cells (MEDAWAR, 1969). Support for this contention is forthcoming from the observations that antilymphocytic antibody fails to suppress the primary humoral response of Balb/c AnN mice to the thymus-independent type III pneumococcus polysaccharide antigen (BAKER *et al.,* 1970a, b) and of other strains of mice to large doses of thymus-dependent antigens (ARGYRIS and PLOTKIN, 1970; LANCE, 1970b).

V. Other Targets of Antilymphocytic Antibodies

A. Effect on Thymus-independent Lymphocytes

In the previous part of this paper evidence has been presented to indicate that antilymphocytic antibodies inactivate thymus-dependent lymphocytes and that this results in the suppression of both cell-mediated and humoral immune responses. At one time it was felt that the effect of antilymphocytic antibody on thymus-dependent lymphocytes was fairly selective, but now there is sufficient evidence to indicate that it may also interfere with the thymus-independent (B) lymphocyte population, and this may be summarized as follows:

(1) Spleen cells from antilymphocytic antibody-treated mice have to be supplemented with both bone marrow and thymus cells before they can restore to normal the immunological responsiveness of cyclophosphamide-treated C57B1/6J mice to sheep erythrocytes (JEEJEEBHOY, 1970).

(2) Prolonged antilymphocytic antibody treatment of mice severely depresses the colony-forming unit activity of their bone marrow, reducing lymphoid as well as erythroid and blastoid elements. This may result from the inactivation within the bone marrow of the common stem cell pre-

cursors or from their removal or inactivation while in transit from the bone marrow to the thymus or spleen (DE MEESTER *et al.,* 1968).

(3) In certain situations antilymphocytic antibody may suppress immune responses which are believed to exhibit a considerable degree of thymus independency. These include the response of CBA mice to low doses of type III pneumococcus polysaccharide (JAMES and MILNE, 1971) and of C57B1 mice to *Salmonella adelaide* flagellin (SHELLAM, 1969).

It is perhaps important to stress that as previously suggested the effect of antilymphocytic antibody on the thymus-independent (B) cell population may only become apparent when large doses of potent antisera are used (JEEJEEBHOY, 1970).

B. Effect on Macrophages

A number of *in vivo* and *in vitro* studies have clearly indicated that certain antilymphocytic sera may strongly cross-react with monocytes (macrophages). As suggested in the introduction, this cross-reactivity may be due to antigenic similarities between lymphocytes and macrophages or to macrophage contamination of the various cell inocula used to prepare these antilymphocytic sera (MARSMAN *et al.,* 1970). On occasions the ability of antilymphocytic sera to cross-react with macrophages could conceivably influence their immunosuppressive activity. It should be stressed, however, that antimacrophage sera, as such, are usually not very effective immunosuppressants.

In general, *in vivo* studies have shown that antilymphocytic antibody severely impairs the clearance of macromolecular substances from the bloodstream (BOAK, 1968; SHEAGREN *et al.,* 1969; GROGAN, 1969; BARTH *et al.,* 1969). The studies of Barth and his colleagues (1969) on the uptake of labled *Salmonella typhi* antigen suggest that antilymphocytic antibody may preferentially impair the uptake of antigen by splenic dendritic macrophages. Furthermore it has been suggested that it is the antigen taken up by these cells which is necessary for the initiation and maintenance of antibody production (BARTH *et al.,* 1969). *In vitro* studies with monocyte monolayers have also confirmed that antilymphocytic antibodies may drastically interfere with the function of these cells.

A number of ways in which antilymphocytic antibody may interfere with macrophage function have been proposed and these are as follows: It may:

(1) have a direct cytotoxic effect on macrophages (WOODRUFF *et al.,* 1966; SHEAGREN *et al.,* 1969; HÜBER *et al.,* 1969);

(2) contain or cause the liberation of macrophage migration inhibitory substances which limit macrophage mobility (CASPARY *et al.,* 1970) and perhaps increase macrophage contact (MACLAURIN and HUMM, 1970);

(3) block IgG and complement receptors on macrophages, thus interfering with antigen processing (HÜBER *et al.,* 1969);

(4) neutralize serum opsonins and complement components, thereby impairing phagocytosis (GROGAN, 1969);

(5) produce large amounts of antigen/antibody complexes which "saturate" the macrophages, thus causing a kind of reticuloendothelial blockade (SHEAGREN *et al.,* 1969).

It should perhaps be stressed that the reduced antigen uptake observed following antilymphocytic antibody treatment may also be due, in part, to an alteration in the transport function of capillary endothelial cells (SHEAGREN *et al.,* 1969) and a decreased perfusion of reticuloendothelial system organs (SHEAGREN *et al.,* 1970).

VI. Noncellular Targets of Antilymphocytic Antibody

In general the immunosuppressive properties of antilymphocytic antibodies are directly attributable to their ability to interact with cell-specific antigens on the surface of lymphocytes, thus causing their lytic destruction or opsonization. It is possible, however, that antilymphocytic antibodies may modify immune responses by interacting with (or releasing) soluble cellular and serum factors which may be involved in the immune processes. For example, certain antilymphocytic antibody preparations may contain antibodies to the following:

A. Thymic Humoral Factor

High levels of these antibodies are claimed to be present when thymus tissue is used in the production of the antilymphocytic antibody. The antibodies to thymic humoral factor are believed to potentiate the immunosuppressive effects of the antilymphocytic sera by causing a marked depletion in output of cells by the thymus and hence a more sustained lymphopenia (NAGAYA, 1970).

B. Lymph Node Permeability Factor

Many of the properties of antilymphocytic antibody are similar to those exhibited by antisera produced against the membrane-free extracts of lymph nodes (TURK and WILLOUGHBY, 1969). For example, they both suppress delayed hypersensitivity to tuberculin, cutaneous hypersensitivity to dinitrochlorobenzene, pertussis hypersensitivity, and allergic thyroiditis. It is feasible, therefore, that in certain instances part of the immunosuppressive activity of antilymphocytic antibody is due to the presence of antibodies against similar soluble constituents of lymph node cells.

C. Lymphokines

Both normal and sensitized lymphoid cells may liberate soluble factors which exhibit a variety of effects on the lymphoid system (DUMONDE,

1970). It is feasible that on certain occasions antilymphocytic antibodies contain antibodies to these components. Alternatively they may cause the release of these substances *in vivo,* and in this connection it has been suggested that antilymphocytic antibody may interfere with macrophage migration by causing the release of macrophage inhibitory factors (CASPARY *et al.,* 1970; see also Section V-B of this article). Further, it has recently been shown that mitogenic substances can release skin-reactive factors from normal guinea pig lymphocytes (PICK *et al.,* 1970). As antilymphocytic antibodies possess mitogenic activity (at least in complement-deficient media) they may cause the release of similar substances.

D. Immunoglobulin Receptors on Antigen-sensitive and Other Cells

Prior to absorption many antilymphocytic sera contain antibodies to immunoglobulins (JAMES *et al.,* 1970). If these were not completely absorbed they could conceivably block the immunoglobulin receptors on antigen-sensitive cells and thus suppress immune responses. In this connection it is interesting to note that antisera to immunoglobulin light chains can inhibit the *in vitro* responses of human lymphocytes to tuberculin and HL-A antigens and mouse spleen cells to sheep erythrocytes (GREAVES, 1970). Furthermore, antilymphocytic sera have been shown to block receptors on macrophages (HÜBER *et al.,* 1969).

E. Complement

WILLOUGHBY *et al.* (1968) have shown that the anti-inflammatory effects of antilymphocytic sera can be reproduced with antibody against the third component of complement. This suggests that certain effects of the antilymphocytic sera may be due to a direct action on complement.

F. Serum Proteins

Many antilymphocytic sera contain antibodies to other serum proteins which could conceivably play an indirect role in immune processes (JAMES *et al.,* 1970).

VII. Concluding Comments

While it is generally agreed that antilymphocytic antibodies exert their major effect upon immune responses by interacting with the long-lived, antigen-sensitive, thymus-dependent, small lymphocytes it is fairly evident, as suggested by TURK and WILLOUGHBY (1969), that they may modify these responses by other mechanisms, *i.e.,* exert a multipotential effect via a number of targets (see Table 1).

This complex effect is to a large extent due to the heterogeneity of the antigens used to prepare antilymphocytic sera. However, it is anticipated that antisera currently being produced using subcellular and soluble

Table 1. The Multipotential Mode of Action of Antilymphocytic Antibodies

---

1. Peripheral inactivation of thymus dependent lymphocytes by cytolysis, opsonization, or other means*
2. Inactivation of thymus independent lymphocytes
3. Blindfolding of homograft or tumour (enhancement)
4. Interference with macrophage function

---

5. Inhibition of thymic humoral factor
6. Inhibition of lymph node permeability factor
7. Inhibition of complement or other serum components
8. Inhibition or activation of lymphokines.

---

* Major mechanism.

components of lymphoid cells will exhibit a much greater specificity (interact with well-defined targets) and that these will be of much greater value as therapeutic agents and as tools for investigating immunological processes.

References

Argyris, B. F., and D. M. Plotkin (1970). Effects of anti-macrophage, anti-thymocyte, anti-lymphocyte and anti-spleen serum on the immune response in mice. *Clin. Exp. Immunol., 7*:551–564.

Bach, J. F. (1970a). *In vitro* assay for anti-lymphocyte serum. *Fed. Proc., 29*:120–122.

—— (1970b). Mechanism and significance of rosette inhibition by anti-lymphocyte serum. *Symp. Ser. Immunobiol. Standard., 16*:189–198.

Baker, P. J., R. F. Barth, P. W. Stashak, and D. F. Amsbaugh (1970). Enhancement of the antibody response to type III pneumococcal polysaccharide in mice treated with anti-lymphocyte serum. *J. Immunol., 104*:1313–1316.

Baker, P. J., P. W. Stashak, D. F. Amsbaugh, B. Prescott, and R. F. Barth (1970). Evidence for the existence of two functionally distinct types of cells which regulate the antibody response to type III pneumococcal polysaccharide. *J. Immunol., 105*:1581–1583.

Barth, R. F., and G. F. Carroll (1970). Immunosuppressive effects of anti-lymphocytic serum on complement-deficient mice: Evidence that immune cytolysis is not essential for ALS activity. *J. Immunol., 104*:522–524.

Barth, R. F., R. L. Hunter, J. Southworth, and R. S. Rabson (1969). Studies on heterologous anti-lymphocyte and anti-thymocyte sera. III. Differential effects of rabbit anti-mouse sera on splenic lymphocytes and macrophages. *J. Immunol., 102*:932–940.

Berenbaum, M. C. (1967). Time-dependent immunosuppressive effects of anti-thymocyte serum. *Nature, Lond., 215*:1481–1482.

Boak, J. L. (1968). Inhibition of circulating lymphocytes by anti-lymphocytic serum. *B. J. Surg., 55*:771–774.

Caspary, E. A., D. Hughes, and E. J. Field (1970). On the mode of action of anti-lymphocytic serum: Experiments on electrophoretic mobility of macrophages in experimental allergic encephalomyelitis. *Clin. Exp. Immunol., 7*:395–400.

De Meester, T. R., N. D. Anderson, and C. F. Shaffer (1968). The effect of heterologous anti-lymphocyte serum on mouse hemopoietic stem cells. *J. Exp. Med., 127*:731–748.

Denman, A. M., E. J. Denman, and P. H. Embling (1968). Changes in the life span of circulating small lymphocytes after treatment with anti-lymphocyte globulin. *Lancet, i*:321–325.

Dumonde, D. C. (1970). "Lymphokines." Molecular mediators of cellular immune responses in animals and man. *Proc. R. Soc. Med., 63*:899–902.

Greaves, M. F. (1970). Biological effects of anti-immunoglobulins: Evidence for immunoglobulin receptors on T and B lymphocytes. *Transpl. Rev., 5*:45–75.

Greaves, M. F., A. Tursi, J. H. L. Playfair, G. Torrigiani, R. Zamir, and I. M. Roitt (1969). Immunosuppressive potency and *in vitro* activity of anti-lymphocyte globulin. *Lancet, i*:68–72.

Grogan, J. B. (1969). Alterations in phagocytic function of rats after treatment with anti-lymphocytic serum. *J. Reticuloendoth. Soc., 6*:411–418.

Hüber, H., G. Michlmayr, and H. H. Fudenberg (1969). The effect of anti-lymphocyte globulin on human monocytes *in vitro*. *Clin. Exp. Immunol., 5*:607–619.

Humphrey, J. H. (1969). General introduction. *Antiobiotica Chemother., 15*:1–6.

James, K. (1967). Some factors influencing the ability of antilymphocytic antibody to suppress humoral antibody formation. *Clin. Exp. Immunol., 2*:685–690.

—— (1969). The preparation and properties of antilymphocytic sera. *Progr. Surg.* (Basel), *7*:140–216.

—— (1970a). Studies with anti-lymphocytic antibody and antibody fragments *Progr. Immunobiol. Stand., 4*:256–264.

—— (1970b). Effect of anti-lymphocyte antibody on humoral antibody formation. *Fed. Proc., 29*:160–166.

James, K., and N. F. Anderson (1967). Effect of anti-rat lymphocyte antibody on humoral antibody formation. *Nature, Lond., 213*:1195–1197.

James, K., and V. S. Jubb (1967). Effect of anti-rat lymphocyte antibody on humoral antibody formation. *Nature, Lond., 215*:367–371.

James, K., and I. Milne (1971). The effect of anti-lymphocytic antibody on the primary immune response of mice to sheep erythrocytes, bovine serum albumin and type III pneumococcus polysaccharide. *Transplantation, 12*:109–113.

James, K., D. M. Pullar, V. S. James, A. Wood, H. B. G. Epps, and L. Rahr (1970). The development and distribution of anti-lymphocytic and other antibodies in horses immunized with human lymphoid antigens. *Transplantation, 10*:208–226.

Jeejeebhoy, H. F. (1970). The effect of heterologous anti-lymphocyte serum on lymphocytes of thymus and marrow origin. *J. Exp. Med., 132*:963–975.

Kalden, J., and K. James (1971). The effect of anti-lymphocytic antibody on the development of experimental thyroiditis in rats induced by the intralymph node injection of rat thyroglobulin in Freund's complete adjuvant. *Immunology, 20*:269–275.

Lance, E. M. (1970a). The selective action of anti-lymphocyte serum on re-circulating lymphocytes: A review of the evidence and alternatives. *Clin. Exp. Immunol., 6*:789–802.

—— (1970b). The effect of heterologous anti-lymphocyte serum (ALS) on the humoral antibody responses to *Salmonella typhi* "H" antigen and bovine serum albumin. *J. Immunol., 105*:108–117.

Leuchars, E., V. J. Wallis, and A. J. S. Davies (1968). Mode of action for anti-lymphocyte serum. *Nature, Lond., 219*:1325–1328.

Levey, R. H., and P. B. Medawar (1967). Further experiments on the action of anti-lymphocytic antiserum. *Proc. Nat. Acad. Sci. U.S.A., 58*:470–477.

Maclaurin, B. P., and J. A. Humm (1970). Macrophage specificity of antiserum against thymus lymphocytes. *Clin. Exp. Immunol., 6*:125–136.

Marsman, A. W. J., M. van der Hart, and J. J. van Loghem (1970). Antigenic differences between macrophages and lymphocytes. *Clin. Exp. Immunol., 6*:899–903.

Martin, W. J. (1969). Assay for immunosuppressive capacity of anti-lympho-cytic serum. I. Evidence for opsonization. *J. Immunol., 103*:979–990.

Martin, W. J., and J. F. A. P. Miller (1968). Cell to cell interaction in the immune response. IV. Site of action of anti-lymphocyte globulin. *J. Exp. Med., 128*:855–874.

Medawar, P. B. (1969). Immunosuppressive agents, with special reference to anti-lymphocyte serum. *Proc. R. Soc. Lond. B., 174*:155–172.

Möller, G., and C. Zukowski (1968). Heterologous antilymphocyte serum: Differential effect on antigen-sensitive and antibody-producing cells. *Surgery, 64*:39–47.

Nagaya, H. (1970). Anti-lymphocyte or anti-thymus serum. *Arch. Intern. Med., 125*:499–502.

Pick, E., J. Krejoi, and J. L. Turk (1970). Release of skin-reactive factor from guinea pig lymphocytes by mitogens. *Nature, Lond., 225*:236–238.

Playfair, J. H. L. (1971). Cell cooperation in the immune response. *Clin. Exp. Immunol., 8*:839–856.

Raff, M. C. (1969). Theta isoantigen as a marker for thymus-derived lympho-cytes in mice. *Nature, Lond., 224*:378–379.

Roitt, I. M., M. F. Greaves, G. Torrigiani, J. Brostoff, and J. H. L. Playfair (1969). Cellular basis of immunological responses. *Lancet, ii*:367–371.

Schlesinger, M., and I. Yron (1969). Antigenic changes in lymph node cells after administration of antiserum to thymus cells. *Science, 164*:1412.

Sheagren, J. N., R. F. Barth, J. B. Edelin, and R. A. Malmgren (1969). Reticuloendothelial blockade produced by anti-lymphocyte serum. *Lancet, ii*:297–298.

—— (1970). Mechanisms of reticuloendothelial system blockade produced by anti-lymphocyte serum. *J. Immunol., 105*:634–641.

Shellam, G. R. (1969). Mechanism of induction of immunological tolerance. VI. Tolerance induction following thoracic duct drainage or treatment with anti-lymphocyte serum. *Immunology, 17*:267–280.

Taub, R. N., and E. M. Lance (1968a). The histopathological effects of heterologous anti-lymphocytic serum in mice. *J. Exp Med., 128*:1281–1296.

—— (1968b). Effects of heterologous anti-lymphocyte serum on the distribution of ^{51}Cr-labelled lymph node cells in mice. *Immunology, 15*:633–642.

Turk, J. L. (1969). *Immunology in Clinical Medicine.* London: Heinemann.

Turk, J. L., and D. A. Willoughby (1967). Central and peripheral effects of anti-lymphocytic serum. *Lancet, i*:249–251.

—— (1969). An analysis of the multiplicity of the effects of anti-lymphocyte serum—a comparison with the action of other immunosuppressive agents in the cell-mediated immune response and non-specific inflammation. *Antibiotica Chemother., 15*:267–294.

Tursi, A., M. F. Greaves, G. Torrigiani, J. H. L. Playfair, and I. M. Roitt (1969). Response to phytohaemagglutinin of lymphocytes from mice treated with anti-lymphocyte globulin. *Immunology, 17*:801–811.

Weitzel, H. K., and K. Rother (1970). Studies on the role of serum complement in allograft rejection and in immunosuppression by anti-thymocyte serum (ATS). *Eur. Surg. Res., 2*:310–317.

Willoughby, D. A., L. Polak, and J. L. Turk (1968). Suppression of contact hypersensitivity and acute inflammation by anti-complement serum. *Nature, Lond., 219*:192–193.

Woodruff, M. F. A., N. F. Anderson, and H. M. Abaza (1966). Experiments with anti-lymphocytic serum. In *The Lymphocyte in Immunology and Haemopoiesis,* pp. 286–291. London: E. Arnold Ltd.

PROPERTIES OF RABBIT ANTISERUM
TO MOUSE THYMOCYTE MEMBRANE*

C. Bron, D. Sauser, and J. Galley

Institute of Biochemistry, University of Lausanne,
Lausanne, Switzerland

I. Introduction

The immune response leads to differentiation of lymphoid cells specialized in the synthesis and secretion of humoral antibodies and of cells responsible in cell-mediated immunity. A distinction between those lymphoid cell populations is apparent in certain strains of mice by the presence of an antigenic marker on the surface of thymus-derived cells in peripheral lymphoid organs—the θ-alloantigen (Reif and Allen, 1964; Schlessinger and Yron, 1969; Raff, 1969; Raff and Wortis, 1970; Raff and Owen, 1971).

Using *in vitro* assay systems, it has been possible to demonstrate two types of effector cells which appear following the initiation of transplantation immunity in mice. The first type represents cells releasing alloantibody which is cytotoxic in the presence of complement. The second type consists of cells displaying a direct cytotoxic activity without addition of complement. Precursors of cytotoxic lymphocytes are present in the thymus, whereas antibody-producing cell precursors are not (Cerottini *et al.*, 1970a, b). Moreover, treatment of sensitized lymphocytes by anti-θ serum, in the presence of complement, abolished cytotoxic activity without affecting alloantibody-forming cells (Nordin *et al.*, 1971; Cerottini *et al.*, 1971).

Using heterologous antiserum, only mouse specific lymphocyte antigen (MSLA) has so far been demonstrated among the surface antigens specific for lymphoid cells (Shigeno *et al.*, 1968). No functional properties have been attributed to its presence although it is expressed at a higher density on thymocytes than on cells from other peripheral lymphoid organs (Boyse and Old, 1969). It was of interest to investigate the cellular mem-

* This work supported by SNSF grant 3-361-70. We wish to thank Dr. J. P. Krahenbühl for electron micrography and Drs. D. Nash and J. C. Cerottini for their helpful discussions and aid in revision of the manuscript.

brane of mouse lymphoid cells, especially thymocytes, in order to see if antigenic structures analogous to the θ-alloantigen could be detected by heterologous antiserum. Should this be the case, two types of effector cells could be demonstrated by the abovementioned *in vitro* assays; also, highly specific heterologous antisera could be used for the isolation of such surface antigens.

Preliminary results of such a study are presented in this paper.

II. Properties of Antisera

A. Isolation of Lymphoid Cell Membranes

Single-cell suspension of thymus, lymph nodes, and spleen from 100 Swiss albino female mice (age 3 to 5 weeks) were obtained by gentle disruption, followed by three washes in Eagle's medium. The cell membranes were obtained by hypotonic lysis in 1 mM NaHCO$_3$ followed by sucrose density gradient centrifugations, a method previously described by NEVILLE (1968) (Figure 1).

B. Preparation of Antisera

A sample of normal serum was obtained from each animal before immunization to serve as controls. Rabbits were immunized with cellular

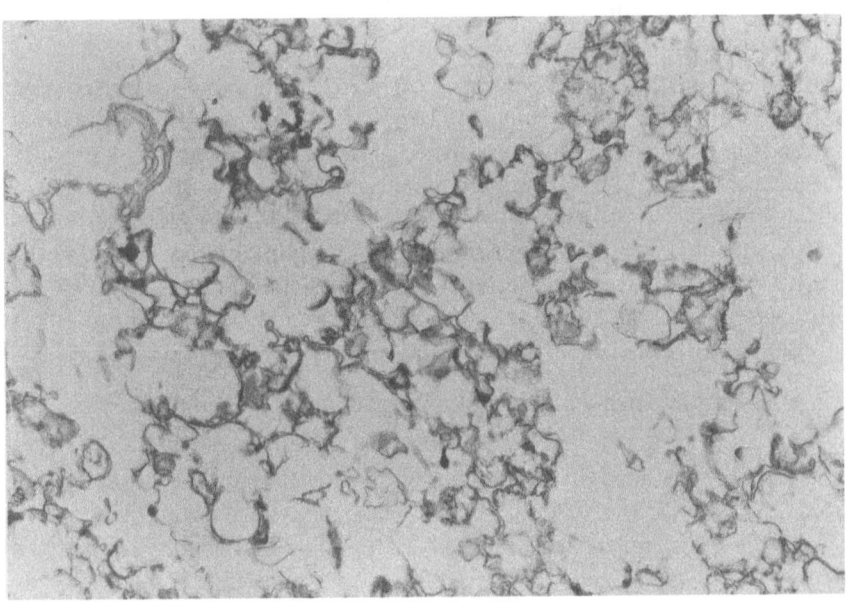

Fig. 1. Electron micrograph of mouse thymocyte membrane enriched fraction prepared by the hypotonic lysis method of Neville Section fixed in glutaraldehyde—OsO$_4$ and stained with uranyl acetate and lead hydroxyde. Magnification: 10,300

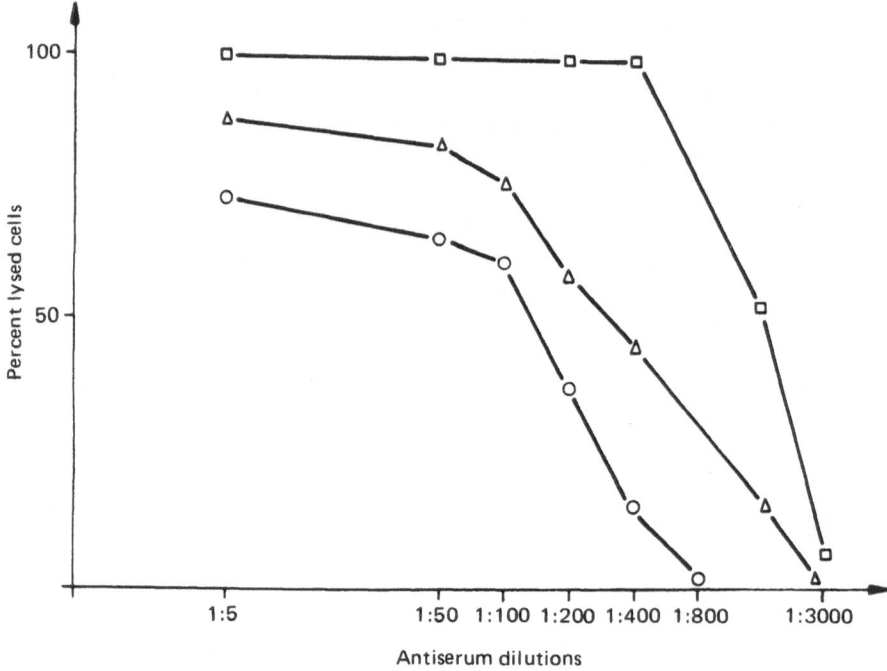

Fig. 2. Cytotoxic activity of rabbit antimouse thymocyte membrane serum against Cr⁵¹ labeled target cells: (○——○) spleen cells, (△——△) lymph node cells, (□——□) thymus cells.

membranes, partly solubilized in 8 *M* urea, and mixed with complete Freund's adjuvant. An average of 2 mg of protein was injected into the foot pads and back muscles every 10 days over a period of 4 to 5 months.

Each antiserum was routinely inactivated and absorbed with mouse and sheep red blood cells as well as with lyophilized mouse serum.

Absorptions were also performed using a glutaraldehyde-polymerized mixture of mouse liver and serum (AVRAMEAS and TERNYNCK, 1969) or with DBA/2 mastocytoma cells (P-815-X2) at 5×10^8 cells per 1 ml of antiserum. The absorbed antiserum will be referred to as AMTS.

C. In Vitro Activity of Rabbit Antimouse Thymocyte Membrane Serum (AMTS)

The cytotoxic activity of antiserum was tested on Cr⁵¹-labeled lymphoid target cells (WIGZELL, 1965). Normal rabbit serum, absorbed at 4°C with 0.25 g/ml of purified agar (BBL, Cockeysville, Maryland), diluted 1:12, was used as a source of complement (COHEN and SCHLESSINGER, 1970). Subcellular fractions of mouse thymocytes and especially membranes are very potent immunogens and thus are capable of inducing highly active heterologous antilymphocyte sera (LANCE *et al.*, 1968; HAYES *et al.*, 1970). Our antisera showed cytotoxic titers of up to 1:3000, whereas antisera raised against whole lymphoid cell homogenates were usually 10 times less active.

Antisera raised against thymocyte membranes always showed a higher cytotoxic activity against thymocytes than against lymph node or spleen cells (Figure 2). This was corroborated using the *in vitro* opsonization assay described by MARTIN (1969). Such antisera diluted 1:500 still opsonized 62% of Cr^{51}-labeled thymocytes. On the other hand, antisera raised against lymph node or spleen cell membranes demonstrated almost no opsonizing capacity under these conditions.

D. Quantitative Absorptions with Lymphoid Cells

The capacity of different lymphoid cell populations to absorb the cytotoxic activity of antisera was determined by incubating several dilutions of AMTS with increasing numbers of lymphoid target cells (COOLEY *et al.*, 1970). Using thymocytes as target cells, the absorptive capacity of lymph node cells was 5 times and that of spleen cells 10 times lower than the capacity of thymocytes to reduce the cytotoxicity by as much as 50% (Figure 3, Table 1).

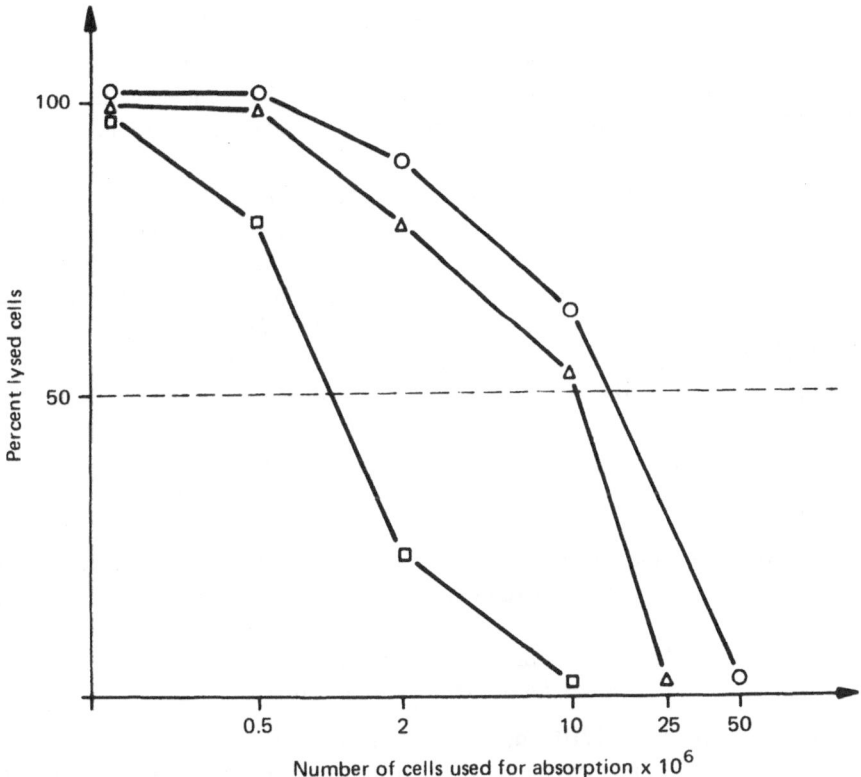

Fig. 3. Quantitative absorption of rabbit antimouse thymocyte membrane serum. Cytotoxic assay using mouse thymocytes as Cr^{51} labeled target cells. Each point represents absorption of 0.5 ml of 1:500 diluted antiserum with the number of mouse cells indicated. (□——□) thymus cells, (△——△) lymph node cells, (○——○) spleen cells.

Table 1. Absorptive Capacity of Mouse Lymphoid Cells for Rabbit Serum Antimouse Thymocytes Membranes

Target cells	Absorptive capacity		
	Thymus cells	Lymph node cells	Spleen cells
Thymus cells	100%*	17, 5%	8, 5%
Lymph node cells	200%	100%*	—
Spleen cells	150%	100%	100%*

* The number of cells required to reduce the cytotoxicity to 50% is expressed in relation to the number of target lymphocytes required for the same reduction.

When lymph node cells or spleen cells were used as the target, the absorptive capacity of thymocytes always remained higher than that of cells from the other lymphoid organs studied (Table 1).

E. Immunofluorescence

Direct and indirect immunofluorescence tests were performed on viable cell populations (CEROTTINI and BRUNNER, 1967). Spleen cells showed weak fluorescence. Lymph node or thymus cells gave intense and uniform staining.

These observations suggest that this antiserum does not show any specificity for one-cell population in peripheral lymphoid organs. Since this antiserum had been rendered specific for lymphoid cells by a variety of absorption procedures (see above), the observed differences in intensity of staining between thymocytes and lymph node cells as opposed to spleen cells can be interpreted as (1) a higher density of the same antigens on the former; or (2) the presence, in lymph nodes, of thymus-derived cells in higher concentration than found in spleen. The latter interpretation seems reasonable in light of the recent results of RAFF and OWEN (1971). It is probable, however, that the heterologous antiserum used here recognizes antigenic determinants, other than θ-alloantigen, common to thymocytes and lymph node lymphocytes that are present in higher concentration in cells of these organs rather than cells found in the spleen.

F. Effect of Rabbit AMTS on Spleen Cells from Mice Immunized with Allografts

Spleen cells of C57BL mice immunized with DBA/2 mastocytoma (P-815-X2) were incubated in the presence of rabbit complement and AMTS diluted 1:100 or normal rabbit serum. The surviving cells were washed and their cytotoxicity against mastocytes, measured in the absence of complement, was tested as a means of detecting cell-mediated immunity (BRUNNER *et al.*, 1968). Simultaneously, the humoral immune response was tested in a plaque assay using mastocytes as target cells (CEROTTINI *et al.*, 1971).

Cytotoxicity was reduced by 70%, whereas 30% of the plaque-forming cells were destroyed. These preliminary results suggest that AMTS might have a preferential action on thymus-derived cells.

To increase the specificity of such antisera, absorptions with lymph node or spleen subcellular fractions will be performed as described, for rat antithymocyte sera, by BACHVAROFF *et al.* (1969) and COOLEY *et al.* (1970).

III. Characterization of Surface Antigens

A. Preparation of Thymocyte Membrane Antigens

An attempt was made to isolate surface antigens, target of antithymocyte serum. The following procedure for cell fractionation was adopted.

Washed lymphoid cells were suspended in an isotonic solution containing 10 mM KCl, 0.25 M sucrose, pH 7.4. Cells were centrifuged for 10 minutes at 1500 $\times$ g and resuspended in a hypotonic solution containing 10 mM KCl, 1 mM MgCl$_2$, 0.25% Triton X-100. After stirring for 3 minutes the isotonicity was restored with 2 M sucrose and the cells were disrupted by 10 strokes of a tight Dounce homogenizer. The homogenate was centrifuged at 10,000 $\times$ g for 20 minutes. The pellet with very low antigenic activity was discarded. The supernatant was centrifuged for 1 hour at 125,000 $\times$ g. The pellet (P) was resuspended in isotonic saline. All steps were carried out at 4°C.

The antigenic activity of soluble or particulate subcellular fractions was analyzed by the microcomplement fixation test recently described by VON FELLENBERG (1971).

Antithymocyte membrane serum diluted 1:1000 was incubated for 20 hours at 4°C in the presence of guinea pig complement with whole cell homogenate, subcellular fraction (S), or fraction (P). Antigen concentrations did not exceed 100 μg protein/ml. As shown in Figure 4, the antigenic activity of cell homogenate, represented by the maximum C fixation for 50 μg of protein antigen, was significantly enhanced in the subfraction P, where the maximum C fixation was reached with 6 μg of protein antigen. On the other hand, the activity of fraction S was significantly reduced since the maximum of C fixation was obtained with 100 μg of protein antigen.

Cytotoxicity of AMTS could be abolished by absorptions with fraction P. These experiments indicate that fraction P contains thymus membrane antigens and will be used in future studies in the characterization and isolation of these antigens.

B. Identification of Thymocyte Surface Components

As many as 35 protein-stainable bands were noted when SDS-solubilized thymocyte fraction P was applied on 15% polyacrylamide gel in

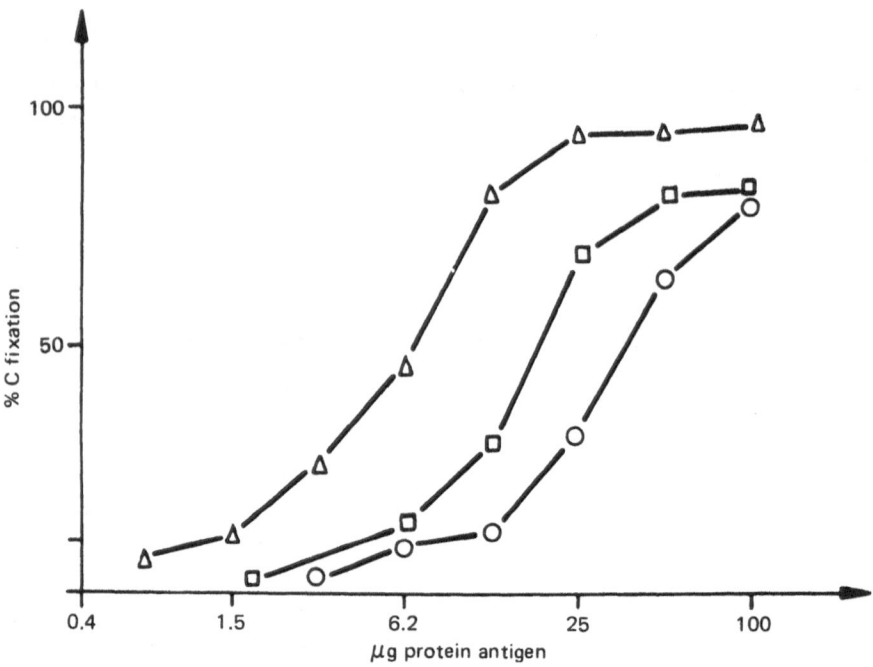

Fig. 4. Complement fixation of whole thymocyte homogenate (□———□), supernatant S (○———○), and pellet P (△———△) AMTS diluted 1:1000.

the presence of 0.1% SDS (LAEMMLI, 1970; FAIRBANKS et al., 1971). To further characterize these various bands, surface proteins of living thymocytes were labeled with I^{125}, using lactoperoxydase as a catalyst of iodination (PHILIPS and MORRISON, 1970). Cell fractionation procedures were then performed as described above. The SDS-solubilized fraction P was electrophoresed on a 15% polyacrylamide gel in SDS. After staining for protein, the gel was cut in 2-mm width discs and counted. Low counts were found at several levels in the gel, but most of the radioactivity was concentrated in one peak, corresponding to a fast-migrating protein band having an M.W. estimated at 10,000.

IV. Conclusion

These results confirm the high immunogenicity of thymocyte membranes as compared to whole cells. Although cytotoxicity, opsonization, and immunofluorescence of AMTS indicate a preferential activity toward thymocytes, no specific marker identifying surface structures related to functional properties of lymphoid cells could so far be detected. Nevertheless, the simultaneous use of two *in vitro* tests, by which both humoral and cellular immune responses can be demonstrated in the same animal with one alloantigen (CEROTTINI *et al.*, 1971), suggests that among the antigens detected by AMTS, one would be either specific or at least in higher density on thymus-derived cells. Further absorptions of AMTS with θ-negative lymphoid cells should increase this specificity. The technique of microcomplement fixation has been shown to be appropriate for tracing antigenic activity in subcellular soluble or particulate fractions (VON FELLENBERG, 1971). This technique, used conjointly with inhibition of cytotoxicity of AMTS and I^{125} labeling of the cell surface, is useful for the identification and characterization of lymphoid cell antigens. The purification of such antigens is currently being attempted as a means of obtaining material for the induction of highly specific AMTS.

References

Avrameas, S., and T. Ternynck (1969). The cross-linking of proteins with glutaraldehyde and its use for the preparation of immunoadsorbants. *Immunochemistry, 6*:53.

Bachvaroff, R., F. Galdiero, and P. Grabar (1969). Antithymus cytotoxicity: identification and isolation of rat thymus-specific membrane antigens and purification of the corresponding antibodies. *J. Immunol., 103*:953.

Boyse, E. A., and L. Old (1969). Some aspects of normal and abnormal cell surface genetics. *Ann. Rev. Genet., 3*:269.

Brunner, K. T., J. Mauel, J. C. Cerottini, and B. Chappuis (1968). Quantitative assay of the lytic action of immune lymphoid cells on Cr^{51}-labelled allogeneic target cells *in vitro:* inhibition by isoantibody and by drugs. *Immunology, 14*:181.

Cerottini, J. C., and K. T. Brunner (1967). Localization of mouse isoantigens on cell surface as revealed by immunofluorescence. *Immunology, 13*:393.

Cerottini, J. C., A. A. Nordin, and K. T. Brunner (1970a). *In vitro* cytotoxic activity of thymus cells sensitized to alloantigens. *Nature, Lond., 227*:72.

—— (1970b). Specific *in vitro* cytotoxicity of thymus-derived lymphocytes sensitized to alloantigens. *Nature, Lond., 228*:1308.

—— (1971). Cellular and humoral response to transplantation antigens. I. Development of allo-antibody-forming cells and cytotoxic lymphocytes in the graft-versus-host reaction. *J. Exp. Med., 134*:553.

Cohen, A., and M. Schlessinger (1970). Absorption of guinea pig serum with agar: a method for elimination of its cytotoxicity for murine thymus cells. *Transplantation, 10*:130.

Cooley, D. G., A. Malakian, and B. H. Waksman (1970). Cellular differentiation in the thymus. II. Thymus-specific antigens in rat thymus and peripheral lymphoid cells. *J. Immunol., 104*:585.

Fairbanks, G., T. L. Steck, and D. F. H. Wallach (1971). Electrophoretic analysis of the major polypeptides of the human erythrocyte membrane. *Biochem., 10*:2606.

Hayes, C. R., L. F. Willard, and R. Wilson (1970). The biological activity of antisera raised against fractional antigens from murine thymocytes and lymphocytes. *Transplantation, 9*:343.

Laemmli, V. K. (1970). Cleavage and structural proteins during the assembly of the head of bacteriophage T4. *Nature, Lond., 227*:680.

Lance, E. M., P. J. Ford, and M. Ruszkiewicz (1968). The use of subcellular fractions to raise antilymphocytic serum. *Immunology, 15*:571.

Martin, W. J. (1969). Assay for the immunosuppressive capacity of anti-lymphocyte serum. *J. Immunol., 103*:979.

Neville, D. M. (1968). Isolation of an organ specific protein antigen from cell-surface membrane of rat liver. *Biochim. Biophys. Acta, 154*:540.

Nordin, A. A., J. C. Cerottini, and K. T. Brunner (1971). The antibody response of mice to allografts as determined by a plaque assay with allogeneic target cells. *Eur. J. Immunol., 1*:55.

Philips, D. R., and M. Morrison (1970). The arrangement of proteins in the human erythrocyte membrane. *Biochem. Biophys. Res. Commun., 40*:284.

Raff, M. C. (1969). Theta isoantigen as a marker of thymus-derived lymphocytes in mice. *Nature, Lond., 224*:378.

Raff, M. C., and J. J. T. Owen (1971). Thymus derived lymphocytes: their distribution and role in the development of peripheral lymphoid tissues of the mouse. *Eur. J. Immunol., 1*:27.

Raff, M. C., and H. H. Wortis (1970). Thymus dependence of tumor-bearing cells in the peripheral lymphoid tissues in mice. *Immunology, 18*:931.

Reif, A. E., and J. M. V. Allen (1964). The AKR thymic antigen and its distribution in leukemia and nervous tissues. *J. Exp. Med., 120*:413.

Schlessinger, M., and I. Yron (1969). Antigenic changes in lymph-node cells after administration of anti-serum to thymus cells. *Science, 164*:1412.

Shigeno, N., U. Hammerling, C. Arpels, E. A. Boyse, and L. J. Old (1968). Preparation of lymphocyte-specific antibody from antilymphocyte serum. *Lancet, 2*:320.

von Fellenberg, R. (1971). Antigens of mouse spleen cells: immunochemical analysis with heterologous rabbit antisera. *Immunology, 20*:963.

Wigzell, H. (1965). Quantitative titrations of mouse H-2 antibodies using Cr^{51}-labeled target cells. *Transplantation, 3*:423.

Part 5

Normal Tissue Antigens

MEMBRANE-ASSOCIATED HISTOCOMPATABILITY ANTIGENS*†

LIONEL A. MANSON

Wistar Institute, Philadelphia, Pennsylvania

I. Introduction

It is a well known fact that tissues exchanged between random members of an outbred population of a single mammalian species will be accepted temporarily and then be vigorously rejected. This rejection has been shown to be due to an immune response by the host to "foreign" antigens that are detected by the host in the graft. Studies with inbred animals have demonstrated that these antigens, the transplantation antigens, are the phenotypic products of a number of complex genetic loci called the histocompatibility genes. These histocompatibility loci have been most carefully delineated in the mouse (LENGEROVÁ, 1969; HILDEMAN, 1970) and recently, have been described in detail for the rat (PALM, 1971). A major effort is being made to describe the histocompatibility systems in the human species, and the results have been summarized in a series of workshops, the latest of which has recently been published (TERASAKI, 1970).

In all the species thus far examined, the transplantation antigens have been found firmly attached to, as an integral part of, the surface membranes of cells. In fact, to be a transplantation antigen, an antigen may have to be expressed on the surface membranes of cells. Nevertheless, detailed investigations of the intracellular distribution of the transplantation antigens, specifically the H-2 transplantation antigens of the mouse, have shown that in some tissues there are significant amounts of this antigen

* I am indebted to my collaborators, C. Hickey, H. Friedman, L. Goldstein, and T. Simmons, who helped in carrying out many of the studies described here. I am especially indebted to my colleague, Dr. Joy Palm, who not only participated in these investigations, but who also helped develop the theoretical concepts upon which this analysis is based. We are grateful for the excellent assistance given us by Mr. G. Vincent Foschi and J. Trachtenberg (deceased) and by Misses R. Crocamo, A. Dalbey, B. van Dyke, and Mrs. Gail Nameroff.

† These studies were supported in part by Public Health Service Research Grants CA-07973, CA-10815 and CA-10097 from the National Cancer Institute, and RR-05540 from the General Research Support Branch.

expressed on internal membranes of the cells as well (PALM and MANSON, 1966; MANSON *et al.*, 1968; HICKEY and MANSON, 1971). The advances that have been made to solubilize the membranes and the chemical properties of fragments containing the antigenic determinants isolated from these membranes have been thoroughly reviewed (REISFELD and KAHAN, 1970) and will therefore not be further mentioned. There has been very little discussion of possible regulatory mechanisms governing the phenotypic expression of the histocompatibility genes in various tissues and among the subcellular organelles within each tissue. In this presentation some of these problems will be developed and our more recent attempts to dissociate cell membranes with detergents will be presented.

II. Subcellular Distribution of Transplantation Antigens

A. Tissue Culture Cell Lines

The data on the distribution of H-2 antigen among the subcellular organelles of two mouse cell lines, L-5178Y lymphoblast (of DBA/2 origin) and L-cell fibroblast (of C3H origin), have been reviewed in detail elsewhere (MANSON and PALM, 1972). By either nitrogen decompression (a vigorous homogenization procedure) or Dounce homogenization (a very gentle procedure), the H-2 antigen could be isolated in a large yield associated with a microsomal lipoprotein fraction (MLP). Careful studies showed that with these tissues, the MLP was not contaminated with surface membrane fragments to any significant extent, and therefore in these two tissues the H-2 antigen was significantly expressed not only on the surface membranes of the cell, but also on the internal membranes of the endoplasmic reticulum. It was also found that the MLP was a more potent source, per milligram of protein, of the H-2 than was surface membrane.

A preliminary study with a human lymphoid line grown in tissue culture, WI-L2, has indicated that the MLP isolated from the cells by nitrogen decompression has HL-A antigens (the major histocompatiblity system in man) that can be detected in a solid-phase radioimmune assay with specific sera (MANSON, 1972).

B. Spleen and Thymus Lymphocytes

In spite of the fact that lymphocytes and thymocytes have very different sensitivities to H-2 cytotoxins, suggesting a qualitative and/or quantitative difference in the surface distribution of the H-2 antigen on the two cell types, the MLP isolated from spleen is qualitatively and quantitatively equal in H-2 content to the MLP isolated from thymus (PALM and MANSON, 1965, 1966). It should be noted, however, that mouse thymocytes yielded only 5.2 mg MLP per gram of tissue, whereas spleen yielded 9.4 mg MLP per gram of tissue. Since the MLP from the two tissues has a similar potency, whereas the surface membranes of the two cell types have quite different antigenic content (BOYSE, 1971), it is not unreasonable to conclude that the H-2 is also expressed on endoplasmic reticular membranes in these cells.

C. Liver and Kidney

In most surveys of tissue distribution of alloantigens (BILLINGHAM and SILVERS, 1963; LENGEROVÁ, 1969) it has been generally noted that liver and kidney are less potent sources of these materials than lymphocytes. This has also been found to be true with lipoprotein fractions isolated from these tissues. At best, in side-by-side comparisons, PALM and MANSON (1966) found that liver MLP or kidney MLP had one-fourth the antibody-combining activity with H-2 antisera than was observed with spleen MLP. Further studies with liver tissue suggested that the surface membrane of liver or kidney, in contradistinction to that of lymphocytes or tissue culture cells, was easily fragmented during nitrogen decompression, being converted to microsome-sized vesicles. A more gentle homogenization of liver tissue showed that the H-2 was found only on surface membrane and perhaps on lysosomes, but not on membranes of the endoplasmic reticulum. This latter conclusion has been confirmed by POPP et al. (1969), EVANS and BRUNING (1970), and WILSON and AMOS (1971). Thus liver and kidney tissue differ drastically from fibroblasts, lymphocytes, and thymocytes in that in the former, the H-2 is not found on membranes of the endoplasmic reticulum, whereas in the latter it is.

The conclusions reached with mouse tissues have recently been extended to rat tissues by Palm (to be published). She has found that rat spleen MLP is 4 to 8 times as potent in absorbing antibodies to products of the Ag-B locus (PALM, 1971) as is rat liver MLP. Cellular localization studies with rat tissues have not yet been carried out. It will be of interest to determine if the Ag-B antigens are expressed on internal membranes in spleen and not in liver or kidney, as in mouse tissue.

III. Immunogenic Capacity of Various Membrane Preparations as Transplantation Antigen

Even in the early attempts to isolate transplantation antigens from tissues, it was noted that liver and kidney tissue appeared to be a poor source as compared to spleen (reviewed by BILLINGHAM and SILVER, 1963). When MANSON et al. (1963) first demonstrated that transplantation antigens could be isolated from spleen in association with the microsomal lipoproteins as potent immunogens, they also reported that liver MLP was inactive as an immunogen, even though H-2 antigen was present in the fraction. This was more carefully documented by PALM and MANSON (1966), who showed that liver MLP had one-fourth the antibody-combining activity, on a protein basis, as spleen MLP, yet the liver MLP would not sensitize for allograft rejection at doses as high as 400 times the minimal sensitizing dose of spleen MLP. Similar observations were made with kidney MLP. A number of laboratories have confirmed the fact that liver or kidney extract have transplantation antigen activity when tested serologically, but little or no immunogenic activity either for the induction of humoral immunity or for an accelerated allograft rejection. In fact, several groups have tried to induce specific immunodepression to allografts with liver extracts, with some moderate success (KELLY *et al.*, 1966; CALNE *et al.*, 1969; BALDWIN and COHEN, 1970). The most recent report that kidney cells are poor immunogens is that of MAIN *et al.* (1971), who found that dissociated rat allogenic skin cells would induce thymidine incorporation in mixed culture with rat peripheral leukocytes, whereas kidney cells were completely inactive as stimulators.

One reason for this lack of immunogenicity of liver and kidney extracts as alloantigens may be the presence in these extracts of a nonspecific blocking factor (PALM and MANSON, 1965; MANSON and PALM, 1968a). This blocking factor was found in the lysosome-rich subcellular particulates of liver and kidney extracts, the fractions which also contained the bulk of the H-2 activity of the homogenates. If it was injected at least one hour before, this factor temporarily suppressed the ability of mice to respond to active H-2 immunogens.

There is another possible explanation, for which there is no direct evidence available, but which takes note of a series of coincidences that have been already hinted at in the previous section. The H-2 is expressed on mouse red blood cells (RBC), yet RBC are notoriously poor immunogens for an H-2 response, either humoral or cellular (see review by BILLINGHAM and SILVERS, 1963). Mouse RBC also have no endoplasmic reticulum. Liver cells have H-2 expressed on their surface membrane, but not on their endoplasmic reticulum, as has been noted previously. When surface membrane preparations of a lymphoblast, L-5178Y, were compared for immunogenic potency to the MLP, which in this cell is derived from the endoplasmic reticulum, they were found to be much less potent immunogens. The differences in the surface expression and the internal expression of the H-2 may be only quantitative. However, there may also be a qualitative difference in the expression of the gene product on the two membranes, a difference that may reside either in the chemical properties of the antigenic determinants themselves or in the other structures in membranes with which the H-2 antigenic determinants are structurally associated. To evaluate both of these possiblities will require the development of a technology for the systematic dissociation of cell membranes into their constituents. Then it will be possible to isolate the units carrying the antigens programmed by the histocompatibility genes from each membrane separately. In this way one may be able to delineate in a rational manner what type of structure carrying an antigenic determinant is an immunogen and what types of associated molecules are desirable for the induction of specific tolerance.

IV. An *In Vitro* Homologue of Allograft Immunity

A necessary tool for the understanding of the biology and molecular biology of the immune response, both inter- and intracellularly, is the development of a variety of systems where the immune response can be studied from beginning to end outside the animal. During the past several years, data have accumulated (DUTTON, 1967; BACH, 1970), suggesting that the interaction termed the "mixed lymphocyte reaction" (MLR) is, in fact, the induction of allograft immunity in the test tube with cells from unprimed animals.

It occurred to us that the sensitizing antigen in both the two-way MLR (where two allogenic lymphocyte populations stimulate each other) and the one-way MLR (where one partner has been treated with either mitomycin C or X-ray to inhibit its replication) might very well be a cell-free transplantation antigen. In a series of investigations we have shown that the MLP can be used as the sensitizing antigen instead of whole or inactivated cells. Specific sensitization was induced at MLP levels of 0.01 to 0.0001 μg protein/ml in lymphocytes cultured for three days in a tissue culture regimen. This could be measured by the enhanced incorporation of tritiated thymidine into the cells (MANSON and SIMMONS, 1969; SIMMONS and MANSON, 1971a) and in three different effector assays for cellular immunity, a plaque reduction test (MANSON *et al.,* 1971), a tumor target test, and a specific radio-labeled, antigen-binding test (MANSON and SIMMONS, 1971). The response that was induced in each case was immunogenetically specific. If higher antigen concentrations were used (1 to 10 μg/ml), unresponsiveness was induced in each test that again was immunogenetically specific. Detailed publications of these data will appear shortly (SIMMONS and MANSON, 1972a, b).

No attempt has yet been made to study the various parameters of the *in vivo* alloimmune response to the *in vitro* homologue. Nor has there been any attempt to use this system to evaluate immunogenicity. Once the variables in this *in vitro* system are understood and other comparable systems devised, a sustained attack on such problems can be undertaken.

V. Use of Detergents to Dissociate Cellular Membranes

Cellular membranes are associations of proteins, lipids, carbohydrates, and nucleic acid components, held together by both covalent and hydrophobic bonds leading to an insoluble product, the membrane. All of the sophisticated tools that are available for the fractionation of natural materials have been designed and depend on the existence of these substances in a dispersed state. Since defined methods for membrane dispersion are still not available, any understanding that we have at present of membrane structure is at least superficial.

The data available with respect to the various attempts to solubilize transplantation antigens from cells have been thoroughly reviewed recently by NATHENSON (1970) and by REISFELD and KAHAN (1970, 1971). The bulk of the information is based on attempts to isolate soluble fragments by autolytic digestion, or deliberate proteolytic digestion of cells or crude membrane fractions of cells. These procedures have yielded relatively small fragments (less than 100,000 daltons in molecular weight) with transplantation antigen activity and in relatively small amounts. These fragments will be useful in determining the chemical structures of the various antigenic determinants and their allelic forms; however, the manner of isolation and the yields obtained are such that little or no information can be found with these approaches as to the manner in which these fragments are integrated into cellular membranes. The newest of such procedures, reviewed by PAPERMASTER (1972), the extraction of whole cells by 3 *M* KCl in the cold, was described by Dr. Mann at this conference as being essentially an autolytic procedure. He presented data which showed that less than 10% of the transplantation antigen activity of washed cell membrane fragments was liberated when incubation was in 3 *M* KCl alone. If the cytosol of the cells was added during incubation, soluble fragments were formed according to the kinetics described by REISFELD and KAHAN (1971). For solubilization it was necessary that the soluble fractions of cells be present during the 16-hour incubation period and that various procedures known to inactivate cellular autolytic enzymes did not also inactivate the soluble fraction. Even though much higher yields of active material can be obtained by this procedure, it is subject to the same criticism that can be applied to any autolytic procedure.

There are very few recorded attempts to extract transplantation antigens from membranes with detergents. The most successful have been those of KANDUTSCH and STIMPFLING (1963) and HILGERT *et al.* (1969). In both procedures Triton nonionic detergents were used. Some active material was obtained, but the fragments were not stable in the absence of detergents.

The approach taken in our laboratory to obtain soluble transplantation antigens depended on the development of methods for dispersing all the protein of the membrane and then subjecting this dispersed mixture to a variety of separation procedures. Since the lymphocyte or lymphoblast MLP was the most potent of the subcellular fractions as a sensitizing antigen for H-2 and non-H-2, it was to be our starting material.

The first attempts (MANSON and PALM, 1968b) at solubilization of the MLP involved the use of butanol to disperse the MLP. An active material was obtained that had transplantation antigen activity both *in vitro* and as an immunogen. The product, however, was still a large aggregate, of approximately 6×10^6 molecular weight. The approach was abandoned when the solubilization procedure could not be made reproducible.

We next attempted to solubilize the MLP with two commonly used anionic detergents, sodium deoxycholate (DOC) and sodium dodecyl sulfate (SDS). The treatment in each case consisted of adding 10 mg of MLP at pH 7 to a 1% DOC or SDS solution, then dialyzing overnight in the cold. The opalescent product was tested for activity in the accelerated allograft rejection test (MANSON et al., 1963). The data obtained indicated that the DOC had not affected the activity of the material, whereas the SDS solution seemed to have yielded an inactive product (Table 1).

Table 1. The Action of Detergents (DOC and SDS) on H-2 Transplantation Activity of L-5178Y MLP

Material	Dosage (μg/animal)	Number of animals	MST (days)
—	—	21	11.1 ± 0.9
MLP	20	9	$8.5 \pm 0.7*$
	100	10	$8.7 \pm 0.6*$
MLP-DOC	25	8	$8.5 \pm 1.0*$
	50	5	$7.9 \pm 0.6*$
	100	10	$7.5 \pm 0.5*$
MLP-SDS	100	10	10.3 ± 0.9
	500	10	10.0 ± 1.1
	1,250	10	10.7 ± 0.7

* Statistically significant depression with a value of $p < 0.05$. Host: C57BL/10 (H-2^b); skin donor: B10.D2 (H-2^d); MLP donor: L-5178Y (H-2^d) grown in tissue culture.

In the early experiments it was thought that it would be possible to dialyze away the detergents, so that the derivatives of the MLP obtained could be assayed in the usual cytotoxin-inhibition serological tests. It soon became evident that such was not to be the case. With detergents such as SDS or sodium decyl sulfate (SDeS) enough detergent was still bound to the dispersed product, even after extensive dialysis, so that target cells were nonspecifically lysed in the absence of complement. Consequently, we have developed two assays in which the direct combination of antigen and antibody can be detected even in the presence of detergent and in which there is no need to rely on a favorable distribution of antigenic sites on some target cells.

A. Solid-phase Radioimmunoassay for Alloantigens and Alloantibodies

Initial reports describing this assay have already appeared (FOSCHI and MANSON, 1970; MANSON, 1970). In this assay, antibody is adsorbed onto disposable plastic tubes, and the tubes are incubated with a preparation of MLP that has been iodinated with $Na^{125}I$. The empty tubes are then counted in a gamma-counter to detect the amounts of antigen that have been specifically bound to the antibody molecules adsorbed to the plastic. The assay has been found to be immunogenically specific and can be used to detect the combination of antigen and antibody for products of the H-2 locus, some non-H-2 loci and also for tumor-specific activities. The assays are very sensitive, *e.g.*, the usual working limiting dilution for H-2 sera is at least 1:50,000 and the specific binding of fractions of nanograms of antigens are detected.

With this method we have been able to evaluate the effects of a number of detergents on the transplantation antigen of the L-5178Y lymphoblast MLP. It was found that this MLP could be completely solubilized by a combination of DOC and SDeS in minutes in the cold. This detergent-treated product, when iodinated, was just as efficient in the assay as the untreated MLP (Table 2), both with a serum that detected the H-2D antigen (C57BL/10) and a serum that detected a tumor-specific antigen (DBA/2).

Table 2. The Ability of Detergent-Treated L-5178Y MLP to Combine with Anti-L-5178Y Antibody in Solid-Phase Radioimmune Assay*

Serum	Dilution	^{125}I-MLP	^{125}I-MLP-Det
C57BL/10	1:10,000	0.65	0.62
	1:100,000	0.19	0.19
DBA/2	1:10,000	0.46	0.45
	1:100,000	0.12	0.08

*The data are averages of three separate experiments. Specific binding greater than 0.05 ng is significant with a value of $p < 0.05$.

Table 3. Comparison of ^{125}I-MLP and ^{125}I-DLP in Radioimmune Assay*

Strain of donor	^{125}I-MLP	^{125}I-DLP
C57BL/10	0.36	0.36
A/J	0.14 ± 0.06	0.11 ± 0.07

*Source of antigen: L-5178Y. Serum used at 1×10^{-4} dilution. Antibody in sera directed against H-2D antigen.

A more stable detergent-treated product was obtained when the lipids of the MLP were extracted, in the presence of DOC and SDeS, by ethanol and ether, according to the method of ALBUTT (1966). This product (DLP), now stable to freezing, was again as efficient when iodinated as the MLP from which it was derived in the radio assay (Table 3). The DLP was also tested to see whether it would induce an accelerated rejection of an allograft and was found to be as active as the MLP from which it was made (Table 4). It was thus apparent that by means of detergent

Table 4. The Activity of L-5178Y MLP and DLP as
Transplantation Antigens

Material	Dosage (μg/animal)	Number of animals	MST (days)
—	—	9	11.9 ± 1.1
MLP	5	10	11.1 ± 0.8
	10	10	10.5 ± 0.5*
	25	10	9.5 ± 0.8*
DLP	5	10	11.3 ± 1.0
	10	10	10.0 ± 1.0*
	25	9	9.4 ± 0.7*
	100	9	9.5 ± 0.7*

* Statistically significant depression with a value of $p < 0.05$.
Host: A_2G strain ($H-2^b$); graft donor: DBA/2 ($H-2^d$); MLP donor: L-5178Y($H-2^d$) grown in tissue culture.

treatment a cellular membrane could be dispersed completely without the destruction of any of its associated transplantation antigen activities, both serologically and as an immunogen *in vivo*.

B. Radioimmunoprecipitation Assay

Recently, we have been able to devise an immune precipitation assay with iodinated detergent-solubilized derivatives of the MLP. In this assay, the dispersed labeled antigen is allowed to react with antibody, and the complex is precipitated with a sheep antimouse immunoglobulin serum. About 2.5 to 3% of the DLP is specifically precipitated with the same C57BL/10 serum as that used in the solid-phase radioimmunoassay (Tables 2 and 3). This assay also appears to be immunogenically specific and will be applicable to a variety of transplantation antigen assays for antibody, as well as for antigen, detection tests. At present this assay seems to be most useful in antigen purification tests, since as inactive material is removed in the fractionation procedures, the amount of the input antigen specifically precipitated by antisera should increase from the original 2.5 to 3% level. The details of this procedure will be published shortly.

VI. Perspectives

From the data reviewed in this presentation and the efforts under way in other laboratories, one can look to the very near future with a good deal of optimism for greater understanding of the nature of the alloimmune response and the biological significance of the transplantation antigens as integrated structures in cellular membranes.

It is now accepted that the mixed lymphocyte reaction, as measured by an enhanced response to thymidine, is an *in vitro* analogue of the alloimmune response. The demonstration that the response can be induced by a cell-free preparation with transplantation antigen activity such as the MLP opens the way to a detailed study of the parameters of the inductive phase of cellular immunity. Since effectors of cellular immunity, the "killer" cells, can be induced as well, this demonstrates that the efferent arm of the cellular immune response is also amenable to detailed investigation. It remains only for the positive demonstration of the formation of humoral antibodies in the *in vitro* system for models such as this to be formally accepted as the test tube equivalent of the *in vivo* response. It should then be possible to systematically dissociate the cellular response from the humoral response and thus delineate that which is common and that which is unique in these two elements of the alloimmune response.

Such studies will permit the immunobiologist to compare the parameters of the alloimmune response with other classical systems, leading to the humoral response or to delayed-type hypersensitivity. There is the distinct possibility that we will truly understand the humoral response to a well known antigen (such as dinitrophenyl bovine serum albumin (DNP-BSA), where the relationship between the carrier and the hapten as an immunogenic complex is very difficult to delineate, since the carrier is an independent immunogen) only by understanding the alloimmune response to MLP, where all except the antigenic determinant of the complex is by definition "self." Perhaps in the latter system it may be easier to define what is required to induce blastogenesis and what role this plays in the immune response. Many more problems will be investigated in such systems—the nature of cell-to-cell interactions during the immune response, the kinetics and dynamics of effector cell formation, among others.

There is another aspect of the phenomenon of cellular and transplantation antigens that is worthy of comment. As has already been pointed out, the transplantation antigens are the phenotypic products of the histocompatibility genes and, as such, are subject to regulation in differentiated tissues. Some aspects of this regulation have been noted, such as the different immunogenic capacities of the mouse H-2 gene product when expressed on surface membranes of cells as compared to that found on intracellular membrane fractions, such as the lymphocyte MLP. Studies along this line should give us greater insight into the regulation of gene expression in mammalian cells during differentiation.

There is a third area in which these antigens will prove useful. Since they are found integrated into cell membranes, they provide another tool for the systematic analysis of membrane structure. The study of cell membranes is at best difficult and complex, and providing scientists with another class of markers for analysis can only lead to a better understanding of this class of subcellular organelles. From studies of the structure of cellular membranes will come hints as to the nature of the biosynthetic mechanisms involved in membrane formation. It is perhaps to this area of molecular biology, membrane biosynthesis, that we should look for a greater understanding of gene regulation and cellular and tissue differentiation.

References

Albutt, E. C. (1966). A study of serum lipoproteins. *J. Med. Lab. Tech.,* *23*:61–82.

Bach, F. H. (1970). Transplantation: Pairing of donor and recipient. *Science,* *168*:1170–1179.

Baldwin, W. M., and N. Cohen (1970). Liver-induced immunosuppression of allograft immunity in urodele amphibians. *Transplantation,* *10*:530–537.

Billingham, R. E., and W. K. Silvers (1963). Sensitivity to homografts of normal tissues and cells. *Ann. Rev. Microbiol., 17*:531–564.

Boyse, E. A. (1971). Selective gene action in the specialization of cell surfaces. *Fed. Proc.* (in press).

Caine, R. Y., R. A. Sells, J. R. Pena, D. R. Davis, P. R. Millard, B. M. Herbertson, R. M. Binns, and D. A. L. Davies (1969). Induction of immunological tolerance by porcine liver allografts. *Nature, Lond., 223*:472–476.

Dutton, R. W. (1967). *In vitro* studies of immunological responses of lymphoid cells. *Adv. Immunol., 6*:254–336.

Evans, W. H., and J. W. Bruning (1970). Studies on the distribution of some histocompatibility antigens in mouse liver plasma membrane and microsomal fractions. *Immunology, 19*:735–741.

Foschi, G. V., and L. A. Manson (1970). Radioimmune assay for histocompatibility antigens. *Nature, Lond., 225*:853.

Hickey, C., and L. A. Manson (1971). *H-2 Immunogenic Activity of Surface Membrane of Mouse Cells.* In press.

Hildeman, W. H. (1970). *Immunogenetics.* New York: Holden-Day, Inc.

Hilgert, I., A. A. Kandutsch, M. Cherry, and O. D. Snell (1969). Fractionation of murine H-2 antigens with the use of detergents. *Transplantation,* *8*:451–461.

Kandutsch, A. A., and J. H. Stimpfling (1963). Partial purification of tissue isoantigens from a mouse sarcoma. *Transplantation, 1*:201–216.

Kelly, W. D., M. F. McKneally, F. Oliveras, C. Martinez, and R. A. Good (1966). Acquired tolerance to skin grafts induced with cell-free antigenic material: further tissue sources, frozen storage, dose-duration requirements. *Transplantation, 4*:489–497.

Lengerová, A. (1969). *Immunogenetics of Tissue Transplantation.* Amsterdam: North Holland Publishing; New York: John Wiley and Sons, Inc.

Main, R. K., K. D. Cochrum, M. J. Jones, and S. L. Kountz (1971). DNA synthesis in mixed cultures of rat leukocytes and allogeneic dissociated skin cells. *Proc. Nat. Acad. Sci. U.S.A., 68*:1165–1168.

Manson, L. A. (1970). A radioimmune assay for transplantation antigens, pp. 469–473. In *Histocompatibility Testing,* 1970, P. Terasaki (ed.), Copenhagen: Munskgaard.

—— (1972). Extraction of transplantation antigens by pressure homogenization. In *Clinical Markers of Biological Individuality,* R. A. Reisfeld and B. Kahan (eds.), New York: Academic Press, Inc. In press.

Manson, L. A., G. V. Foschi, and J. Palm (1963). An association of transplantation antigens with microsomal lipoproteins of normal and malignant mouse tissue. *J. Cell. Comp. Physiol., 61*:109–118.

Manson, L. A., C. A. Hickey, and J. Palm (1968). H-2 alloantigen content of surface membrane of mouse cells, pp. 93–103. In *Biological Properties of the Mammalian Surface Membrane,* L. A. Manson (ed.), Philadelphia: Wistar Press.

Manson, L. A., and J. Palm (1968a). Mouse transplantation antigens. A blocking factor present in mouse liver extracts. *Transplantation, 6*:667–669.

—— (1968b). The solubilization of mouse transplantation antigens, pp. 301–304. In *Advances in Transplantation,* J. Dausset, J. Hamburger and G. Mathé (eds.), Copenhagen: Munksgaard.

—— (1972). Intracellular distribution of transplantation antigens. In *Clinical Markers of Biologic Individuality,* R. A. Reisfeld and B. Kahan (eds.), New York: Academic Press, Inc. (in press).

Manson, L. A., and T. Simmons (1969). Induction of the alloimmune response in mouse lymphocytes by cell-free transplantation antigens *in vitro:* enhancement of DNA synthesis and specific sensitization. *Transpl. Proc., 1*:498–501.

—— (1971). An *in vitro* model for the study of the development of the alloimmune response, pp. 235–240. In *Cellular Interactions in the Immune Response,* S. Cohen, G. Cudkowicz and R. T. McCluskey (eds.), Basel: S. Karger.

Manson, L. A., T. Simmons, L. Mills, and H. Friedman (1971). *In vitro* induction of immunity: Detection by an antibody plaque reduction assay, pp. 193–206. In *Proceedings of the Fourth Annual Leucocyte Culture Conference,* O. Ross MacIntyre (ed.), New York: Appleton/Century/Crofts.

Nathenson, S. G. (1970). Biochemical properties of histocompatibility antigens. *Ann. Rev. Genet., 4*:69–90.

Palm, J. (1971). Immunogenetic analysis of Ag-B histocompatiblity antigens in rats. *Transplantation, 11*:175–183.

Palm, J., and L. A. Manson (1965). Tissue distribution and intracellular sites of some mouse isoantigens, pp. 21–35. In *Isoantigens and Cell Interactions,* J. Palm (ed.), Philadelphia: Wistar Press.

—— (1966). Immunogenetic analysis of microsomal and nonmicrosomal lipoproteins from normal and malignant mouse tissues for histocompatibility-2 (H-2) antigens. *J. Cell Physiol., 68*:207–220.

Papermaster, B. W., V. M. Papermaster, R. A. Reisfeld, M. A. Pellegrino, S. Ferrone, B. D. Kahan, P. I. Terasaki, M. Takasugi, and E. A. Albert (1972). HL-A antigen characterization and isolation from continuous cultured human lymphocyte cell lines. This volume, p. 186.

Popp, R. A., D. M. Popp, M. G. Anderson, and L. H. Elrod (1969). Use of the zonal centrifuge to separate particles containing transplantation antigen. *Biochim. Biophys. Acta, 184*:625–633.

Reisfeld, R. A., and B. D. Kahan (1970). Transplantation antigens. *Adv. Immunol., 12*:117–200.

—— (1971). Extraction and purification of soluble histocompatibility antigens, *Transpl. Rev., 6*:81–112.

Simmons, T., and L. A. Manson (1971). *In vitro* induction of allograft immunity with mouse microsomal lipoprotein transplantation antigen. *Transpl. Proc., 3*:253–256.

—— (1972a). Induction of the alloimmune response *in vitro.* I. Antigen-induced lymphocyte proliferation. In press.

—— (1972b). Induction of the alloimmune response *in vitro.* II. Formation of effector cells. In press.

Terasaki, P. (1970). *Histocompatibility Testing.* Copenhagen: Munksgaard.

Wilson, L., D. B. Amos, and W. Boyle. The subcellular location of antigens. This volume, p. 263.

SURFACE ANTIGENS OF
MAMMALIAN CELLS*

F. Milgrom and K. Kano

*Department of Microbiology, School of Medicine,
State University of New York at Buffalo,
Buffalo, New York*

Studies on tissue and species specificity have been performed by and large on saline-soluble antigens in body fluids and/or tissue extracts. Analysis of cell-membrane antigens, most of which are not saline-extractable, has been largely neglected, except for investigations on erythrocytes. The traditional procedure for study of cell-surface antigens, the agglutination test, requires stable cell suspension, and this could not be achieved with most nucleated cells.

The procedure which we have extensively employed for studies on cell-surface antigens has been the mixed agglutination test with cell cultures, described by FAGRAEUS and EPSMARK (1961). (An excellent review of mixed agglutination procedures has been recently published by COOMBS and FRANKS (1969).) We have used established and primary cell culture monolayers of various species origin. The cultures were incubated with a serum studied for antibodies to cell-surface antigens. Following washing of unbound proteins, the binding of immunoglobulins to the monolayer was detected by indicator erythrocytes. Human or sheep erythrocytes were sensitized by antierythrocyte antibodies from the same species as the serum tested against the cell culture. The sensitized erythrocytes were agglutinated by an excess of the proper antigamma globulin serum. If antibodies in the tested serum were bound to the cell cultures, indicator erythrocytes would adhere to the monolayer; otherwise they would remain free. The results were assessed by inspection of the monolayer with a microscope at low magnification. The details of the procedure are described elsewhere (MILGROM et al., 1964).

The results of our initial studies (MILGROM et al., 1964) clearly showed that the predominant antigens of mammalian cell surface are de-

* Supported by USPHS research grant AI-06754 from the National Institute of Allergy and Infectious Diseases.

Table 1. Mixed Agglutination with Cell Cultures of Bovine
Adrenal Medulla

Antiserum dilution 1:	Antiserum against bovine				
	Adrenal suspension	Brain suspension	Liver suspension	Liver extract	Serum
10	++++	++++	++++	±	++
100	++++	++++	++++	±	+
1000	++++	++++	+++	−	−
10,000	++++	+++	+++	−	−
100,000	++	++	+	−	−
1,000,000	−	±	−	−	−

void of tissue specificity. It may be seen in Table 1 that cell cultures of bovine adrenal medulla gave equally strong reactions with rabbit antisera to bovine adrenal, brain, and liver suspensions. It should be noted that the antibodies participating in these reactions were directed against insoluble antigens. This is evidenced by the fact that antiserum to a bovine tissue extract gave virtually negative results and antiserum to bovine serum gave a very weak reaction, which might have been produced by antibodies formed as a result of the stimulus exerted by small fragments of lymphocytes or platelets not removed from the serum used for immunization. The procedure was characterized by very impressive sensitivity since antiserum titers reached and surpassed 100,000. Lack of tissue specificity in the described reactions was further ascertained by demonstrating that the serologic activity of antiadrenal serum against adrenal cell cultures can be readily removed by absorption not only with adrenal suspension but also with suspensions of other bovine organs. Antisera to erythrocytes would also react with cultures of nucleated cells, but these reactions would be much weaker. Antisera to tissue suspensions could be extensively absorbed with erythrocytes without appreciably losing their activity against cell cultures. This indicated that a relatively small proportion of surface antigens of nucleated cells is shared by erythrocytes.

Similar results were obtained in mixed agglutination studies with cell cultures of other species origin including man, pig, rabbit, guinea pig, hamster, rat, and mouse. Antisera raised in a foreign species by immunization with crude tissue suspensions or suspensions of any nucleated cells detected cell-surface antigens which were not tissue-specific. On the other hand, these reactions exhibited a high degree of species specificity, as exemplified by the results presented in Table 2. Antisera to bovine, hamster, and human antigens gave by far the strongest reactions with bovine, hamster, and human cell cultures, respectively. The interspecies cross-reactions could be readily removed by proper absorptions which rendered the antisera

Table 2. Mixed Agglutination with Bovine, Hamster, and
Human Cell Cultures

		Titer of mixed agglutination with antisera against suspension of:		
		Bovine brain	Hamster kidney	HeLa cells
Cell culture	Bovine adrenal medulla	100,000	100	1000
	Hamster kidney	100	100,000	1000
	HeLa	100	1000	1,000,000

species-specific. No extensive studies on the physicochemical nature of the antigens under study were performed. Evidence was obtained, however, that they can be destroyed by heating at 80°C or by treatment with ethanol.

Another convenient way of detecting reactions between cell-surface antigens and their antibodies has been cytolysis in agar gel (FUJI *et al.,* 1971; JUJI *et al.,* 1971). In this procedure a suspension of nucleated cells was incorporated into agar and spread on a microscope slide. Droplets of the serum under investigation were placed on the agar surface and allowed to diffuse. Thereafter, the slides were incubated with complement. (In most instances fresh, normal rabbit serum was used as a source of complement.) After fixation of slides with ethanol, clear circular zones were observed wherever cytolytic reactions took place.

In studying reactions between various nucleated cells and antisera to tissue suspensions raised in foreign species, we obtained, in cytolysis in agar, results similar to those of mixed agglutination with cell cultures (JUJI *et al.,* 1971). In our studies on antithymocyte sera, we employed both the mixed agglutination and cytolysis tests (ABEYOUNIS *et al.,* 1971). It may be seen in Table 3 that horse antimouse thymocyte serum gave positive results in mixed agglutination tests with murine fibroblast cultures, "L" cells, and in cytolysis tests with murine thymocytes. Absorption of the antiserum with nonlymphoid cells reduced considerably the mixed agglutination titer but it had little effect on thymocytolysis. On the other hand, absorption with thymus cells not only decreased the mixed agglutination titer but also resulted in spectacular disappearance of thymocytolytic activity of the antiserum. These results can be explained by postulating that antithymocyte serum contains antibodies to two types of antigens. One antigen has a high density on lymphoid and nonlymphoid cells. Antibodies

Table 3. Effect of Absorption on the Titer of Horse Antimouse Thymocyte Serum*

	Mixed agglutination with "L" cells	Cytolysis of murine thymus cells
Unabsorbed antiserum	100,000	1000
Antiserum absorbed with:		
"L" cells	<1000	500
Liver cells	<1000	500
Thymus cells	<1000	<10

* The titer of mixed agglutination is always much higher than the titer of cytolysis. Therefore, only the figures in columns but not in rows can be compared.

to this antigen participate in mixed agglutination but not thymocytolysis. The other antigen has a high density on lymphoid cells but on nonlymphoid cells, it has a low density or it does not appear on these cells at all. Antibodies to this antigen are responsible for lysis of thymus cells.

The biological role of antibodies to cell-surface antigens raised in a foreign species is of considerable interest. The immunosuppressive activities of antilymphocyte and antithymocyte sera have been studied extensively by many investigators and recently reviewed by JAMES (1969). It would appear that the antibody which has immunosuppressive activity combines with antigens which have high density on lymphoid cells, especially on thymus-derived cells.

The pathogenic role of antibodies to species-specific cell-surface antigens has been a subject of our recent investigation (KANO and MILGROM, 1971). Rabbit antisera were produced by immunization with rat brain or rat epithelial cell culture. Following injection of these antisera into rats, proteinuria ranging from 20 to 150 mg per day was noticed in many rats. Immunofluorescent staining of kidney sections within 2 hours after injection revealed localization of rabbit gamma globulin along the capillary walls of glomerulus in a linear pattern. Light microscopy performed 8 days after injection showed swelling and proliferation of endothelial and mesangial cells. Significantly, these sera had a high titer of antibodies combining in mixed agglutination tests with rat cell cultures. Absorption of the antisera with nonrenal tissues to the point where the mixed agglutination reactions became negative abolished also their nephrotoxicity.

The cell-surface antigens which have been most extensively studied in the past decade are undoubtedly the transplantation antigens. Our own studies on transplantation antigens and antibodies have primarily involved mixed agglutination tests with cell cultures. These studies were initiated by examination of sera of C57BL mice which rejected skin grafts of C3H origin (ABEYOUNIS et al., 1964). These sera were tested either against primary cell cultures of C3H origin or against established cell cultures of

Table 4. Mixed Agglutination with "L" Cells and C57BL Anti-C3H
Transplantation Sera

	Number of sera positive at a dilution of 1						
	<90	270	810	2430	7290	21870	Total
Sera after first graft	5	8	14	3	1	0	31
Sera after second graft	0	0	3	6	4	2	15
Normal C57BL sera	24	1	0	0	0	0	25

"L" fibroblast cells. As may be seen in Table 4, the skin-grafted mice formed antibodies readily detectable by mixed agglutination. Similar results were obtained in studying sera of skin graft recipients of other species, including rat (MILGROM et al., 1965a), rabbit (McDONALD et al., 1965), dog (MILGROM et al., 1966), and man (MILGROOM et al., 1965b).

Subsequently, we studied sera of 11 human recipients of renal homografts (MILGROM et al., 1966). The observation period varied from 1 to 24 months and 9 to 35 serum samples from each recipient were examined. Since the donor tissues were not available, the tests were performed with a panel of established human cell cultures. This was done in anticipation that at least some cell cultures would share transplantation antigens with the graft donor and accordingly give reactions with the recipient serum. However, we obtained negative results in all these studies. We suspected that these negative results might be due to the removal of transplantation antibodies by the functioning renal graft. This hypothesis was supported by results obtained with sera of patients who had rejected their grafts and from whom the grafts were surgically removed (MILGROM et al., 1966; WILLIAMS et al., 1968b). In most instances, sera of such patients contained transplantation antibodies. The thesis about the removal of transplantaton antibodies by functioning renal grafts was further supported by elution studies. We found that by heat elution procedures, transplantation antibodies can be recovered from excised renal grafts. This study included not only rejected grafts but also the grafts which functioned well to the death of the recipients of nonrenal diseases (KANO and MILGROM, 1969).

The role of humoral transplantation antibodies in the rejection process was investigated in connection with hyperacute rejection of renal homografts. This is a clinical catastrophe in which the graft is rejected within a few minutes or a few hours after vascular anastomoses. Our first study was performed on 4 patients of the Medical College of Virginia, who rejected hyperacutely 6 renal grafts matched for ABO groups (WILLIAMS et al., 1968a). All these patients had antibodies in their pretransplantation sera, which could be shown by means of mixed agglutination tests with cell cultures from the rejected grafts or cell cultures of established human cell lines. Furthermore, transplantation antibodies could be recovered from rejected grafts. Our material on hyperacute rejections of renal grafts encompasses now 12 patients, 10 of whom had antibodies in their pretransplantation sera, which were formed as a result of previous grafts, transfusions or pregnancies. In contrast, of 73 patients who did not reject their grafts hyperacutely, only one had antibodies in his pretransplantation serum (MILGROM *et al.*, 1971b).

The role of humoral antibodies in the rejection of renal grafts could be further assessed in rabbit experiments (KLASSEN and MILGROM, 1969; MILGROM and KLASSEN, 1971). It was shown that most recipients of the first renal grafts reject them in 1 to 3 weeks with the morphologic hallmark of infiltration by mononuclear cells. In contrast, recipients of second and third renal grafts frequently rejected their grafts within 48 hours and sometimes even within a few hours. The morphologic hallmarks of this rejection were accumulation of polymorphonuclear neutrophils in glomerular and peritubular capillaries and, if the graft was left *in situ* long enough, renal cortical necrosis. This form of rejection could be clearly related to humoral antibodies detectable by mixed agglutination with donor cell cultures. In one experiment, 20 rabbits were sensitized by repeated skin grafts, and transplantation antibodies were demonstrated in all of them. Renal grafts were selected in such a way that 12 of them contained antigens of specificity corresponding to the recipients' antibodies, and 8 did not have antigens corresponding to the recipients' antibodies. In the first group, rejection was characterized by its rapid tempo, and histologic examination revealed polymorphonuclear neutrophils in glomerular and peritubular capillaries and/or cortical necrosis. In the second group there was good cortical perfusion at the time of assessment 48 hours after grafting. Histologically, rejection was characterized by mononuclear cell infiltration.

As mentioned previously, we have good reason to believe that all or most all human recipients of renal homografts form antibodies and that these antibodies are absorbed by functioning renal grafts. This situation certainly differs from the situation in hyperacute rejection. In the latter, the graft is exposed to immediate action of very large amounts of antibodies, whereas in the former, the graft is exposed to relatively small amounts of antibodies over a long period of time. This may explain why the graft placed into an unsensitized recipient does not suffer from any dramatic damage. Still, the possibility remains that the chronic exposure to humoral antibodies may also be deleterious to the graft. This should be considered as one of the possible mechanisms involved in the formation of late glomerular lesions which are responsible for deterioration of many human renal grafts occurring several months or even more than one year after transplantation (MILGROM *et al.*, 1971a).

Our studies on human transplantation raised the question of how antigens detected by mixed agglutination compare to the histocompatibility antigens detected by standard lymphocytotoxicity tests. There is general agreement that lymphocytotoxicity tests detect primarily antigens of the HL-A system. We have reason to believe that mixed agglutination is also capable of detecting these antigens. This may be supported by the reactions of HEp-2 cell cultures with HL-A typing sera. It may be seen in Table 5 that positive reactions were obtained with anti-HL-A2 sera but not with

Table 5. Mixed Agglutination with HEp-2 Cell Cultures and HL-A Typing Sera

	Reactions of sera at dilutions of 1:			
	100	300	900	2700
Anti-HL-A2 sera				
Jochum	++++	+++	++	±
Piquard	++++	+++	+	−
Pinquett	+++	+++	+	−
Revillard	+++	++	−	−
Antisera to other HL-A specificities*	−	−	−	−

* The tests were performed with four anti-HL-A1 sera, five anti-HL-A3 sera, four anti-HL-A9 sera, one anti-HL-A11 serum, three anti-HL-A5 sera, four anti-HL-A7 sera, four anti-HL-A8 sera, two anti-HL-A12 sera, and one anti-HL-A13 serum.

antisera of other HL-A specificities. It should be noted that the titer of the reaction with some antisera was around 1000, which is rather impressive in view of the fact that the same antisera had only a titer of 8 in lymphocytotoxicity. We also showed that absorption of an anti-HL-A2 serum with buffy-coat leukocytes of HL-A2 individuals would remove its activity against HEp-2 cells. Similar absorption with buffy coats of individuals lacking the HL-A2 antigen did not affect the reaction of the antiserum with HEp-2 cells. Conversely, HEp-2 cells were capable of removing the lymphocytotoxic properties of anti-HL-A2 sera without affecting similar properties of antisera with other specificities. Concerning the detection of non-HL-A antigens by means of mixed agglutination, we may state that we studied several transplantation sera which were completely negative by lymphocytotoxicity tests against panel lymphocytes but which reacted with cell cultures in mixed agglutination. Significantly, these sera reacted with each of four different cell cultures examined, and the titer of their reactions was higher than 1000. Although this is not proof, it supports the thesis that these sera reacted with specificities other than HL-A. We will return to this discussion later.

The presence of HL-A2 antigen on HEp-2 cells offered an opportunity for studying the relation of species-specific antigens to HL-A antigens. The experiment, the results of which are depicted in Table 6, was performed in the following way. Multiple tubes with HEp-2 cultures were incubated with a human anti-HL-A2 serum at various dilutions. Following washing, the cultures were submitted to a second incubation with a rabbit antiserum to human species-specific antigens. Thereafter indicator erythrocytes which

Table 6. Mixed Agglutination with HEp-2 Cells and Anti-HL-A2 Serum. Replacement by Rabbit Antiman Serum

First incubation	Second incubation							
	Rabbit antiman serum at a dilution of 1:							
	Diluent	50	100	200	400	800	1600	3200
Anti-HL-A2 serum at a dilution of 1:								
50	+++	−	−	±	++	+++	+++	+++
100	+++	−	−	−	−	++	+++	+++
200	+++	−	−	−	−	+	+++	+++
400	+++	−	−	−	−	−	+	+++
800	++	−	−	−	−	−	+	++
1600	−	−	−	−	−	−	−	−

detected binding of human immunoglobulins were added. It may be seen that rabbit antiserum at dilutions of 1:50, 1:100, and 1:200 displaced completely human HL-A antibodies. At dilutions of 1:400 and 1:800 partial displacement was noted. This would indicate that HL-A antigens are located in very close proximity to species-specific antigens and therefore binding of antibodies to the latter antigens interferes sterically with the binding of antibodies to the former antigens. The reverse experimental design was also followed. HEp-2 cell cultures were first incubated with a rabbit antiserum to human tissues and thereafter with antiHL-A2 serum. This did not have any appreciable effect on the subsequent reaction with indicator erythrocytes which detected binding of rabbit antibodies. This would indicate that not all sites containing species-specific antigens are accompanied by HL-A antigens.

In the past few years we have been interested in the study of cell-surface antigens on inter-species hybrid cells obtained by cell fusion mediated by Sendai virus (KANO *et al.*, 1969). The first hybrid which we studied was obtained by fusion of human cell culture KL with murine cell culture cl 1D. Hybrid cells along with parental cells were studied for human and

Table 7. Mixed Aggluination Test with Man-Mouse
Hybrid Cells

| | Parents | | Hybrid 133-5 |
	KL	cl 1D	
Antiman serum at dilution of 1:			
300	++++	−	++++
900	++++	−	++++
2700	+++	−	++
8100	+	−	−
Antimouse serum at dilution of 1:			
300	−	++++	++++
900	−	++++	++++
2700	−	++++	+++
8100	−	+++	++

murine species-specific antigens (Table 7). Rabbit antiserum to HeLa cell suspension was extensively absorbed with murine tissues and served as an "antiman serum." Rabbit antiserum to pooled murine spleen absorbed with human tissues served as an "antimouse serum." It may be seen that both these antisera reacted with hybrid cells whereas they gave only the predicted species-specific reactions with parental cultures. Analogous results were obtained with man-rat and man-hamster hybrids.

The man-mouse hybrid cells were also studied for the presence of human and murine transplantation antigens. In our initial studies (KANO et al., 1969), we used sera of C57BL mice which rejected skin grafts from C3H mice and human sera with undefined multispecific transplantation antibodies in order to detect murine and human transplantation antigens, respectively. We showed that all hybrid cells studied reacted with the murine antiserum. These reactions could be anticipated since the cl 1D cells were originally derived from a C3H mouse. We also observed positive reactions with human antisera and obtained evidence that most clones of the hybrid contained human transplantation antigens.

In continuing this line of investigation, we fused murine cell cultures IT with peripheral blood leukocytes of three human donors whose HL-A types were known (KANO et al., 1971). The hybrid cells were tested for species-specific antigens and were shown to contain both human and murine antigens. Subsequently, mixed agglutination tests were performed with HL-A typing sera. It may be seen in Table 8 that hybrid cells originating from human parent PL gave definite positive reactions with antisera to HL-A specificities 1, 9, and 8, which corresponded very well with the

Table 8. Mixed Agglutination with Hybrid Cells

Serum dilution	IT × PL							IT × KK						
	Typing sera to HL-A							Typing sera to HL-A						
	1	2	3	9	5	7	8	1	2	3	9	5	7	8
1: 100	+++	-	+	++	-	++	+++	+	+	-	++	+	+	++
300	++	-	-	+++	-	-	+++	-	++	-	+++	++	-	-
900	+	-	-	+++	-	-	+++	-	+	-	+++	++	-	-
2700	-	-	-	-	-	-	-	-	-	-	-	-	-	-
LCT*	+++	-	-	+++	-	-	+++	-	++	-	+++	++	-	-

* Standard cytotoxicity tests with lymphocytes of PL and KK donors.

phenotype of the leukocyte donor. In a similar fashion on the second hybrid, antigens 2, 9, and 5 were detected, which were also present on leukocytes of the donor KK. Results of tests with other antisera were either negative or very weakly positive. These tests were repeated, using four sera for each specificity and results were very similar.

Further studies were performed on clones obtained from the hybrid IT × KK (Table 9). It may be seen that all clones but clone 3 retained

Table 9. Mixed Agglutination with Hybrid IT × KK and Its Clones

		Original hybrid	Clones				
			2	4	6	7	3
Antiman serum	100	+++	+++	+++	+++	+++	±
	300	+++	++	+++	+++	+++	−
	900	+++	−	++	+++	+++	−
	2700	++	−	−	+	−	−
Anti-HL-A9 serum	100	+++	+++	+++	+++	+++	−
	300	+++	++	++	+++	+++	−
	900	++	−	+	+	±	−
Anti-HL-A2 serum	100	+++	−	−	−	−	−
	300	+++	−	−	−	−	−
	900	+	−	−	−	−	−
Anti-HL-A5 serum	100	+++	−	−	−	−	−
	300	+	−	−	−	−	−
	900	−	−	−	−	−	−

human species-specific antigens as evidenced by the reactions with antiman serum. Of 3 HL-A antigens present on the original hybrid, HL-A2 and 5 were lost from all clones, whereas HL-A9 was retained by all clones except clone 3. Our interpretation of these findings is the following: Upon propagation of hybrid cells human chromosomes tend to be deleted. This has been substantiated by karyotypic studies. The fact that clones 2, 4, 6, and 7 retained HL-A9 antigen but lost HL-A2 and 5 antigens, may be explained by the loss of one of the two HL-A chromosomes. This would infer that the KK donor has one chromosome carrying the information for HL-A2 and 5 and another chromosome controlling HL-A9, which is consistent with the most likely genotype of this individual, established on the basis of the haplotype frequency in the population. Clone 3 must have lost all chromosomes controlling the appearance of human species-specific antigens. It may be said that clone 3 lost all the "manness" returning to pure "mouseness"; it obviously retained murine species-specific antigens but, for the sake of brevity, this has not been shown in Table 9. The fact that the loss of human species-specific antigens was accompanied by the loss of HL-A antigens is consistent with the previously discussed experiments, which demonstrated that the HL-A specificity is located in close proximity to the species specificity. This could be also construed into an assumption that the HL-A chromosome controls also the appearance of some species-specific antigen on the cell surface.

The third hybrid (IT × BK) gave virtually negative results with all HL-A typing sera (Table 10), pointing to the loss of HL-A antigens by this hybrid. Interestingly, this hybrid gave definite positive results with some human transplantation sera containing multispecific antibodies of undefined specificity. This would indicate that besides HL-A antigens there are other human histocompatibility antigens detectable by mixed agglutination and that the loss of HL-A antigens need not be accompanied by the loss of other histocompatibility antigens.

Since procedures for isolation and purification of cell-surface antigens are not readily available, we were interested in developing serological, analytical procedures for studies of these antigens. These studies were, by and large, limited to erythrocyte antigens. Erythrocyte stromata were prepared by hemolyzing red blood cells in distilled water. The stromata were further fragmented by supersonic vibration. By using very soft agarose gel at a concentration of 0.5% we succeeded in obtaining double diffusion reactions between stroma preparations and corresponding antibodies (MILGROM and LOZA, 1969). In several instances, the reaction was composed of two lines. This may be illustrated by the reactions of Rh antigens with their antibodies. It may be seen in Figure 1 that anti-Rh serum containing anti-D and anti-G antibodies gave two lines of reaction with CD stromata. Both these lines merged with a single line formed by cD stromata. Only one line merged with the reaction line formed by Cd stromata; the second line continued "unimpressed" by the neighboring reaction with Cd. The interpretation of this finding is as follows: The CD erythrocytes contain antigen G at high density and antigen D at low density. Following fragmentation of stromata, some particles contain only G antigen, whereas other particles contain both G and D. The G particles have apparently higher concentration and therefore are the first to form a reaction line with anti-G. This line is freely crossed by anti-D antibodies which react with particles containing both antigens and this results in formation of the second reaction line which is closer to the well containing CD stromata. cD erythrocytes apparently contain G and D specificities equally distributed. Therefore, all reactive particles combine with both antibodies which results in one reaction line merging with the two lines formed by CD stromata. The only reacting particles in Cd stromata are those containing the G antigen. Therefore, it is not surprising that the line formed in reaction with these stromata merged with the G line formed by CD stromata and did not affect the D line formed by these stromata.

Table 10. Mixed Agglutination with Hybrid IT × BK

Serum dilution 1 to:	Typing sera to HL-A							Multispecific sera		
	1	2	3	9	5	7	8	NC	CR	SW
100	±	–	–	–	–	–	–	+	++	+++
300	–	–	–	–	–	–	–	+++	++	+++
900	–	–	–	–	–	–	–	+++	–	+++
2700	–	–	–	–	–	–	–	–	–	++
LCT with lymphocytes BK	+++	+++	–	–	–	–	+++	+++	+++	n.d.

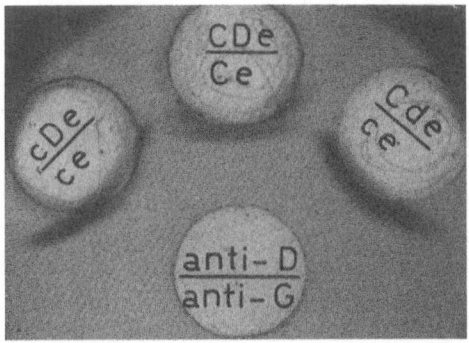

Fig. 1. Double diffusion in gel. Upper wells: stromata of human erythrocytes with the indicated Rh groups. Lower well: anti-Rh serum containing anti-D and anti-G antibodies.

Summary

Mixed agglutination with cell cultures, cytolysis in agar gel, and double diffusion in gel with cell-membrane particles were presented as procedures for the study of surface antigens of mammalian cells. Predominant surface antigens of nucleated cells show no tissue specificity but have pronounced species specificity. However, surface antigens apparently restricted to lymphoid cells were demonstrated. Antibodies to cell-surface antigens raised in a foreign species may have immunosuppressive or nephrotoxic properties. Humoral transplantation antibodies combining with cell-surface antigens are responsible for the hyperacute rejection of the renal graft. Human HL-A transplantation antigens are closely associated with some species-specific antigens. In man-mouse hybrid cells, species-specific and transplantation antigens of both species coexist. Deletion of human HL-A antigens may be partial, involving antigens controlled by one of the pair of HL-A chromosomes. Deletion of species-specific antigens is accompanied by deletion of histocompatibility antigens. Agar diffusion tests performed with cell-membrane particles permit distinguishing antigens with patchy distribution from those with high density.

References

Abeyounis, C. J., F. Milgrom, and E. Witebsky (1964). Homotransplantation antibodies detected by mixed agglutination with cell culture monolayers. *Nature, Lond., 203*:313–314.

Abeyounis, C. J., J. J. Trentin, and F. Milgrom (1971). Serologic studies on antilymphocytic sera. *Fed. Proc., 30*:2754.

Coombs, R. R. A., and D. Franks (1969). Immunological reactions involving two cell types. In *Progress in Allergy*, P. Kallos and B. H. Waksman (eds.), *13*:174–272, Basel: S. Karger.

Fagraeus, A., and A. Epsmark (1961). Use of a "mixed haemadsorption" method in virus-infected tissue culture. *Nature, Lond., 190*:370–371.

Fuji, H., M. Zaleski, and F. Milgrom (1971). Immune response to alloantigens of thymus studied in mice with plaque assay. *J. Immunol., 106*:56–64.

James, K. (1969). The preparation and properties of anti-lymphocytic sera. In *Progress in Surgery*, M. Allogöwer, S.-E. Bergentz and R. Y. Calne (eds.), *7*:140–216, New York: S. Karger.

Juji, T., K. Kano, and F. Milgrom (1971). Antibody-induced lysis of nucleated cells in agar gel. *Int. Arch. Allergy Appl. Immunol., 41*:739–753.

Kano, K., W. Baranska, B. B. Knowles, H. Koprowski, and F. Milgrom (1969). Surface antigens on interspecies hybrid cells. *J. Immunol., 103*:1050–1060.

Kano, K., B. Knowles, H. Koprowski, and F. Milgrom (1971). HL-A antigens on man-mouse hybrid cells. *Fed. Proc., 30*:2769.

Kano, K., and F. Milgrom (1969). Relation of anti-γ-globulin antibodies to transplantation antibodies in human renal allograft recipients. *Transplantation, 7*:281–289.

—— (1971). Nephrotoxicity of antibodies to species-specific antigens. *Int. Arch. Allergy Appl. Immunol., 40*:470–480.

Klassen, J., and F. Milgrom (1969). The role of humoral antibodies in the rejection of renal homografts by rabbits. *Transplantation, 8*:566–575.

McDonald, J. C., F. Milgrom, C. J. Abeyounis, and E. Witebsky (1965). Mixed agglutination with cell cultures in rabbit homotransplantation. *Proc. Soc. Exp. Biol. Med., 118*:397–399.

Milgrom, F., C. J. Abeyounis, J. C. McDonald, and E. Witebsky (1965a). Studies on antibodies accompanying homograft rejection. *Bibl. Haemat., 23*:155–162.

Milgrom, F., K. Kano, A. L. Barron, and E. Witebsky (1964). Mixed agglutination in tissue cultures. *J. Immunol., 92*:8–16.

Milgrom, F., K. Kano, and E. Witebsky (1965b). The mixed agglutination test in studies on human transplantation. *J.A.M.A., 192*:845–848.

Milgrom, F., and J. Klassen (1971). Immunopathologic studies on allograft rejection. In *Cellular Interactions in the Immune Response*, S. Cohen, G. Cudkowicz and R. T. McCluskey (eds.), pp. 290–296, Basel: S. Karger.

Milgrom, F., K. Klassen, and H. Fuji (1971a). Immunologic injury of renal homografts. *J. Exp. Med., 134*:193–207.

Milgrom, F., J. Klassen, K. Kano, and H. Fuji (1971b). Renal graft rejection by humoral antibodies. *Immunopathology*. P. A. Miescher (ed.) pp. 236–249, Basel, Schwabe & Co.

Milgrom, F., B. I. Litvak, K. Kano, and E. Witebsky (1966). Humoral antibodies in renal homograft. *J.A.M.A., 198*:226–230.

Milgrom, F., and U. Loza (1969). Immunodiffusion tests with Rh antigens and antibodies. *Vox. Sang., 16*:470–477.

Williams, G. M., D. M. Hume, R. P. Hudson, Jr., P. J. Morris, K. Kano, and F. Milgrom (1968a). "Hyperacute" renal-homograft rejection in man. *New Eng. J. Med., 279*:611–618.

Williams, G. M., D. M. Hume, K. Kano, and F. Milgrom (1968b). Transplantation antibodies in human recipients of renal homografts. *J.A.M.A., 204*:119–122.

THE SUBCELLULAR LOCATION
OF ANTIGENS

L. A. Wilson, D. B. Amos, and W. Boyle

*Duke University Medical Center, Durham,
North Carolina, 27710; University of Melbourne,
Parkville, Victoria, Australia*

I. Introduction

It is evident from various serological assays such as agglutination and cytotoxicity that histocompatibility antigens are expressed on the cell surface. Studies with fluorescein and ferritin-labeled antibodies have variously shown the external location of these antigens to be in patches (Aoki *et al.,* 1969; Cerottini and Brunner, 1967) or evenly distributed (Möller, 1961; Davis and Silverman, 1968) on plasma membrane, depending on the cells studied and the methods used (Davis *et al.,* 1971). This evidence indicates that antigens are located on the surface, but does not exclude the possibility of an additional internal location. Haughton (1966) tried to rule out internal location by exclusion. Essentially, his method involved quantitation of the antigenic activity of whole cells, equated to the total activity of lysed cells. An increase in total activity of lysed cells would indicate a contribution of antigen from internal sources. In fact, there was little increase in antigen and the conclusion was drawn that at least 80% of the histocompatibility antigens are located on the cell surface. Schwartz and Nathenson (1970) have reported the release of 60% of the total available antigen from the surface of intact, viable cells by papain treatment. These percentages indicate that most of the antigen is exposed on the plasma membrane, but the experiments still do *not* eliminate subcellular location for *some* of the antigen.

According to the previously mentioned studies, intact cells are not permeable to labeled antibodies, and there would appear to be a percentage of antigen unaccounted for. Therefore, one must turn to subcellular fractionation studies to pinpoint the internal location, if any, of additional sources of histocompatibility antigens. Successful fractionation studies would provide several advantages. The availability of membrane preparations of the cytoplasmic organelles, as well as of plasma membrane, would permit a comparative examination of their antigenic determinants, and provide a method of obtaining information on antigen synthesis and turnover.

The methods and problems of cell fractionation have been previously reviewed by WALLACH (1967). However, all of the procedures have the following features in common: homogenization; separation of subcellular organelles, identification of the subcellular fractions and analysis of their purity; and quantitation of antigen.

Aside from the early reports of a nuclear location of H-2 antigens, which was later disproven, HERTZENBERG and HERTZENBERG (1961) described histocompatibility antigens to be limited to plasma membrane. Others have also reported antigen on endoplasmic reticulum (DUMONDE et al., 1963), on lysosomes (BASCH and STETSON, 1963) and on various combinations of the three organelles. The only consensus among investigators has been the lack of antigenic activity contributed by mitochondria.

The recent publications of several investigators are reviewed here as being representative of the diverse methodology, terminology, and results reported for the subcellular distribution of histocompatibility antigens in the mouse and rat systems. Little work has been done in the human HL-A system which is the subject of part of this report. Attention is also directed to other reviews containing information on this subject (PALM and MANSON, 1965; BOYLE, 1967; NATHENSON, 1970). H-2 antigenic activity has been shown to be associated with a microsomal lipoprotein (MLP) fraction derived from a nitrogen decompression homogenization of mouse lymphoblasts by MANSON et al. (1968). The fraction was isolated by a sucrose flotation method. The MLP was characterized by a comparison of sialic acid content of the fraction with whole cells and fluorescein mercuric acetate-prepared plasma membranes. Antigenic activity was assayed by graft survival time. They concluded that the microsomal fraction was internally derived and, therefore, that H-2 antigenic determinants were not limited to the surface membrane. It was also proposed that different tissues vary in the amount of intracellular H-2 antigen. Lymphoblasts, spleen, thymus, and fibroblasts were reported by these investigators to have considerable intracellular antigen. In contrast, they felt that liver H-2 antigens were limited to the plasma membrane.

OZER and WALLACH (1967) have reported that the H-2 antigenic determinants 19, 3, and 8 from a mouse lymphoma are associated exclusively with the plasma membrane. H-2,4 was found to be in both plasma membrane (PM) and endoplasmic reticulum (ER) fractions. Their method of homogenization was also by nitrogen decompression. They contended that this method contributed to the formation of small vesicles of plasma membrane, which are not easily separated from endoplasmic reticulum by frequently used methods. However, they accomplished the separation of vesicular plasma membrane and endoplasmic reticulum by centrifugation on a Ficoll barrier of specific gravity 1.09. WALLACH (1967) had shown earlier that PM and ER localized isopycnically at 1.06 and 1.12, respectively. The ratios of one enzyme membrane marker to another were calculated in order to quantitate contamination of their fractions. Assayed by this method, an enrichment of twofold or more was reported for both PM and ER fractions.

The association of H-2 antigenic activity with mouse liver plasma membranes, endoplasmic reticulum, and probably with lysosomes has been described by HERBERMAN and STETSON (1965). Subcellular fractions were prepared by various differential centrifugation methods. They also employed enzymatic activity as markers for identification of each fraction; however, no attempt was made to check for plasma membrane contamination in each antigen-containing fraction.

EVANS and BRUNING (1970) found H-2 antigens on the PM of mouse liver, although some antigenic activity was evident in their microsomal ER fraction. Enzyme markers and cytoxicity inhibition were used to assay the membrane fractions. They prepared two subfractions derived from large plasma membrane fragments, a "light" vesicular form, and remaining "heavy" fragments of PM. The specific activity of both antigen and enzyme marker increased in the light PM fraction, suggesting to the authors that H-2 antigens may be present at higher concentrations in certain areas of the liver plasma membrane.

Inhibition of the mixed agglutination test for histocompatibility activity in rat-spleen and mouse-liver subcellular fractions derived from differential centrifugations was reported by ABEYOUNIS et al. (1968). Purity of the fractions was assayed by electron microscopy (EDEBO et al., 1968). The results indicated to the authors that most, if not all, histocompatibility antigenic activity was located on the plasma membrane. The mouse-liver PM fraction had twice the antigen content of the next highest absorbing fraction, which was ER. Plasma membrane derived from rat liver had 16 times more absorptive capacity than any other subcellular fraction. Absorption of sera by fractions other than PM was attributed to contamination by PM, but was not confirmed.

POPP et al. (1969) separated large amounts of small subcellular particles from homogenized mouse spleens, using a zonal centrifugation technique. These particles had high concentrations of transplantation antigens and isolated at a density corresponding to cell membrane. Localization of the subcellular fractions with high antigenic content coincided with regions of ATPase acid phosphatase, and nonspecific esterase activity. Antigenic activity within the fractions was quantitated by hemagglutination inhibition assays. Immunogenicity of the material was assayed by its ability to induce antibody formation. Mitochondrial and microsomal fractions were found to be devoid of antigenic activity.

II. Experimental Approach

This brief review indicates that different investigators have found histocompatibility antigens in a wide variety of subcellular organelles. The variation in interpretation of subcellular antigen location may be largely attributed to the multiplicity of techniques and tissues employed; however, some general conclusions can be drawn. It has already been established that histocompatibility antigens are located on the cell surface or plasma membrane. Our concern has been to establish whether the quantities of histocompatibility antigens observed in subcellular fractions other than plasma membrane are truly associated with that fraction or are the result of subcellular cross-contamination with plasma membrane. It is apparent that histocompatibility antigens remain associated with the plasma membrane following homogenization. Thus the association of a known surface-located antigen with subcellular organelles, other than plasma membrane, is meaningless unless the *extent* of plasma membrane contamination can be ascertained (BOYLE, 1968). We felt that two criteria must be met to study successfully the subcellular topology of any known cell surface antigen. First, an assay, other than that of antigen, must be established for detecting the fate of the major portion of plasma membrane after homogenization. Second, specific markers are needed for each subcellular organelle for measuring the possible contamination of preparations by plasma membrane fragments. In an attempt to fulfill these conditions we have used the following enzyme markers, which correspond to specific subcellular organelles (WEAVER and BOYLE, 1969): 5′ nucleotidase for plasma membrane; glucose-6-phosphatase for endoplasmic reticulum; acid phosphatase for lysosomal membranes; and succinic dehydrogenase for mitochondria. Each zonal and Ficoll fraction of subcellular membranes (Figure 1) was profiled quantitatively for all enzymes. Protein concentrations were also established in order to compare specific activities serologically and enzymatically for each fraction.

All the standard methods for enzyme determination were time-consuming and required large quantities of material. To circumvent this problem, we developed a semiautomated microtechnique, employing the Technicon Autoanalyzer. This technique enabled us to make the 160 determinations required for each run in one tenth of the time with more than enough material left for serologic assays (WILSON and BOYLE, 1970).

Figure 1 is a flow diagram of the cell fractionation process. A simple rate zonal centrifugation was used for rat and human liver. There were minor differences for human liver because of the tougher connective tissue (WILSON and AMOS, 1971). The nuclear fraction was used because it contained the larger plasma membrane sheets described by others (NEVILLE, 1960; WEAVER and BOYLE, 1968, 1969). This procedure is essentially a combination of the methods previously described (WILSON and BOYLE, 1970). Nitrogen decompression homogenization was required for tissue

CELL FRACTIONATION PROCEDURE

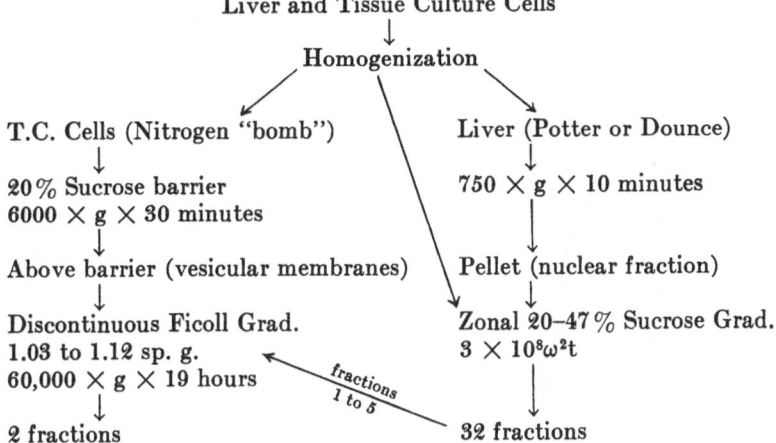

Liver and Tissue Culture Cells
↓
Homogenization

T.C. Cells (Nitrogen "bomb")
↓
20% Sucrose barrier
6000 × g × 30 minutes
↓
Above barrier (vesicular membranes)
↓
Discontinuous Ficoll Grad.
1.03 to 1.12 sp. g.
60,000 × g × 19 hours
↓
2 fractions

Liver (Potter or Dounce)
↓
750 × g × 10 minutes
↓
Pellet (nuclear fraction)
↓
Zonal 20–47% Sucrose Grad.
3 × 10⁸ω²t
↓
32 fractions

fractions 1 to 5

Protein, 5′nucleotidase, G-6-Phosphatase
Acid Phosphatase, Succinic Dehydrogenase
↓
Cytotoxicity Inhibition Assay

Fig. 1. A flow-diagram of the homogenization of liver and tissue culture cells with the procedures for separation of subcellular organelles. All fractions from zonal or isopycnic separations were characterized by enzyme markers, protein, and antigen content.

culture cells followed by a crude isolation of vesicular membranes and then a Ficoll isopycnic separation of PM and ER. The diagonal route indicates that in some cases whole homogenates were placed on the zonal followed by the Ficoll separation for processing small vesicular membrane fractions (WILSON and AMOS, 1971). Zonal fractions 1 to 5 which contained vesicular plasma membrane, endoplasmic reticulum, and lysosomes were pooled for the Ficoll separation.

A. Detection of a Species-Specific Cell Surface Antigen

Rat liver was chosen as a model system because the tissue is (1) easy to homogenize, (2) more readily available than human liver, and (3) larger than mouse liver. For these preliminary studies we attempted to define the location of a species-specific surface antigen in rat liver homogenates (WILSON and BOYLE, 1971). A typical run for rat liver is presented in Figure 2. There are two major peaks of protein with only one entering the gradient. Figure 3 shows the relationship of the phosphorus-splitting enzymes to each other. The endoplasmic reticulum, or G-6-Pase, peaks sharply at fraction 5 with a little activity at 6 and 7. Lysosomes represented

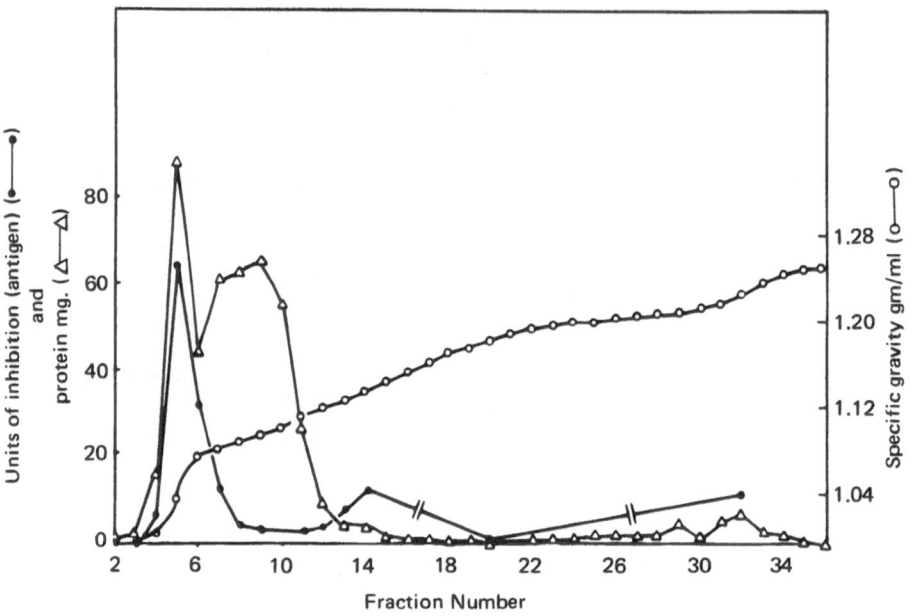

Fig. 2. A profile of units inhibition, protein, and specific gravity of zonal fractions from a rat liver fractionation. (WILSON and BOYLE, 1971).

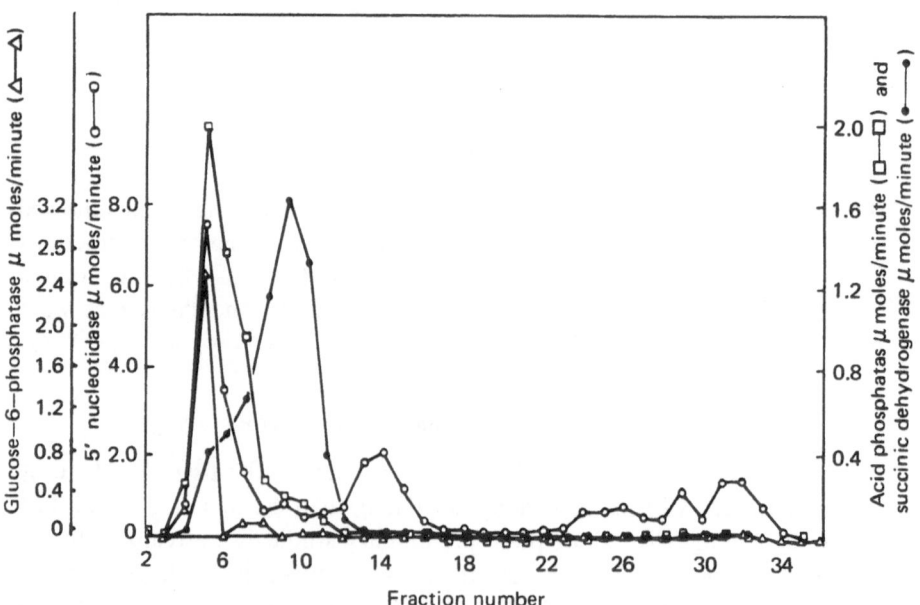

Fig. 3. Enzyme profiles for the rat liver zonal fractions in Figure 1. Subsequent zonal runs at greater total centrifugal force tend to spread existing peaks over a greater portion of the gradient, but did not improve separation of the subcellular organelles. (WILSON and BOYLE, 1971).

by acid phosphatase peak at fraction 5 but trail into the gradient slightly. A single peak of mitochondria is shown in fraction 9. 5′ nucleotidase, or plasma membrane, is distributed in two major peaks at fractions 5 and 14. Fractions 13, 14, and 15, which are essentially uncontaminated by other organelles, contain the large fragments of membrane described first by NEVILLE (1960).

To assay for antigen content, selected fractions were used to absorb the heteroantiserum (Figure 4). The absorbed serum was then tested for its ability to release ^{51}Cr from labeled rat spleen lymphocytes in the presence of guinea pig complement (BOYLE, 1968). Units of antigen were calculated to be the reciprocal of that dilution of fraction which inhibited at 30%.

A control for nonspecific absorption was included, using antilymphocyte serum against a human tissue culture line. Only fractions 5 and 9

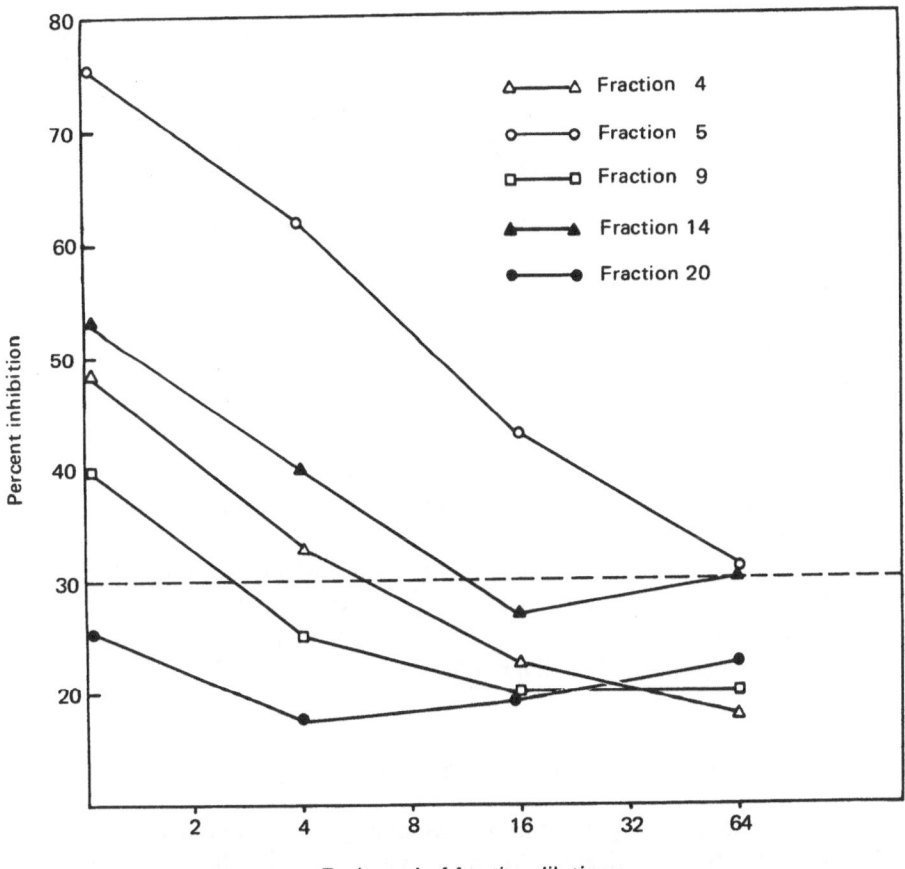

Fig. 4. Titration curves for key rat liver fractions in Figures 1 and 2 against rabbit anti-rat liver plasma membrane serum. (WILSON and BOYLE, 1971).

show significant absorption and then only at very low dilutions of the fraction. The correlation of enzyme activity and protein with antigen content is shown in figures 2 and 3. The peak antigen activity is limited to fractions 5 and 14 and is thus not related to protein content (Figure 2). The only enzyme that peaks in fractions 5 and 14 is 5' nucleotidase, the plasma membrane marker (Figure 3). To relate enzyme activity with antigen content, a ratio of enzyme to units of antigen was calculated (Table 1). Note

Table 1. Ratio of Enzyme Activity and Protein to Units of Inhibition (Rat Liver Zonal Fractions) *

Fraction Number	5'nucleo-tidase	G-6-Phos-phatase	A. Phos-phatase	Suc. De-hydrogenase	Protein
3	−†	−†	−†	NT	−†
4	.12	.07	.07	.01	2.00
5	.12	.04	.03	.01	0.69
6	.11	−†	.04	.02	0.69
7	.13	.01	.08	.05	2.50
8	.14	.04	.07	.29	7.80
9	.18	−†	.07	.54	10.80
11	.20	.01	.03	.14	4.53
12	.17	−†	.01	.02	1.15
13	.23	.001	.002	NT	0.26
14	.17	.001	.002	.001	0.15
20	−†	−†	−†	−†	−†
32	.12	.001	.001	−†	0.29

* WILSON and BOYLE (1971).
† No value as either divisor or dividend, equal to zero.
‡ NT Not tested.

the constant ratio of 5' nucleotidase to units of inhibition among the fractions. Other markers have ratios of 10- to 70-fold variation between fractions. Differences in ratios of the slow-moving PM in fractions 4 to 8 and the larger fragments of PM in fractions 12 to 15 is most likely due to differences in either antigen or enzyme content of the specialized portion (large PM fragments) of the liver cell.

B. Subcellular Distribution of HL-A Antigens

Human liver was processed in essentially the same manner for the detection of HL-A antigens with the enzyme profiles shown in Figure 5 (WILSON and AMOS, 1971). 5' nucleotidase, the PM marker, peaks sharply at fraction 4, but trails into the gradient through fraction 19. The

Human liver fractionation

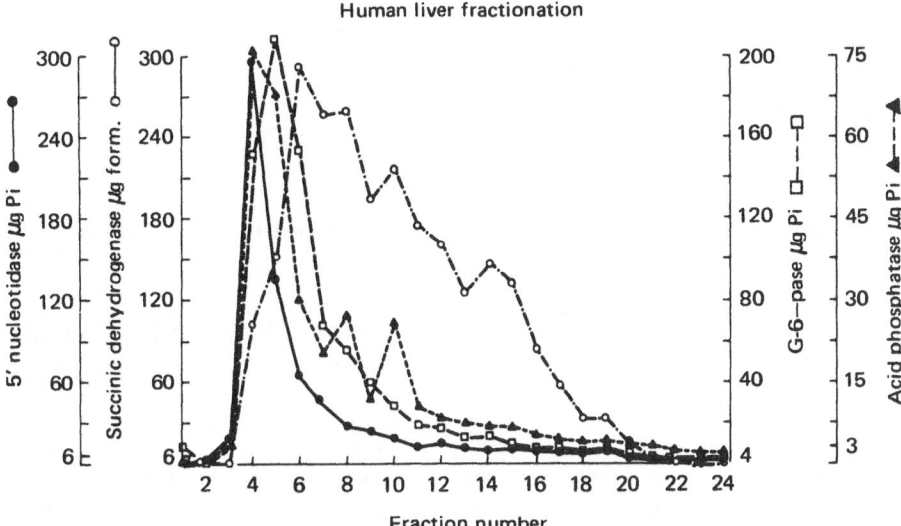

Fig. 5. Enzyme profiles of zonal fractions from human liver. Units of activity are total μg of inorganic phosphorus released by the fraction in 15 minutes, or in the case of succinic dehydrogenase, total μg of formazan produced by the fraction in the same time. (WILSON and AMOS, 1971).

lack of a second peak is probably due to over homogenization and the fracture of the large membrane sheets into vesicles. Acid phosphatase peaks at 4, but stays very high in fraction 5, which indicates some entrance into the gradient. Endoplasmic reticulum peaks in fraction 5, separate from either the plasma membrane in fraction 4 or the mitochondria in fraction 6. This separation was not evident in any other liver preparation, human or rat. The same profile was repeated in separate homogenizations and zonal runs from the same liver. A comparison of the serology with enzyme activity and protein is shown in Table 2. This patient was typed as being HL-A 2, 9, and 7, or FJH. The peaks of greatest activity are shown for each column by a star. Notice that the peak antigenic activity corresponds only with 5′ nucleotidase and acid phosphatase activity in fraction 4. Enzyme activity and protein to unit inhibition ratios are shown in Table 3. Again only the plasma membrane correlates directly with the antigenic content of each fraction by having a constant ratio similar to that of the rat experiments. Variation in other markers, including acid phosphatase, is quite large.

To determine if human tissue culture cells derived from a malignancy and from peripheral lymphocytes could be assayed in a similar manner, IMI* and RPMI 7249† lines were processed by zonal centrifugation (Fig-

* Courtesy of Dr. D. L. Mann.
† Supplied by Dr. B. W. Papermaster.

ure 1; WILSON and AMOS, 1971). The results of the IMI fractionation are shown in Figure 6. With the exception of mitochondria, all the sub-cellular membranes peaked in fraction 5. Also, there was no real second peak of plasma membrane, indicating again the lack of large fragments. The gradual decrease of 5′ nucleotidase with each fraction may indicate

Table 2. Human Liver Enzymes and Protein with Units of Inhibition*

	Units Inhibition			5′nucl.	G-6-Pase	A. Phos.	Suc. Dehy.	Prot.
Fraction	RA	KH	ED	μg Pi	μg Pi	μg Pi	μg Form.	mg.
3	8	4	>64	20	12	4	0	1
4	128†	64†	2048†	296†	152	76†	104	17
5	64	32	1024	136	208†	68	152	9
6	32	16	512	66	154	30	296†	11
7	16	8	>64	42	68	20	256†	8
9	8	4	64	24	41	12	196	6
14	4	2	32	11	14	7	148	3
19	4	1	32	10	5	4	33	2

* WILSON and AMOS (1971).
† Peaks of activity for units inhibition and enzymes.

Table 3. Human Liver Ratios of Enzymes and Protein to Units of Inhibition*

Fractions	5′nucleo-tidase	G-6-Phosphatase	A. Phosphatase	Suc. Dehydrogenase	Protein
3	2.5†	1.5†	0.50†	—†	0.13†
4	2.3	1.2	0.59	0.81	0.13
5	2.1	3.3	1.10	2.40	0.14
6	2.1	4.8	0.94	9.30	0.34
7	2.6	4.3	1.30	16.00	0.50
9	3.0	5.1	1.50	24.50	0.75
14	2.8	3.5	1.80	37.00	0.75
19	2.5	1.3	1.00	8.25	0.50

* WILSON and AMOS (1971).
† Enzyme or protein/units of inhibition.

a range of vesicle or fragment size. A scan of antigenic activity was performed, using two target cells with appropriate antisera. Peak antigenic activity was found in fraction 5. To separate PM from ER, another homogenate of IMI was prepared. A 20% sucrose barrier was used to separate the PM and ER vesicular membranes from most of the mitochondria, nuclei, and damaged cells. The material above the barrier was placed

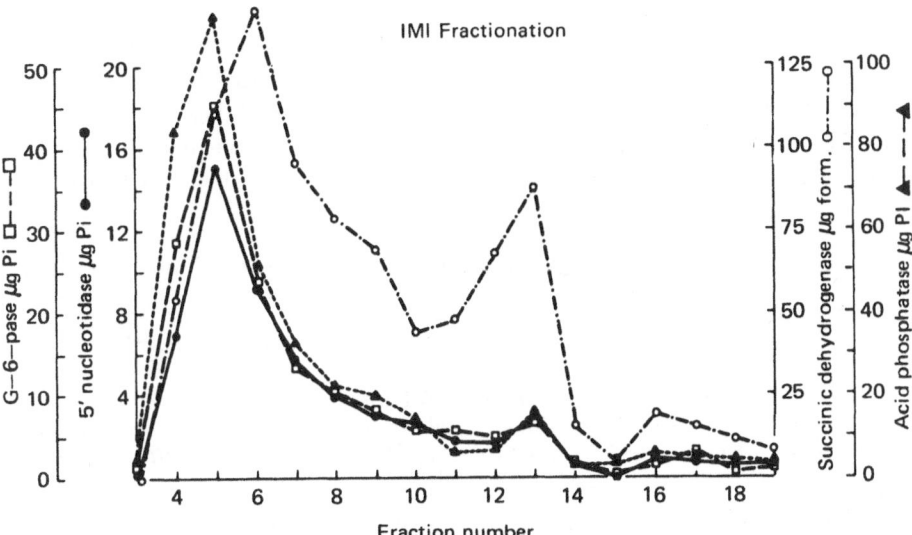

Fig. 6. Enzyme profiles of zonal fractions from IMI tissue culture cell homogenate. Units of activity are the same as in Figure 5. All enzymes except succinic dehydrogenase peak in fraction 5 necessitating further separation on Ficoll. (WILSON and AMOS, 1971).

on discontinuous Ficoll gradients. Fractions were isolated at an approximate specific gravity of 1.06, representing PM, and at a greater than 1.12 pellet, representing ER. Units of inhibition and enzyme activities were calculated (Table 4). The ratio of PM to ER enzyme and antigenic activity was equivalent for only 5' nucleotidase. In preliminary experiments, RPMI 7249 cells were processed in the same manner as the IMI cells with one exception. Zonal fractions 1 through 5 were pooled and placed on the

Table 4. IMI Ficol Separation*

					Enzyme and protein units/fraction			
		Units of Inhibition		5'nucl.	G-6-Pase	A. Pase	Suc. Dehy.	
Fraction	KH	BC	DAL	μg Pi	μg Pi	μg Pi	μg Form.	Prot. αg
PM	192	128	112	35	7	20	5	8
	(2.4)†	(2.0)	(2.8)	(2.50)	(0.46)	(0.54)	(0.38)	(1.30)
ER	80	64	40	14	15	37	13	6

* WILSON and AMOS (1971).

† Numbers in parentheses indicate the ratio of PM to ER for units inhibition, enzymes, and protein.

Ficoll gradients (Figure 1). The results were similar to the IMI studies using the 20% barrier material, with good correlation between plasma membrane content and antigenic activity (Table 5). However, the antigen ratio of PM to ER is not the same as the 5′ nucleotidase ratio of PM to ER. There is less antigenic activity in the ER fraction than one would expect from the nucleotidase content. This discrepancy does not favor the

Table 5. RPMI 7249 Ficol Separation*

| | Units of inhibition | | Enzyme and protein units/fraction | | | | |
| | | | 5′nucleo-tidase | G-6-Phos-phatase | A. Phos-phatase | Suc. Dehydros. | Prot. |
Fraction	ED	BM	μg Pi	μg Pi	μg Pi	μg Form	mg
PM	8,192	1024	45	8	46	1	11
	(16.0)†	(16.0)	(2.0)	(0.21)	(0.45)	(0.50)	(1.0)
ER	512	64	22	38	102	2	11

* Wilson and Amos (1971).
† Numbers in parentheses indicate the ratio of PM to ER for units inhibition, enzymes, and protein.

supposition that ER, lysosomes, or mitochondria contribute to antigenic activity. On the contrary, it would appear that the greater amount of lysosomal enzymes in the ER fraction could be responsible for antigen degradation during the prolonged contact of the membranes in the pellet of the isopycnic separation. An alternative explanation might be that there is greater aggregation of membranes in the ER pellet as opposed to the PM fraction, which remains dispersed at the interface of 1.06 specific gravity Ficoll during centrifugation, thus inhibiting antibody absorption.

III. Conclusions

A model system has been developed to investigate the subcellular location of surface antigens, such as histocompatibility, tumor-specific, species-specific, and organ-specific antigens. Quantitative markers are provided for all the major subcellular organelles, which allows quantitation of an antigen on subcellular organelles by enzyme marker ratios without the necessity of trying to obtain uncontaminated fractions of each organelle. Aggregation of membranes by repetitive sedimentation, used in many methods for the preparation of subcellular organelles, has been restricted in this system to allow better antibody absorption for antigenic quantitation. The procedure is applicable to the liver of several species and, with the additional Ficoll separation, to tissue culture cells. The results of this study would indicate that the surface antigens detected here are exclusive to the plasma membrane within the limits of the system for quantitation. Certainly, no large amounts of antigen would appear to be intracellular for either the species-specific rat antigen, or for the 7 HL-A determinants tested. This does not mean that all the HL-A determinants are exclusively plasma-membrane located, but the determinants tested do represent both segregant series of the HL-A locus. In addition, diverse types of tissue cells were used with comparable results.

We are presently investigating the possibility of detecting the intracellular location of a common "core" structure (LYO) or precursor of HL-A with a heteroantiserum, which has been shown to block HL-A alloantisera from the surface of cells, as well as papain-derived soluble antigen. Preliminary results indicate the absence of association of LYO with any intracellular membranes in human liver, spleen, or tissue culture cells.

In an attempt to obtain some information on HL-A synthesis and turnover, we are in the process of fractionating cells, stripped of most surface HL-A by papain digest, just prior to the reappearance of HL-A or LYO antigen on the surface (SCHWARTZ and NATHENSON, 1970). It is our belief that this procedure may lead to the stimulation of the synthesis of either "core" structure and/or HL-A antigens, which might then be detectable intracellularly.

References

Abeyounis, C. J., L. Edebo, and F. Milgrom (1968). Localization and characterization of mouse and rat tissue isoantigens, pp. 305–309. In *Advances in Transplantation*, J. Dausset, J. Hamburger and G. Mathé (eds.). Copenhagen: Munksgaard.

Aoki, T., U. Hammerling, E. deHarven, E. A. Boyse, and L. J. Old (1969). Antigenic structure of cell surfaces. An immunoferritin study of the occurrence and topography of H-2 and TL alloantigens on mouse cells. *J. Exp. Med., 130*:979–1001.

Basch, R. S., and C. A. Stetson (1963). Quantitative studies on histocompatibility antigens of the mouse. *Transplantation, 1*:469–480.

Boyle, W. (1967). Immunochemistry of mammalian cell membranes, pp. 243–258. In *Immunity, Cancer and Chemotherapy*, E. Mihich (ed.). New York: Academic Press.

Boyle, W. (1968a). In *Biological Properties of the Mammalian Surface Membrane*, p. 113. L. Manson (ed.). Philadelphia: Wistar Institute Press.

—— (1968b). An extension of the ^{51}Cr-release assay for the estimation of mouse cytotoxicity. *Transplantation, 6*:761–764.

Cerottini, J. C., and K. T. Brunner (1967). Localization of mouse isoantigens on the cell surface as revealed by immunofluorescence. *Immunology, 13*:395–403.

Davis, W. C., M. A. Alspaugh, J. H. Stimpfling, and R. L. Walford (1971). Cellular surface distribution of transplantation antigens: Discrepancy between direct and indirect labeling techniques. *Tissue Antigens, 1*:89–93.

Davis, W. C., and L. Silverman (1968). Localization of mouse H-2 histocompatibility antigen with ferritin-labeled antibody. *Transplantation,* 6:535–543.

Dumonde, D. C., S. Al-Askari, H. S. Lawrence, and L. Thomas (1963). Microsomal fractions as transplantation antigen. *Nature, Lond.,* 198:598–612.

Edebo, L., C. J. Abeyounis, H. Themann, and F. Milgrom (1968). Preparation of cell membranes for studies on transplantation antigens. *Int. Arch. Allergy,* 34:257–268.

Evans, W. H., and J. W. Bruning (1970). Studies on the distribution of some histocompatibility antigens on mouse liver plasma membrane and microsomal fractions. *Immunology, 19:735–741.*

Haughton, G. (1966). Transplantation antigen of mice: cellular localization of antigen determined by the H-2 locus. *Transplantation, 4:238–244.*

Herberman, R., and C. A. Stetson (1965). The expression of histocompatibility antigen on cellular and subcellular membranes. *J. Exp. Med., 121:533–549.*

Herzenberg, L. A., and L. Z. Herzenberg (1961). Association of H-2 antigens with the cell membrane fraction of mouse liver. *Proc. Nat. Acad. Sci. U.S.A., 47:762–767.*

Manson, L. A., C. A. Hickey, and J. Palm (1968). H-2 alloantigen content of surface membrane of mouse cells, pp. 93–103. In *Biological Properties of Mammalian Surface Membrane,* L. A. Manson (ed.). Philadelphia: Wistar Institute Press.

Moller, G. (1961). Demonstration of mouse isoantigens at the cellular level by the fluorescent antibody technique. *J. Exp. Med., 114:415–431.*

Nathenson, S. (1970). Biochemical properties of histocompatibility antigens. *Ann. Rev. Genet., 4:69–90.*

Neville, D. M., Jr. (1960). The isolation of a cell membrane fraction from rat liver. *J. Biophys. Biochem. Cytol., 8:413–422.*

Ozer, J. H., and D. F. H. Wallach (1967). H-2 components and cellular membranes. Distinction between plasma membrane and endoplasmic reticulum governed by the H-2 region in the mouse. *Transplantation, 5:652–667.*

Palm, J., and L. A. Manson (1965). Tissue distribution and intracellular sites of some mouse isoantigens, pp. 21–33. In *Isoantigens and Cell Interactions,* J. Palm (ed.). Philadelphia: Wistar Institute Press.

Popp, R. A., D. M. Popp, N. C. Anderson, and L. H. Elrod (1969). Use of the zonal centrifuge to separate particles containing transplantation antigen. *Biochim. Biophys. Acta, 184:625–633.*

Schwartz, B., and S. G. Nathenson (1970). The regeneration of transplantation antigen on mouse cells. *Third Inter. Congr. Transpl. Soc.,* p. 275.

Wallach, D. F. H. (1967). Isolation of plasma membranes of animal cells, pp. 129–163. In *The Specificity of Cell Surfaces,* B. D. Davis and L. Warren (eds.). Englewood Cliffs, N.J.: Prentice-Hall, Inc.

Weaver, R., and W. Boyle (1968). Isolation of plasma membranes from human liver. In *Abstracts of the Conference on Viral, Tumor, and Transplantation Antigen Isolation, Transplantation, 6:676–677.*

—— (1969). Purification of plasma membranes of rat liver. Application of zonal centrifugation to isolation of cell membranes. *Biochim. Biophys. Acta,* 173:377–388.

Wilson, L. A., and D. B. Amos (1971). Subcellular location of HL-A antigens. *Tissue Antigens.* In press.

Wilson, L. A., and W. Boyle (1970). Rapid micro procedure for scanning distribution of subcellular components in density gradient fractions of a tissue homogenate. *Anal. Biochem., 35*:466–474.

—— (1971). A model system for the study of subcellular distribution of cell surface antigens. *J. Immunol., 108*:460–466.

Part 6

Malignant Tissue Antigens

CELLULAR IMMUNITY TO TUMOR ANTIGENS—POSSIBLE CLINICAL USEFULNESS OF THE FINDINGS SO FAR OBTAINED*

KARL ERIK HELLSTRÖM and INGEGERD HELLSTRÖM

*Departments of Pathology and Microbiology,
University of Washington Medical School,
Seattle, Washington*

Lymphocytes from animals as well as from human patients having cancer (or having had cancer) can destroy cultivated neoplastic cells *in vitro* (HELLSTRÖM and HELLSTRÖM, 1967; HELLSTRÖM *et al.*, 1969a; HELLSTRÖM and HELLSTRÖM, 1969a; HELLSTRÖM *et al.*, 1970a, 1968, 1971a, 1969b, 1971b; BALDWIN and EMBLETON, 1971; BUBENIK *et al.*, 1970; SJÖGREN and BORUM, 1971; DATTA and VENDEPUTTE, 1971; JAGARLAMOODY *et al.*, 1971). Sera from individuals with actively growing tumors can specifically block the destructive lymphocyte effect, while sera from individuals whose tumors have been successfully removed cannot do that (HELLSTRÖM *et al.*, 1969b, 1971b; BUBENIK *et al.*, 1970; SJÖGREN and BORUM, 1971; DATTA and VANDEPUTTE, 1971; JAGARLAMOODY *et al.*, 1971). There are recent data which indicate that there also exist—in certain immune sera—"unblocking" (or "de-blocking") antibodies capable of abrogating the blocking activity of sera from tumor-bearers.

All these findings have been presented in detail, and have been repeatedly reviewed (HELLSTRÖM and HELLSTRÖM, 1969b, 1970a, 1970b). We shall, therefore, not review them once more. Instead, we shall center this discussion on what implications the recent observations on cell-mediated antitumor immunity may have with respect to tumor diagnosis, therapy, and prevention. The discussion will have to be speculative, since conclusive evidence as to the ultimate feasibility of the three goals is scarce.

* The authors' work, on which this summary was based, has been supported by Grants CA-10188 and CA-10189 from the National Institutes of Health, by Contract NIH 69-2061 from the National Institutes of Health, and by Grant T-453 from the American Cancer Society.

281

I. Aspects in Relation to Tumor Diagnosis

GOLD and FREEDMAN (1965) showed that human colonic carcinomas have an antigen, which is absent from adult human tissues but present in fetal colon. This is the so-called carcino-embryonic antigen (CEA). It has been demonstrated in the serum of patients with colonic carcinomas, but according to GOLD (1970), not in sera from patients having other types of tumors or in normal sera. Therefore, it has been suggested that screening of sera for CEA may make it possible to diagnose colonic carcinomas.

Recent work by MARTIN and MARTIN (1970) indicates that normal adult colon has a small amount of CEA as well. The quantitative differences between normal and neoplastic colon may be large enough to make diagnosis of the neoplasms possible. Nevertheless, it is not settled as to what extent screening of sera for CEA will be discriminative enough to provide new means for diagnosing colonic carcinomas.

Our own work supports the notion that colonic carcinomas share an antigen, absent from other human tumors (HELLSTRÖM et al., 1971a) and that there is (at least to some degree) cross-reactivity between colonic carcinomas and normal fetal gut epithelium (HELLSTRÖM et al., 1970b). We cannot conclude, however, that the antigen involved in our tests as the "trigger" of target cell destruction by immune lymphocytes is the same as Gold's CEA (GOLD, 1970).

The recent demonstration of a large additional number of tumor groups, within (but not between) which there is antigenic cross-reactivity, suggests the possibility of an immunological diagnosis of many types of human tumors. Most of this information has been obtained by studying lymphocyte-mediated reactions against tumor-associated antigens. Some groups within which cross-reactions have been detected are formed by the following types of neoplasms: Neuroblastomas (HELLSTRÖM et al., 1970a), malignant melanomas; carcinomas of the breast, cervix, endometrium, kidney, bladder; seminomas; sarcomas (HELLSTRÖM et al., 1971b). Whether immunological assays can be helpful in diagnosing these types of tumors may depend upon whether practically simple techniques can be worked out; the colony inhibition assay and the microcytotoxicity tests are too cumbersome for clinical screening of a large number of human subjects to search for those few with a given type of tumor. When trying to develop assays useful for this purpose, it may be important to ascertain whether any soluble tumor antigens can be detected in the patient's sera.

II. Aspects in Relation to Tumor Therapy

The finding that lymphocytes from individuals who have cancer can destroy cultivated neoplastic cells *in vitro* offers the hope that a situation may be created whereby the same thing happens *in vivo*. For this to be feasible, one needs to know how best to stimulate the lymphocytes to kill the tumor cells and how to prevent the tumor cells escaping from the immunological attack.

The discovery that serum from individuals with growing tumors can block the cytotoxic effect of immune lymphocytes (HELLSTRÖM *et al.,* 1969b; HELLSTRÖM and HELLSTRÖM, 1970a) is relevant in this context. If it were known how to remove the blocking serum factors, how to abrogate the formation of more such factors, and how to cancel the blocking effect of the factors present, it might then be possible for the already present immune lymphocytes to destroy the tumor cells. This point of view gains support from recent data showing a high correlation between progressive tumor growth *in vivo* and blocking serum activity *in vitro.* Such a correlation has been seen when individual cases of human cancer patients have been followed (HELLSTRÖM *et al.,* 1971b). It has also been observed when the immune status of mice (HELLSTRÖM *et al.,* 1970c) and rats (SJÖGREN and BANSAL, 1971a) with transplanted tumors has been monitored. The most extensive of the animal experiments have been done on rats with polyoma tumors. The rats were followed for different parameters of tumor immunity *in vitro* and for tumor resistance *in vivo,* as monitored by rejection (or acceptance) of small doses of inoculated syngeneic polyoma tumor cells (SJÖGREN and BANSAL, 1971a).

Theoretically, the most simple approach may be the physical removal of the blocking factors. Attempts to remove the blocking factors may involve plasmaphoresis, or—even better—thoracic duct cannulation combined with removal of the lymph plasma and reinfusion of the cells (L. Rose, personal communication); the latter procedure should be combined with splenectomy so that no antibodies are produced from the spleen. One could also consider the use of immunoadsorbents capable of clearing serum from its blocking factors, the immunoadsorption, for example, to be combined with plasmaphoresis. Although experiments aiming at the physical removal of blocking factors are highly worthwhile, they may be hampered by difficulties in keeping pace with the continuous formation of more blocking factors.

Another point of attack would be to try decreasing either the formation or the way of action of the blocking factors. Precise knowledge as to the nature of the blocking factors seems to be needed in order to achieve this goal, and it is not yet available. Existing evidence suggests, however, that the blocking factors may be complexes between tumor-associated antigens, released from the neoplastic cells, and antibodies formed by the host in response to the antigenic stimulus (SJÖGREN *et al.,* 1971). If this notion is correct, decreases in the blocking serum activity may follow upon procedures leading either to a decreased release of antigen or to a decreased formation of the antibody portion of the complexes. One may speculate that certain drugs might serve the latter purpose by being more toxic to cell clones forming the antibodies than to lymphocytes mediating the cell-mediated immunity. Of course, chemotherapeutic drugs may also decrease the blocking effect by destroying tumor cells that would otherwise release antigen.

The most promising approach to inhibit the action of blocking factors seems to be offered by the discovery of "unblocking" antibodies, which can abrogate the blocking effect of sera from tumor-bearing individuals, when mixed with them *in vitro* (HELLSTRÖM and HELLSTRÖM, 1970c; BANSAL and SJÖGREN, 1971). The mechanism of action of the unblocking antibodies is unknown. If the blocker is an antigen-antibody complex, the addition of unblocking antibodies may sufficiently change the nature of the complex so as to abrogate its blocking ability. The finding that inoculation of sera with unblocking activity could induce regression of primary Moloney sarcomas in mice (HELLSTRÖM *et al.*, 1969c) and of transplanted polyoma tumors in rats (BANSAL and SJÖGREN, 1971) urges further investigations as to the immunotherapeutical value of inoculation of unblocking antibodies. Primary tumors of spontaneous origin, such as mouse mammary carcinomas, should preferably be used, since they would yield information more relevant with respect to human tumors. When such immunotherapeutical studies are performed, it seems fundamental to ensure first, that the sera used are indeed unblocking, and second, that they are given in amounts sufficient for the blocking activity of sera from the inoculated animals to shift into an unblocking one. If this is not checked, tumor growth may be enhanced because of the creation of more blocking complex, composed of antigen released from the tumor and antibodies supplied from outside.

When considering the different approaches to immunotherapy, it is well to recognize that there is a balance between tumor growth on the one hand and the immunological response to the tumors on the other, and that the latter consists of several components, including a cell-mediated immunity (which can destroy the neoplastic cells), blocking factors (which can abrogate the lymphocyte effect), unblocking antibodies (which can cancel out the blocking activity), and cytotoxic antibodies (which can destroy tumor cells together with complement); other factors may be involved as well. A therapeutic effect may be expected from procedures increasing the cell-mediated antitumor immunity (specifically or nonspecifically) if they do not also increase the blocking activity. It is probable that bacillus Calmette-Guérin (BCG) may increase the cell-mediated immunity, but immunotherapeutic schemes based upon that effect should also take into account possible changes in the blocking serum activity (SJÖGREN and BANSAL, 1971b). The cell-mediated immunity may be also improved by adoptive transfer of more immune cells or by inoculation of transfer factor; the last possibility deserves more study. As already stated, inoculation of unblocking antibodies, or of tumor antigens capable of stimulating the formation of such antibodies, may be therapeutically beneficial by decreasing the blocking effect; this is the approach which appears most promising to us at the present time.

Not much can be offered for immunotherapy of human cancer today. However, there are, as mentioned, promising leads for animal experimentation, to be followed up by more animal tests. If these leads prove to be useful, rational procedures for immunotherapeutical trials in man are likely to evolve.

It appears likely that already existing modes of therapy (surgery, radiation, drugs, hormones) influence the various immunological parameters. Studies on the nature of this influence may yield information which could lead to a better administration of conventional cancer therapy. Techniques are available by which this can be investigated. Obviously, information on the tumor-specific response is more relevant than, for example, the question of whether or not lymphocytes will transform well when exposed to phytohemagglutinin, following a particular therapeutical procedure. Such *in vitro* assays may also be helpful when monitoring individual patients during treatment. The appearance of unblocking antibodies (as well as increases in cell-mediated immunity) might be a good clinical sign, while the appearance of blocking factors might indicate either an incomplete tumor removal or a tumor recurrence.

III. Aspects in Relation to Tumor Prevention

The most obvious goal of cancer research is to find ways by which tumor formation can be prevented. Claims were raised more than 40 years ago that, indeed, tumors could be prevented from occurring in experimental animals. Unfortunately, however, these claims were based on transplantation experiments performed in non-inbred animals, so that the resistance detected was primarily one against normal alloantigens.

The fact that tumors can induce an oncocidal type of immune response and that neoplasms of the same morphological type cross-react antigenically, offers realistic possibilities for an immunological prevention of certain tumors. Experiments planned to achieve such prevention should be performed with the goal of establishing a strong cell-mediated immunity in the absence of blocking serum activity. A major concern at the present time is whether such an immunity can be induced and whether it would influence the frequency of primary tumors. No valid conclusions can be made before it is known whether the "tissue-type-specific" tumor antigens identified in human neoplasms can, indeed, serve as tumor-specific transplantation antigens *in vivo*. The easiest way to approach this question may be to search for "tissue-type-specific" tumor antigens in animal neoplasms, *e.g.,* in bladder carcinomas in rats, and then to investigate whether they can act as transplantation antigens. If this is so, one should study whether tumor formation can be prevented by immunizing animals against the "tissue-type-specific" tumor antigens before they are exposed to the respective mode of chemical carcinogenesis.

If it should, indeed, be possible to build up a strong immunity to common antigens of histologically similar animal tumors, and if animals with such an immunity (but no blocking serum activity) should have an increased resistance to carcinogenesis (in the respective organ), the groundwork may have been laid for a future approach to "vaccination" against certain human cancers. Even then, however, extreme caution is necessary. Otherwise, the "vaccination" procedures may lead to more tumors by stimulating the formation of enhancing antibodies or by inducing autoimmunity caused by inoculated tumor antigens that cross-react with antigens occurring normally (but in smaller doses), *e.g.*, during certain periods of the normal cell cycle. The present *in vitro* assays of tumor immunity may serve an important function in the development of vaccination procedures by making it possible to follow different aspects of the immune response.

IV. Conclusion

The recently acquired knowledge on cell-mediated immunity to tumor-associated (specific?) antigens of neoplastic cells, in animals as well as in man, may have implications with respect to tumor diagnosis, therapy, and prevention. The full extent to which this may be the case cannot be determined at the present time.

References

Baldwin, R. W., and M. J. Embleton (1971). Demonstration by colony inhibition methods of cellular and humoral immune reactions to tumor-specific antigens associated with aminoazo-dye-induced rat hepatomas. *Int. J. Cancer, 7*:17–25.

Bansal, S. C., and H. O. Sjögren (1971). Manuscript in preparation.

Bubenik, J., P. Perlmann, K. Helmstein, and G. Moberger (1970). Cellular and humoral immune responses to human urinary bladder carcinomas. *Int. J. Cancer, 5*:310–319.

Datta, S. K., and M. Vandeputte (1971). Studies on cellular and humoral immunity to tumor-specific antigens in polyoma virus-induced tumors of rats. *Cancer Res., 31*:882–889.

Gold, P. (1970). The model of colonic cancer in the study of human tumor-specific antigens, pp. 131–142. In *Carcinoma of the Colon and Antecedent Epithelium*, W. J. Burdette (ed.), Springfield: Charles C Thomas.

Gold P., and S. O. Freedom (1965). Specific carcinoembryonic antigens of the human digestive system. *J. Exp. Med., 122*:467–481.

Hellström, I., and K. E. Hellström (1967). Cell-bound immunity to autologous and syngeneic mouse tumors induced by methylcholanthrene and plastic discs. *Science, 156*:981–983.

—— (1969a). Studies on cellular immunity and its serum mediated inhibition in Moloney virus induced mouse sarcomas. *Int. J. Cancer, 4*:587–600.

Hellström, K. E., and I. Hellström (1969b). Cellular immunity against tumor antigens. *Adv. Cancer Res., 12*:167–223.

—— (1970a). Immunological enhancement as studied by cell culture techniques. *Ann. Rev. Microbiol., 24*:373–398.

—— (1970b). Immunological defenses against cancer. *Hospital Practice,* 5:45–61.

—— (1970c). Colony inhibition studies on blocking and non-blocking serum effects on cellular immunity to Moloney sarcomas. *Int. J. Cancer,* 5:195–201.

Hellström, I., K. E. Hellström, G. E. Pierce, and J. P. S. Yang (1968). Cellular and humoral immunity to different types of human neoplasms. *Nature, Lond., 220*:1352–1354.

Hellström, I., C. A. Evans, and K. E. Hellström (1969a). Cellular immunity and its serum mediated inhibition in Shope virus induced rabbit papillomas. *Int. J. Cancer, 4*:601–607.

Hellström, I., K. E. Hellström, C. A. Evans, G. Heppner, G. E. Pierce, and J. P. S. Yang (1969b). Serum mediated protection of neoplastic cells from inhibition by lymphocytes immune to their tumor specific antigens. *Proc. Nat. Acad. Sci. U.S.A., 62*:362–369.

Hellström, I., K. E. Hellström, G. E. Pierce, and A. Fefer (1969c). Studies on immunity to autochthonous mouse tumors. *Transpl. Proc., 1*:90–94.

Hellström, I., K. E. Hellström, A. H. Bill, G. E. Pierce, and J. P. S. Yang (1970a). Studies on cellular immunity to human neuroblastoma cells. *Int. J. Cancer, 6*:172–188.

Hellström, I., K. E. Hellström, and T. H. Shephard (1970b). Cell-mediated immunity against antigens common to human colonic carcinomas and fetal gutepithelium. *Int. J. Cancer, 6*:346–351.

Hellström, I., K. E. Hellström, and H. O. Sjögren (1970c). Serum mediated inhibition of cellular immunity to methylcholanthrene induced murine sarcomas. *Cell Immunol., 1*:18–30.

Hellström, I., K. E. Hellström, H. O. Sjögren, and G. A. Warner (1971a). Demonstration of cell-mediated immunity to human neoplasms of various histological types. *Int. J. Cancer, 7*:1–16.

Hellström, I., H. O. Sjögren, G. A. Warner, and K. E. Hellström (1971b). Blocking of cell-mediated tumor immunity by sera from patients with growing neoplasms. *Int. J. Cancer, 7*:226–237.

Jagarlamoody, S. M., J. C. Aust, R. H. Tew, and C. A. McKhann (1971). *In vitro* detection of cytotoxic cellular immunity against tumor-specific antigens by radioisotope technique. *Proc. Nat. Acad. Sci. U.S.A., 68*: 1346–1350,

Martin, F., and M. S. Martin (1970). Demonstration of antigens related to colon cancer in the human digestive system. *Int. J. Cancer, 6*:352–360.

Rose, L. Personal communication.

Sjögren, H. O., and S. C. Bansal (1971a). Manuscript in preparation.

—— (1971b). Manuscript in preparation.

Sjögren, H. O., and K. Borum (1971). Tumor-specific immunity in the course of primary polyoma and Rous tumor development in intact and immunosuppressed rats. *Cancer Res., 31*:890–900.

Sjögren, H. O., I. Hellström, S. C. Bansal, and K. E. Hellström (1971). Suggestive evidence that the "blocking antibodies" of tumor-bearing individuals may be antigen-antibody complexes. *Proc. Nat. Acad. Sci. U.S.A., 68*: 1372–1375.

SIGNIFICANCE OF THE CELLULAR ANTIGENS IN BURKITT'S LYMPHOMA*

Eva Klein and George Klein

*Department of Tumor Biology, Karolinska Institute,
Medical School, Stockholm, Sweden*

I. Introduction

Viruses have obviously evolved so that they enter a productive, permissive interaction with cells or, if they enter a temporarily nonproductive interaction, they do so without annihilating the environment of their host cell. Most oncogenic virus-cell interactions are nonpermissive, however, and if successful in producing a tumor, they kill their host. Many of them occur in hosts other than the natural species to which the virus has adapted during its evolution. In oncogenic transformation a range of virus-cell interactions exist, from the permissive cell (containing all components, or antigens, of a full viral cycle) to the nonpermissive (where only some of the early virally induced products are made). By immunological means the T and the transplantation antigen have been regularly identified in such cells.

Antigenic changes regularly associated with neoplastic and/or transformed cells induced by the small oncogenic DNA viruses are of considerable interest, since the virus-contributed genetic information is only sufficient to code for a small number of proteins. Any virus-determined function consistently associated with transformation has, therefore, a relatively high probability of being an essential determinant of transformation (LURIA, 1962).

The transplantation antigen expresses an alteration of the cell membrane. SJÖGREN's (1964) experiments showed that the polyoma antigen was very stable and resisted prolonged negative selection by serial passage in preimmunized mice. We know that exposure of tumor cells derived from

* This study was conducted under USPHS Contract NIH-69-2005 within the Special Viral Cancer Program of the National Cancer Institute. Grants were also received from the Swedish Cancer Society, the Medical Research Council, the Damon Runyon Memorial Fund (DRG-1064), Lotten Bohman's Fund, Harald and Greta Jeansson's Foundation and Åke Wiberg's Foundation.

H-2 heterozygotes to similar selection against one H-2 complex has led, on the other hand, to the regular isolation of antigenic loss variants that failed to reacquire the missing H-2 phenotype when returned to the permissive host (HELLSTRÖM, 1961; KLEIN, 1961; KLEIN *et al.,* 1960). A second selection step against the remaining single complex was inefficient, however, indicating that the cell must maintain one H-2 complex, presumably to preserve the integrity of its membrane. It can be assumed therefore that the virus-induced membrane change reflected by the polyoma antigen is an essential determinant of neoplastic behavior. Conceivably, it may represent a modification of some membrane-associated element involved in growth regulation. The fact that tumor-associated transplantation antigens are specific for a given viral etiological group of tumors that may include a variety of histological types strongly suggests that they are coded by the viral genome. Polyoma tumors in mice include, for instance, osteosarcoma, thymoma, mammary carcinoma, etc., but they all share the same transplantation antigen. The converse is also true: tumors of the same histogenetic origin but induced by different viral agents, *e.g.,* the thymic lymphomas induced by the Gross and the Moloney virus, respectively, carry different transplantation antigens (KLEIN and KLEIN, 1964). There is, in other words, a conspicuous lack of tissue specificity and a strong virally associated common group specificity even for tumors induced in different species (GIRARDI, 1965; SJÖGREN, 1965).

In contrast to the virally induced transplantation antigens that show strict group specificity, membrane receptors detected by certain phytoagglutinins are expressed on all kinds of transformed cells, irrespective of etiology (BURGER, 1969; INBAR and SACHS, 1969). They also appear on the surface of normal, untransformed cells during mitosis, and can be "unmasked" on normal interphase cells by trypsin. Their presence on transformed cells may be a consequence, rather than a cause, of transformation. In fact transformation-defective polyoma mutants do not induce the agglutinin sites either (BENJAMIN and BURGER, 1970).

II. Cell-Surface Localized Immunoglobulins

When it became obvious that virus-induced animal tumors carried common membrane-associated antigens, we searched for such group-specific antigens in tumors in order to obtain clues about their possible viral etiology. Our attention was turned to Burkitt's lymphoma (BL), both because it was postulated to have a viral etiology and because it indicated the existence of immunological host defense in this disease (BURKITT, 1963; BURKITT, 1967; CLIFFORD, 1966; NGU, 1965). Based on the experience obtained in animal systems, especially in murine leukemia, we exposed BL biopsy cells to the sera of BL patients. Afterwards, we looked for reactive antibodies on the cells by the indirect membrane fluorescence technique (KLEIN *et al.*, 1966; KLEIN *et al.*, 1967c; KLEIN *et al.*, 1967a). We found that sera of BL patients reacted more frequently with the cells than did African control sera, also from donors with other neoplastic or non-neoplastic diseases. It was found that the most regularly positive sera were from patients whose tumors regressed after chemotherapy.

While this was encouraging, further studies on the specificity of the reaction were hampered by the use of different biopsy cells as targets which varied in their strength of reactivity. Also it was found that in a number of cases there was already immunoglobulin on the cell surface. This resulted in reactivity with the conjugated anti-immunoglobulin reagents. Both IgM and/or IgG could be detected (KLEIN *et al.*, 1968a; KLEIN *et al.*, 1969). In the cases where the cell surface reacted with anti-IgM conjugates, this occurred on almost 100% of the cells. A number of such biopsies reacted also with anti-kappa serum, but never with anti-lambda serum. When established lines were developed, their IgM on the cell membrane was maintained, but the cells did not secrete IgM into the medium (VAN FURTH *et al.*, personal communication). Characterization of the IgM molecule (ESKELAND and KLEIN, 1971) present on one cell line revealed that 7S subunits are integrated into the cell membrane. Conceivably, this is the neoplastic variety of a normal lymphoid cell that incorporates immunoglobulin molecules into its plasma membrane. Lymphoid cells of this type are postulated to play an important role in immunological memory and/or delayed hypersensitivity (SINGHAL and WIGZELL, 1971). The lymphoma cells may represent the neoplastic variant in the same way as myeloma cells project normal immunoglobulin-secreting plasma cells into a magnified, neoplastic image. The phenomenon is not exclusive for BL; the majority of chronic lymphatic leukemia cases have the same cellular characteristics (JOHANSSON and KLEIN, 1970). Comparisons between different lymphomas showed that they carry different amounts of IgM on their membrane. Although an extensive search has been made on more than 60 biopsies and derived lines, so far no line has been found that would carry other than IgM kappa types of membrane-associated immunoglobulins.

The membrane-associated immunoglobulins carried by the cells have always behaved as cell markers. If present on the cells of a given tumor, they were maintained unchanged in the course of repeated biopsies; if absent, they remained absent. On the other hand, the IgG demonstrable on the biopsy cells was rarely present on untreated BL biopsy cells, but tended to appear if the tumor persisted in spite of treatment.

In contrast with the membrane-associated IgM, the changing amount of IgG on repeated biopsies and its failure to persist on derived *in vitro* lines (KLEIN *et al.*, 1968a; NADKARNI *et al.*, 1969), indicates that it is due to coating (KLEIN *et al.*, 1969; CLIFFORD *et al.*, 1968). Its role in the host-tumor relationship could not be established yet.

The significance of the finding of membrane-bound IgM on both the

Burkitt lymphoma and chronic lymphatic leukemia cells in spite of the dissimilarity in clinical course and age of onset, is that it indicates a common cell type as a target of neoplastic transformation. This cell is probably the bone marrow-derived lymphocyte, as immunoglobulin has been demonstrated on its surface. The tumors provide, therefore, a good tool for studies on the properties and synthesis of this immunoglobulin molecule which is considered to function as an antigen receptor (UNANUE *et al.,* 1971).

III. The Epstein-Barr Virus—(EBV)—Associated Antigen System

A number of BL-derived lymphoblastoid cell lines were tested against BL sera that reacted with BL biopsy cells, and were found to be free of demonstrable isoantibodies (KLEIN *et al.,* 1967b). Four BL-derived lines reacted with the reference serum "Mutua" (derived from a BL patient in long-term regression), whereas three BL-derived lines were negative. Eight control lines derived from various leukemias and, in one case, from a normal donor, were also negative.

When these results were compared with the reactivity of the same cell lines in the Henle test (HENLE and HENLE, 1966) known to detect EB virus capsid antigen (VCA), it was found that the four membrane-positive lines contained EBV-VCA in more than 1% of the cells, whereas the membrane-negative lines were either VCA-negative or contained a very small frequency of positive cells (less than 1%).

This suggested that the membrane antigen (MA) detected by this reference serum may be determined by the EBV genome. More conclusive evidence was obtained in a prospective study (KLEIN *et al.,* 1968b). Fourteen new lines were established from biopsies received from Nairobi, and the frequency of VCA-positive and of membrane-reactive cells was determined in parallel, on coded specimens. The same relationship was found as in the preliminary retrospective study: only the lines that carried a relatively high "EBV load" showed a positive membrane-antigen reactivity. In the reactive lines, the frequency of membrane-positive cells was approximately 10 times higher than the frequency of VCA-positive cells. The biopsies were VCA-negative, and reactivity appeared during the first week in culture, suggesting that production of the viral nucleocapsid antigen is suppressed in the tumor cell *in vivo*. Parallel lines from the same patient, derived from successive biopsies, led to lines with fairly similar VCA levels, whereas lines derived from different patients varied (NADKARNI *et al.,* 1970). This suggests that the viral "load" per cell, or the virus activation potential, or both, are characteristic for the individual tumor. Since the membrane-associated IgM marker, mentioned above, and another study wth G6PD-isozyme markers (FIALKOW *et al.,* 1970) strongly indicate that the BL process has a clonal origin, this may reflect the virus-cell relationship that characterizes a particular clone. This is also suggested by the closely similar levels of EBV-DNA hybridizable cellular DNA in repeated biopsies taken from multiple tumors from the same BL patient (ZUR HAUSEN *et al.,* 1970).

The postulate that the MA is determined by EBV was directly confirmed when it was found that is can be induced to appear in negative lines after infection with EBV (KLEIN *et al.,* 1967d; HENLE and HENLE, 1969; GERGELY *et al.,* 1971a). In the infected cells, MA appeared after 20 to 24 hours. DNA-inhibitors such as cytosine arabinoside (Ara C) or IUDR did not prevent its appearance, whereas puromycin inhibited it completely. It behaves, in other words, like an "early" product of the viral genome, not requiring viral DNA synthesis. In this respect, it shows certain parallels with MA and T antigens found in experimental oncogenic DNA virus systems (RAPP *et al.,* 1965; MEYER *et al.,* 1969; DEICHMANN, 1969).

Recently, HENLE *et al.* (1970b) have described yet another EBV-associated antigen, designated as EA (early antigen), detected in EBV-infected Raji or 6410 cells. The sera of some but not all EB-VCA-positive donors contained antibodies against it. EA precedes the appearance of VCA during the infectious cycle; its appearance is prevented by puromycin but not by DNA-inhibitors (GERGELY *et al.,* 1971a). The behavior of EA is thus not unlike what has been mentioned above for MA, with one important exception: whereas MA is compatible with continued cell growth and DNA synthesis, EA inhibits host macromolecular synthesis, as shown by a combination of immunofluorescence and autoradiography (GERGELY *et al.,* 1971b), and thus probably signals the entry of the cell into the lytic cycle. In contrast to EA, VCA is dependent on DNA synthesis and therefore probably represents a "late" viral product.

Although the relationship between EBV and MA was clarified by these studies, this applies only to the EBV-carrier cultures *in vitro* and it must be kept in mind that similar compelling evidence is lacking about the connection between the membrane antigens detected in the biopsy cells and the virus.

The antigenic components entering the EBV-determined membrane and the intracellular nucleocapsid complex, respectively, are not identical. It was possible to remove the membrane-reactive antibodies by absorption with large numbers of viable MA-positive cells, with only a minor reduction in the anti-VCA titer (PEARSON *et al.,* 1969). Moreover, a few sera could be found with antibodies either against the membrane antigen or the VCA antigen. Although such "discordant" sera were in the minority (20%), their existence was evidence of the immunological distinctness of the two antigen(s) which could also be directly visualized by two-color fluorescence examination (KLEIN *et al.,* 1971).

Further analysis of the two antigen systems must be regarded as antigen complexes, with several distinct subcomponents (SVEDMYR *et al.,* 1970).

The nature of the MA, particularly its specification by the viral or the cellular genome remains to be clarified. Recently, indirect evidence has accumulated suggesting that it may represent a viral envelope component. The ability of different sera to neutralize an artificial infection of EBV-negative culture lines (such as Raji or 6410) was related to the titer of membrane-reactive antibody, and not to the anti-VCA titer (PEARSON *et al.*, 1970). This was particularly apparent with sera that were discordant with regard to their anti-VCA and membrane reactivity. Also, carrier cultures were a high frequency of MA-positive cells absorbed virus-neutralizing antibody better than cultures with a low frequency of such cells, both containing approximately equal numbers of VCA-positive cells (GERGELY *et al.*, 1971b). MA thus may represent viral envelope components, inserted in the outer plasma membrane, perhaps in analogy with the new membrane glycoproteins that appear in the membrane of herpes simplex-infected cells (ROIZMAN and SPRING, 1967). Immunoferritin staining of EBV-carrier cells also showed that MA is present on the outer envelope of the virus, but not on naked virus particles (SYLVESTRE *et al.*, 1971).

IV. Disease-Associated Serological Patterns

Mean anti-EBV-VCA and anti-MA titers are much higher in groups of Burkitt lymphoma patients than in healthy African controls, or in African patients with neoplastic diseases other than BL and nasopharyngeal carcinoma (NPC) (GUNVÉN *et al.*, 1970; HENLE *et al.*, 1969). The same is true for immunoprecipitating antibodies against a soluble antigen extracted from an EBV-carrying Burkitt line (OLD *et al.*, 1968). NPC was unique among the head and neck tumors so far investigated, in showing high EBV-associated antibody levels in the anti-MA, anti-VCA and immunoprecipitin tests as well (HENLE *et al.*, 1970a; OLD *et al.*, 1968; DE SCHRYVER *et al.*, 1969). It was particularly remarkable that hypopharyngeal and oropharyngeal carcinomas were quite different serologically; their EBV-associated antibody pattern resembled that of the controls (DE SCHRYVER *et al.*, 1969).

Among other lymphoid malignancies, the lymphocyte-poor, sarcomatous cases of Hodgkin's disease had high anti-VCA and anti-MA levels, comparable to BL and NPC (JOHANSSON *et al.*, 1970). The lymphocyte-rich, paragranulomatous form that has a better prognosis was characterized, on the other hand, by low anti-VCA and anti-MA titers, similar to those of the controls. The granulomatous form also was intermediate, from the histological and the serological points of view.

Forty percent of the sera from chronic lymphocytic leukemia patients with a high circulating tumor cell count showed relatively high serological reactivities. On the other hand, patients with lymphocytic lymphoma had regularly high EBV-associated serological reactivity (JOHANSSON *et al.*, 1970).

There is strong evidence that EBV is involved in the etiology of infectious mononucleosis (IM) (HENLE *et al.*, 1968). Prospective studies showed that anti-EBV seropositive young adults are protected from IM, while a substantial fraction of seronegatives acquire the disease in the same environment (NIEDERMAN *et al.*, 1970). It is also clear, however, that EBV is widespread and ubiquitous. Infection in early childhood apparently causes seroconversion only, but does not appear to be associated with any recognized clinical syndrome.

The relationship of EBV to BL can be considered according to three main alternatives that may be referred to as the *cofactor* hypothesis, the *multiple-virus* hypothesis and the *passenger* hypothesis.

The *cofactor* hypothesis implies that EBV is the causative agent of IM and of BL as well, but that the malignant proliferation is due to cofactor(s).

The *multiple-virus* hypothesis implies that there are different variants of EBV in nature. Some of these may induce benign and other malignant disease. Attempts to show antigenic differences between the virus associated with IM, BL, and NPC, respectively, have so far failed to reveal any differences, but few such attempts have been made and the present methodology is unable to resolve minor differences in type.

The *passenger* hypothesis departs from the fact that EBV is an inhabitant of human lymphoid tissues and is widespread and ubiquitous. When lymphomas or other lymphocyte-rich tumors arise and proliferate, EBV would be carried, along with a corresponding increase in antigen load and antibody response. This is possible but, while it cannot be excluded, its likelihood has been somewhat reduced by the following facts:

(1) Although 10 to 15% of the children in the high-endemic areas and within the ages at risk are seronegative, all African histologically confirmed BL cases are seropositive and most of them have high antibody titers. If the causation of the disease is entirely unrelated to EBV, at least some seronegative cases would be expected.

(2) Other lymphomas and other malignant diseases of the lymphoreticular system would be expected to show also a high serological reactivity. However, this is by no means the case (HENLE *et al.*, 1969; JOHANSSON *et al.*, 1970; OLD *et al.*, 1968; DE SCHRYVER *et al.*, 1969). Within the Hodgkin's disease group, for example, it is the sarcomatous type, poorest in lymphoid elements, that shows a high EBV-associated serological reactivity, whereas the lymphocyte-rich paragranuloma is similar to the controls (JOHANSSON *et al.*, 1970).

(3) In NPC, other tumors of the same or closely adjacent anatomical regions should show a similar serological pattern, if the passenger hypothesis were correct. Among a variety of tumors localized to the nasopharynx showing low anti-MA and anti-VCA reactivity were reticulum cell sarcoma, Hodgkin's disease, craniopharyngioma, salivary gland tumor, etc. (DE SCHRYVER *et al.*, 1969). Hypo- and oropharyngeal carcinomas did

not show a high EBV-associated serological pattern either (HENLE *et al.,* 1970a; DE SCHRYVER *et al.,* 1969). In contrast, Chinese, American, Swedish, African, and French cases of NPC all had high EBV-associated reactions in the anti-VCA, anti-MA and the immunoprecipitin tests (HENLE *et al.,* 1970a; OLD *et al.,* 1968; DE SCHRYVER *et al.,* 1969). Serological uniformity, uninfluenced by geographic distance, is one of the few criteria that may be usefully applied when trying to distinguish between haphazard agents and viruses more intimately associated with certain tumors.

A difference in the intimacy of the EBV-tumor association in BL and NPC as contrasted to other neoplasms arising in EBV-seropositive patients, is also indicated by recent nucleic acid hybridization studies (ZUR HAUSEN *et al.,* 1970). BL and NPC biopsies contained DNA that hybridized specifically with purified EBV-DNA, whereas tumors of other kinds, arising in EBV-seropositive patients, contained no detectable hybridizable DNA. In 13 Burkitt biopsies, the approximate number of EBV-genome equivalents per cell varied between 2 and 26. In 10 NPC biopsies, the number of EBV-genome equivalents varied between 1 and 19. Interestingly, 3 BL patients from whom double biopsies were taken showed closely similar genome equivalent values in their geometrically distant tumors (2-2, 7-8, and 21-25, respectively, in the 3 cases). This is reminiscent of the different and characteristic numbers of SV40 genome copies in different clones of SV40 transformed cells (WESTPHAL and DULBECCO, 1968).

While the nucleic acid hybridization studies merely confirm that EBV is more closely associated with BL and NPC than with a number of other tumors, this evidence is at least consistent with the behavior of known oncogenic DNA virus systems.

It may be recalled in this connection that the agents of at least three neoplastic diseases, Marek's neurolymphomatosis in the chicken (CHURCHILL, 1968), Lucke's carcinoma in the frog (MIZELL *et al.,* 1964), and a simian lymphoma (HUNT *et al.,* 1970) were recently identified as herpes-type viruses.

References

Benjamin, T. L., and M. M. Burger (1970). Absence of a cell membrane alteration function in non-transforming mutants of polyoma virus. *Proc. Nat. Acad. Sci. U.S.A., 67*:929–934.

Burger, M. M. (1969). A difference in the architecture of the surface membrane of normal and virally transformed cells. *Proc. Nat. Acad. Sci. U.S.A., 62*:994–1001.

Burkitt, D. (1963). A lymphoma syndrome in tropical Africa. *Int. Rev. Exp. Pathol., 2*:67–138.

—— (1967). Chemotherapy of jaw tumors, pp. 94–101. In *Treatment of Burkitt's Tumor,* J. H. Burchenal, (ed.), UICC Monograph Series, vol. 8. Heidelberg: Springer-Verlag.

Churchill, A. E. (1968). Herpes-type virus isolated in cell cultures from tumors of chickens with Marek's disease. I. Studies in cell culture. *J. Nat. Cancer Inst.*, *41*:939–950.

Clifford, P. (1966). Further studies in the treatment of Burkitt's lymphoma. *E. African Med. J.*, *43*:179–199.

Clifford, P., N. Gripenberg, E. Klein, E. M. Fenyö, and G. Manolov (1968). Treatment of Burkitt's lymphoma. *Lancet, ii*:517–518.

Deichmann, G. I. (1969). Immunological aspects of carcinogenesis by deoxyribonucleic acid tumor viruses. *Adv. Cancer Res.*, *12*:101–136.

de Schryver, A., S. Friberg, G. Klein, W. Henle, G. Henle, G. de Thé, P. Clifford, and H. C. Ho (1969). Epstein-Barr virus associated antibody patterns in carcinoma of the post-nasal space. *Clin. & Exp. Immunol.*, *5*:443–459.

Eskeland, T., and E. Klein (1971). Isolation of 7S IgM and kappa chains from the surface membrane of tissue culture cells derived from a Burkitt lymphoma. *Immunology*, in press.

Fialkow, P. J., G. Klein, S. M. Gartler, and P. Clifford (1970). Clonal origin for individual Burkitt tumors. *Lancet, ii*:384–386.

Gergely, L., G. Klein, and I. Ernberg (1971a). Appearance of EBV-associated antigens in infected Raji cells. *Virology*, in press.

—— (1971b). Host cell macromolecular synthesis in cells containing EBV induced early antigens, studied by combined immunofluorescence and radioautography. *Virology*, in press.

Girardi, A. J. (1965). Prevention of SV40 virus oncogenesis in hamsters. I. Tumor resistance induced by human cells transformed by SV40. *Proc. Nat. Acad. Sci. U.S.A.*, *54*:445–451.

Gunvén, P., G. Klein, G. Henle, W. Henle, and P. Clifford (1970). Antibodies to EBV-associated membrane and viral capsid antigens in Burkitt lymphoma patients. Nature, Lond., *228*:1053–1056.

Hellström, K. E. (1961). Studies on the mechanism of isoantigenic variant formation in heterozygous mouse tumors. II. Behavior of H-2 antigens D and K: cytotoxic tests on mouse lymphomas. *J. Nat. Cancer Inst.*, *27*: 1095–1105.

Henle, G., and W. Henle (1966). Immunofluorescence in cells derived from Burkitt's lymphoma. *J. Bacteriol.*, *91*:1248–1256.

Henle, W., and G. Henle (1970). Evidence for a relation of Epstein-Barr virus to Burkitt's lymphoma and nasopharyngeal carcinoma. *Bibl. Haemat.*, *36*:706–713.

Henle, G., W. Henle, P. Clifford, V. Diehl, G. W. Kafuko, B. G. Kirya, G. Klein, R. H. Morrow, G. M. R. Manube, P. Pike, P. M. Tukei, and J. L. Ziegler (1969). Antibodies to Epstein-Barr virus in Burkitt's lymphoma and control groups. *J. Nat. Cancer Inst.*, *43*:1147–1158.

Henle, G., W. Henle, and V. Diehl (1968). Relation of Burkitt's tumor-associated herpes-type virus to infectious mononucleosis. *Proc. Nat. Acad. Sci. U.S.A.*, *59*:94–101.

Henle, W., G. Henle, H. C. Ho, P. Burtin, Y. Cachin, P. Clifford, A. de Schryver, G. de Thé, V. Diehl, and G. Klein (1970a). Antibodies to Epstein-Barr virus in nasopharyngeal carcinoma, other head and neck neoplasms, and control groups. *J. Nat. Cancer Inst.*, *44*:225–231.

Henle, W., G. Genle, B. A. Zajec, G. Pearson, R. Waubke, and M. Scriba (1970b). Differential reactivity of human sera with early antigens induced by Epstein-Barr virus. *Science, 169*:188–190.

Hunt, R. D., L. V. Meléndez, N. W. King, C. E. Gilmore, M. D. Daniel, M. E. Williamson, and T. C. Jones (1970). Morphology of disease with features of malignant lymphoma in marmosets and owl monkeys inoculated with herpes virus saimiri. *J. Nat. Cancer Inst., 44*:447–465.

Inbar, M., and L. Sachs (1969). Interaction of the carbohydrate binding protein concanavalin A with normal and transformed cells. *Proc. Nat. Acad. Sci. U.S.A., 63*:1418–1425.

Johansson, B., and E. Klein (1970). Cell surface localized IgM-kappa immunoglobulin reactivity in a case of chronic lymphocytic leukemia. *Clin. & Exp. Immunol., 6*:421–428.

Johansson, B., G. Klein, W. Henle, and G. Henle (1970). Epstein-Barr virus (EBV)-associated antibody patterns in malignant lymphoma and leukemia. I. Hodgkin's disease. *Int. J. Cancer, 6*:450–462.

Klein, G. (1961). Studies on the mechanism isoantigenic variant formation in mouse tumors. I. Behavior of H-2 antigens D and K: quantitative absorption tests on mouse sarcomas. *J. Nat. Cancer Inst., 27*:1069–1093.

Klein, G., P. Clifford, G. Henle, W. Henle, L. J. Old, and L. Geering (1969). EBV-associated serological patterns in a Burkitt lymphoma patient during regression and recurrence. *Int. J. Cancer, 4*:416–421.

Klein, E., P. Clifford, G. Klein, and C. A. Hamberger (1967a). Further studies on the membrane immunofluorescence reaction of Burkitt lymphoma cells. *Int. J. Cancer, 2*:27–36.

Klein, G., P. Clifford, E. Klein, R. T. Smith, J. Minowada, F. M. Kourilsky, and J. H. Burchenal (1967b). Membrane immunofluorescence reactions of Burkitt lymphoma cells from biopsy specimens and tissue cultures. *J. Nat. Cancer Inst., 39*:1027–1044.

Klein, G., P. Clifford, E. Klein, and J. Stjernswärd (1966). Search for tumor specific immune reactions in Burkitt lymphoma patients by the membrane immunofluorescence reaction. *Proc. Nat. Acad. Sci. U.S.A., 55*:1628–1635.

—— (1967c). Search for tumor specific immune reactions in Burkitt lymphoma patients by the membrane immunofluorescence reaction, pp. 209–232. In *Treatment of Burkitt's Tumor*, J. H. Burchenal (ed.). UICC Monograph Series, vol. 8. Heidelberg: Springer-Verlag.

Klein, G., L. Gergely, and G. Goldstein (1971). Two-color immunofluorescence studies on EBV-determined antigens. *Clin. & Exp. Immunol., 8*:593–602.

Klein, E., and G. Klein (1964). Antigenic properties of lymphomas induced by the Moloney agent. *J. Nat. Cancer Inst., 32*:547–568.

Klein, G., E. Klein, and P. Clifford (1967d). Search for defenses in Burkitt lymphoma: membrane immunofluorescence tests on biopsies and tissue culture lines. *Cancer Res., 27*:2510–2520.

Klein, E., G. Klein, and K. E. Hellström (1960). Further studies on isoantigenic variation in mouse carcinomas and sarcomas. *J. Nat. Cancer Inst., 25*:271–294.

Klein, E., G. Klein, J. J. Nadkarni, J. S. Nadkarni, H. Wigzell, and P. Clifford

(1968a). Surface IgM-kappa specificity on a Burkitt lymphoma cell *in vivo* and in derived culture lines. *Cancer Res., 28*:1300–1310.

Klein, G., G. Pearson, J. S. Nadkarni, J. J. Nadkarni, E. Klein, G. Henle, W. Henle, and P. Clifford (1968b). Relation between Epstein-Barr viral and cell membrane immunofluorescence of Burkitt tumor cells. I. Dependence of cell membrane immunofluorescence on presence of EB virus. *J. Exp. Med., 128*:1011–1020.

Luria, S. E. (1962). Bacteriophage genes and bacterial functions. *Science, 136*:685–692.

Meyer, M. G., F. Birg, and M. H. Bonneau (1969). Cinetique de l'antigène de membrane dans le système virus polyome-hamster. *C.R. Acad. Sci.,* Paris, *268*:2848–2849.

Mizell, M., I. Toplin, and J. J. Isaacs (1969). Tumor induction in developing frog kidneys by a zonal centrifuge purified fraction of the frog herpes-type virus. *Science, 165*:1134–1137.

Nadkarni, J. S., J. J. Nadkarni, P. Clifford, G. Manolov, E. M. Fenyö, and E. Klein (1969). Characteristics of new cell lines derived from Burkitt lymphomas. *Cancer, 23*:64–79.

Nadkarni, J. S., J. J. Nadkarni, G. Klein, W. Henle, G. Henle, and P. Clifford (1970). EB viral antigens in Burkitt tumor biopsies and early cultures. *Int. J. Cancer, 6*:10–17.

Ngu, V. A. (1965). The African lymphoma (Burkitt tumors): Survivals exceeding two years. *Brit. J. Cancer, 19*:101–107.

Niederman, J. C., A. S. Evans, L. Subrahmanyan, and R. W. McCollum (1970). Prevalence, incidence and persistence of EB-virus antibodies in young adults. *New Eng. J. Med., 282*:361–365.

Old, L. J., E. A. Boyse, G. Geering, and H. F. Oettgen (1968). Serological approaches to the study of cancer in animals and in man. *Cancer Res., 28*:1288–1299.

Pearson, G., F. Dewey, G. Klein, G. Henle, and W. Henle (1970). Correlation between antibodies to Epstein-Barr virus (EBV)-induced membrane antigens and neutralization of EBV infectivity. *J. Nat. Cancer Inst., 45*:989–997.

Pearson, G., G. Klein, G. Henle, W. Henle, and P. Clifford (1969). Relation between Epstein-Barr viral and cell membrane immunofluorescence in Burkitt tumor cells. IV. Differentiation between antibodies responsible for membrane and viral immunofluorescence. *J. Exp. Med., 129*:707–718.

Rapp, F., J. S. Butal, L. A. Feldman, T. Kitahara, and J. L. Melnick (1965). Differential effects of inhibitors on the steps leading to the formation of SV40 tumor and viral antigens. *J. Exp. Med., 121*:935–944.

Roizman, B., and S. B. Spring (1967). Alteration in immunologic specificity of cells infected with cytolytic viruses, pp. 85–197. In *Proceedings of the Conference on Cross Reacting Antigens and Neoantigens.* Baltimore: Williams and Wilkins Co.

Silvestre, D., F. Kourilsky, J. P. Levy, and G. Klein (1971). Relationship between the EBV-associated membrane antigen on Burkitt lymphoma cells and the viral envelope, demonstrated by immuniferritin labeling. *Int. J. Cancer,* in press.

Singhal, S. K., and H. Wigzell (1971). Cognition and recognition of antigen by cell associated receptors. *Progr. Allergy,* in press.

Sjögren, H. O. (1964). Studies on specific transplantation resistance to polyoma-virus-induced tumors. IV. Stability of the polyoma cell antigen. *J. Nat. Cancer Inst., 32*:661–666.

—— (1965). Transplantation methods as a tool for detection of tumor specific antigens. *Progr. Exp. Tumor Res., 6*:289–322.

Svedmyr, A., A. Demissie, G. Klein, and P. Clifford (1970). Antibody patterns in different human sera against intracellular and membrane antigens and neutralization of EBV infectivity. *J. Nat. Cancer Inst., 44*:595–610.

Unanue, E. R., H. M. Grey, E. Rabellino, P. Campbell, and J. Schmidtke (1971). Immunoglobulins on the surface of lymphocytes. II. The bone marrow as the main source of lymphocytes with detectable surface-bound immunoglobulin. *J. Exp. Med., 6*:1188–1198.

van Furth, R., E. Klein, J. J. Nadkarni, and J. S. Nadkarni. Personal communication.

Westphal, H., and R. Dulbecco (1968). Viral DNA in polyoma- and SV40-transformed cell lines. *Proc. Nat. Acad. Sci. U.S.A., 59*:1158–1165.

zur Hausen, H., H. Schulte-Holthausen, G. Klein, W. Henle, G. Henle, P. Clifford, and L. Santesson (1970). EBV-DNA in biopsies of Burkitt tumors and anaplastic carcinomas of the nasopharynx. *Nature, Lond., 228*:1056–1058.

CROSS-REACTIVITY BETWEEN SV40-TRANSFORMED CELL SURFACE ANTIGEN AND EARLY EMBRYO ANTIGEN*

P. Koldovsky, V. Sawicki, and H. Koprowski

The Wistar Institute, Philadelphia, Pennsylvania 19104

I. Introduction

Comparisons from various points of view of malignant and embronic tissues have appeared for more than half a century. Thus it is not surprising that attempts to find cross-reacting antigens between these two tissues have been proceeding during the same period of time. These cross-reacting antigens were found to differ in many ways: *i.e.,* in methods of identification, localization in the cell, biological properties, origin of tumor or embryonic tissue, and stage of embryonic development at which they can be detected. Presently, these carcino-embryonic antigens (CEA) can be divided into three groups. Groups 1 and 2 differ biochemically, and are detectable by various serological reactions, but the immunity against them does not affect the viability of tumor or embryonic cells. Group 3 is characterized by production of cytotoxic antibodies and/or transplantation type of reaction.

In the first group belongs an antigen found in mouse hepatomas by Abelev (1963), which was defined as α-globulin. This antigen was also found in new-born mice. Immunization with this antigen has no effect on growth of hepatomas. Similar antigen was found in human liver carcinomas and liver disorders (Masopust *et al.,* 1968).

The second antigen was described as CEA of the digestive tract by Gold and Freedman (1965). It is present in all malignant tumors of the endodermally derived epithelium of the GI tract and in fetal gut, liver and pancreas at 2 to 6 months of gestation. No clinical evidence of the relationship between antibody against CEA and cause of disease was observed. These two antigens differ in physicochemical properties, histogenesis, and duration during embryogenesis. They are similar in that they are

* The authors express their gratitude for the expert technical assistance of Mrs. Marguarite Solomon.

excreted products of cell rather than direct components of the cell membrane.

The third type of CEA is less clearly defined. It can be detected by the transplantation reaction or by reactions detecting cell surface antigens. FURUSAWA *et al.* (1965) showed that guinea pigs immunized against Ehrlich ascites tumor agglutinate Ehrlich tumor cells and erythrocytes of 14-day-old mice embryos in high titers, but erythrocytes of adult mice are agglutinated only in low titers or not at all. SEDALLIAN and JACOB (1967) immunized guinea pigs and rabbits with mouse embryos. The resulting sera were divided into three categories: (1) those which only agglutinate Ehrlich ascitic cells even in the presence of complement; (2) those which exhibit complement dependence of cytotoxic effect; and (34) those which have a direct cytolytic effect which is complement-dependent. The most virogous sera were obtained after immunization with embryos less than 12 days old. immunization with embryos less than 12 days old.

Even though human tumor tissue can grow on immunologically (cortisone-treated) suppressed animals (TOOLAN, 1951, 1957), the animals will produce antibodies with an inhibitory effect (TOOLAN, 1958). Preimmunization with the same tumor can prevent its growth in cortisone-treated animals (BUTTLE *et al.*, 1962). One of the established Toolan tumors (HS_1) can be inhibited by various human tumor tissues, but not by immunization with serum or spleen of adult persons. Adult muscle inhibits slightly the growth of the tumor. Inhibition equal to that produced by malignant tissues was achieved with placenta, fetal muscle, and fetal spleen. Women who repeatedly abort have in their serum an inhibitory factor against Toolan's tumor.

In experiments with non-inbred rat and mouse tumors and in preimmunization with rat or mouse embryonic tissue, respectively, some protective effect against various tumors was achieved. When genetically pure lines of rats and mice were used in similar experiments, however, no effect on tumor growth was observed (BUTTLE *et al.*, 1964; BUTTLE and FRAYN, 1967). More recently, BLAIR (1970) failed to find any cross-reacting antigens between virus-induced mammary tumors and mouse embryonic tissue. TING (1968) failed to induce transplantation resistance against polyoma tumor in mice immunized with syngeneic embryonic tissue. On the other hand, PEARSON and FREEMAN (1968) found cross-reactivity between antigens present in polyoma-transformed hamster cells and in hamster cells from a 12-day-old embryo. COGGIN *et al.* (1970) induced immunity against SV40-induced tumors in hamsters using either hamster or mouse fetal tissue. DUFF and RAPP (1970) found that sera from pregnant hamsters gave a positive indirect membrane immunofluorescence reaction with SV40-transformer hamster cells.

Certainly the most interesting question is whether CEA of the transplantation type is present in some human tumors and, if so, whether the patient is capable of an immunologic reaction against such antigen.

It seems that the CEA described by Gold (GOLD and FREEDMAN, 1965) in tumors of the digestive tract and these antigens are two separate components. This hypothesis is supported by the findings of HELLSTRÖM *et al.* (1970), who found that blood lymphocytes from human patients with adenocarcinoma of the colon inhibited colony formation of cells from colon carcinomas, fetal gut and liver epithelial cells, but not of cells from adult colon mucosa.

In the experiments which we would like to discuss today, we compared the antigenicity of unfertilized mouse eggs with various somatic cells of normal and malignant mouse, human, monkey, and hamster origin. Heterologous sera were prepared in guinea pigs, and cytolytic, cytotoxic, and growth inhibition tests were performed.

II. Results

The serum produced in guinea pigs against C57BL/6 eggs (referred to hereafter as AE serum) was cytotoxic for unfertilized eggs obtained from either BALB/c, C57BL/6 or Swiss mice. When titrated against C57BL/6 eggs, the end point of the cytotoxic titer was over 1:100. No cytotoxic reaction was observed when either rat or Syrian hamster unfertilized eggs were exposed to AE serum.

The cytotoxic reaction was apparently complement-dependent because it was abolished after inactivation of the serum by heat, but restored by the addition of fresh guinea pig serum.

AE serum was not cytotoxic for methylcholanthrene tumors of two strains of mice; nor was it cytotoxic for 3T3 cells originating from mouse embryo. No cytotoxicity was observed in the reactions between AE serum and lymph node cells obtained from normal adult C57BL/6 mice.

In contrast, AE serum showed the highest cytotoxic activity against SV40-transformed mouse cells (S15 and PF-1), somewhat less activity against Moloney leukemia virus-infected cells (YAC) and even less against adenovirus 12- (E1) and polyoma virus-transformed (Py-A1/n) cells (Table 1).

Table 1. Cytotoxic Effect of AE Guinea Pig Serum on Mouse Cells of Various Origins

| Serum | Highest dilution of serum showing cytotoxic effect against at least 50% of cells | | | | | |
	PF-1	S15	E1	YAC	MR8	PyAv/n
Immune	1:96	1:96	1:12	1:24	<1:3	1:12
Control	<1:3	<1:3	<1:3	<1:3	<1:3	<1:3

The results in both the cytotoxicity and the colony inhibition test (Table 2) indicate that only the SV40-transformed mouse cells (PF-1) were susceptible to the cytotoxic effect of AE serum, whereas SV40-transformed cells of hamster, monkey, or human origin were not.

Table 2. The Effect of AE Guinea Pig Serum on SV40-Transformed Cells of Various Species Origin

Cells			Assays	
Origin	Type	Serum	Colony Inhibition†	Cytotoxicity†
Mouse	PF-1	I	3, 0, 9	1:96
		C	182, 185, 200	<1:3
Hamster	BTH	I	63, 67, 91	<1:3
		C	87, 53, 60	<1:3
Monkey	cl 2A-1	I	43, 37, 55	<1:3
		C	48, 39, 29	<1:3
Human	W126Va₁	I	N.T.	<1:3
		C	N.T.	<1:3

* Number of colonies growing out of 300 cells seeded, for each of three petri dishes.

† Highest dilution of serum showing cytotoxic effect against at least 50% of cells.

NT: Not tested.

AE sera were absorbed with cells of various origin and tested for their cytotoxic effect on mouse eggs. Only after absorption with the SV40-transformed S15 cells was the cytotoxicity titer of AE serum for eggs considerably reduced (Table 3). Absorption of the AE serum with cells transformed by either polyoma or adenovirus had no effect on the cytotoxicity of the serum for eggs. Equally without effect was absorption of AE serum with Moloney virus-infected cells, malignant mouse cells induced by chemical carcinogen (MC57G), normal thymus cells or SV40-transformed hamster cells (BTH).

We also prepared antisera in guinea pigs against SV40-transformed PF-1. The anti-PF-1 guinea pig serum was absorbed either with PF-1 cells or with normal spleen cells and tested for its cytotoxic effect on cells. The results (Table 4) indicate that absorption of anti-PF-1 serum with PF-1 cells completely eliminates its cytotoxic effect for any of the cells tested. In contrast, absorption of anti-PF-1 serum with mouse spleen cells does not reduce the cytotoxic effect of the serum for any of the cells tested, with the exception of spleen cells.

On the other hand, sera produced in guinea pigs against unfertilized single-cell mouse eggs were cytotoxic for mouse eggs and PF-1 cells, serum

Table 3. Cytotoxicity of AE Serum for Mouse Eggs after Absorption of the Serum with Cells of Various Origins

AE serum absorbed† with cells	Ratio of eggs destroyed after exposure to AE serum in dilutions:				
	1:10	1:30	1:60	1:100	1:200
None	10/10	10/10	10/10	10/10	10/10
S15	10/10	9/10	0/10	0/10	—
BTH	—	—	8/9	—	9/9
PyAv/n	—	—	8/9	—	9/9
YAC	—	—	9/9	—	9/9
E1	—	—	9/9	—	9/9
G1	—	—	8/9	—	9/9
MC57G	10/10	9/10	10/10	7/10	—
Thymus (C57BL/6)	—	10/10	10/10	6/10	—
Absorbed with various mouse organs†	25/31	—	—	27/29	—

* The sera were absorbed in dilution 1:10 with 10^7 cells per 0.4 ml.

† This serum was absorbed by Dr. L. J. Old with various organs of inbred mice representing all major H-2 locus-controlled antigens.

against PF-1 cells was cytotoxic for PF-1 cells and mouse eggs, while sera against normal mouse spleen cells were cytotoxic for spleen cells and PF-1 cells only (Table 5).

These results indicate that the anti-C57BL/6 spleen serum, which showed much higher cytotoxicity for C57BL/6 spleen cells than for the SV40-transformed PF-1 cells, had no effect on mouse unfertilized eggs. In contrast, the anti-PF-1 serum, which in dilutions of 1:48 and 1:24 was cytotoxic for PF-1 cells and normal spleen cells, respectively, was cytotoxic for mouse eggs and dilutions higher than 1:100.

Table 4. Cytotoxic Effect of Anti-PF-1 Serum Absorbed with PF-1 or Mouse Spleen Cells on SV40-Transformed Cells

Anti-PF-1 serum absorbed with cells	Highest dilution of serum showing cytotoxic effect against at least 50% of cells				
	PF-1	BTH	cl 2A1	Mouse spleen	Mouse egg
None	1:48	1:12	1:24	1:96	>1:100
C57 spleen	1:48	1:12	1:24	<1:10	>1:100
PF-1	<1:10	<1:10	<1:10	<1:10	<1:10
Normal guinea pig serum	<1:3	<1:3	<1:3	<1:3	<1:3

Table 5. Comparative Effects of AE, Anti-PF-1, and
Antimouse-Spleen Guinea Pig Sera on EGG,
PF-1 and Spleen Cells

| | | | Highest dilution of serum showing cytotoxic effect against at least 50% of cells | |
| | | | | |
Serum	PF-1	Spleen	Mouse eggs	Hamster eggs
Antispleen	1:24	1:512	<1:3	N.T.
Anti-egg	1:96	<1:10	>1:120	<1:3
Anti-PF-1	1:48	1:24	>1:100	N.T.

N.T.: Not tested.

The AE serum, which was not cytotoxic for spleen cells, was highly cytotoxic for PF-1 cells. Interestingly, the AE serum showed no cytotoxic effect on hamster eggs.

III. Discussion

The results obtained indicate that unfertilized single-cell mouse eggs show the presence of an antigen which induces in guinea pigs formation of an antibody cytotoxic not only for eggs of the donor strain but also for eggs obtained from a strain of mice which differs from the donor strain at the H-2 locus. The anti-egg antibody was cytotoxic neither for rat nor hamster unfertilized single-cell eggs and may, therefore, be considered as species-specific. Furthermore, the egg antigen is probably not in H-2 antigen since serum containing H-2B antibody, which destroyed allogeneic and C57BL/6 lymph node cells, was not cytotoxic for C57BL/6 eggs. This finding seems to confirm the observation of HEYNER *et al.* (1969) who found no evidence for the presence of H-2 antigens in eight-cell embryos obtained from fertilized mice.

What is, perhaps, most interesting was the cross-reactivity of AE serum with the SV40-transformed mouse PF-1, 3T3-SV, and S15 cells.

In contrast, AE serum was not cytotoxic for hamster, human, or monkey cells transformed by SV40. It seems, therefore, that the antigen which the mouse eggs have in common with the SV40-transformed cells is species-specific. This antigen is apparently unrelated either to the structural or the functional proteins of the virus itself, or to any "new" cellular antigens induced by infection with SV40. The species specificity of the reaction could be checked easily in experiments on cross-reactivity between immune serum produced against eggs of species other than mice and SV40-transformed cells of homologous species.

The ability of the hamster to react against SV40-transformed syngeneic cells after immunization with hamster embryo cells (FURUSAWA et al., 1965) renders this system different from the SV40-mouse egg system. The immunogenic reactivity of the hamster may, perhaps, be explained by the fact that 9- and 12-day-old hamster embryos were used for immunization. Old mouse embryos might also differ immunogenically and immunosensitively from single-cell unfertilized eggs. In the light of the negative results obtained in cross-reactivity tests between AE serum and adult mouse tissue (spleen) on the one hand, and antispleen serum and mouse eggs on the other, it will be interesting to determine when in the prenatal or postnatal period the reactivity of the tissues against AE serum becomes suppressed.

Taking into consideration the fact that: (1) AE serum is not toxic for SV40-transformed cells of other species; (2) guinea pig anti-PF-1 (SV40-transformed cells) serum is cytotoxic for SV40-transformed cells of *other* species in addition to the mouse; and (3) absorption of guinea pig anti-PF-1 serum with normal spleen cells does not affect its cytotoxicity for SV40-transformed cells, we may postulate the existence of at least two types of antigens in SV40-transformed cells, as was suggested earlier (DUFF and RAPP, 1970), one coded directly by the SV40 genome which is the true TSTA and which is responsible for species cross-reaction and the second, the embryo antigen itself, as a result of derepression of the cell genome by SV40.

References

Abelev, G. I., S. D. Perova, N. I. Khramkova, Z. A. Postnikova, and I. S. Irlin (1963). Production of embryonal α-globulin by transplantable mouse hepatomas. *Transplantation, 1*:174–180.

Blair, P. B. (1970). Search for cross-reacting antigenicity between mammary tumor virus-induced mammary tumors and embryonic antigens: Effect of immunization on development of spontaneous mammary tumors. *Cancer Res., 30*:1199–1202.

Buttle, G. A. H., J. L. Eperon, and E. Kovacs (1962). An antigen in malignant and in embryonic tissues. *Nature, Lond., 194*:780.

Buttle, G. A. H., J. L. Eperon, and D. N. Menzies (1964). Induced tumor resistance in rats. *Lancet, ii*:12–14.

Buttle, G. A. H., and A. Frayn (1967). Effect of previous injection of homologous embryonic tissue on the growth of certain transplantable mouse tumors. *Nature, Lond., 215*:1495–1497.

Coggin, J. H., K. R. Ambrose, and N. G. Anderson (1970). Fetal antigen capable of inducing transplantation immunity against SV40 hamster tumor cells. *J. Immunol., 105*:524–526.

Duff, R., and F. Rapp (1970). Reaction of serum from pregnant hamsters with surface of cells transformed by SV40. *J. Immunol., 105*:521–523.

Furusawa, M., M. Kotani, H. Takeuchi, and S. Asayama (1965). Some antigen similarities between mouse erythrocytes and Ehrlich ascites tumour cells. *Nature, Lond., 207*:1204.

Gold, P., and S. O. Freedman (1965). Specific carcinoembryonic antigens of the human digestive system. *J. Exp. Med., 122*:467–481.

Hellström, I., K. E. Hellström, and T. H. Shepard (1970). Cell mediated immunity against antigens common to human colonic carcinomas and fetal gut epithelium. *Int. J. Cancer, 6*:346–351.

Heyner, S., R. L. Brinster, and J. Palm (1969). Effect of iso-antibody on pre-implantation mouse embryos. *Nature, Lond., 222*:783–784.

Masopust, J., K. Kithier, J. Rádl, J. Koutecký, and L. Kotál (1968). Occurrence of fetoprotein in patients with neoplasms and non-neoplastic diseases. *Int. J. Cancer, 3*:364–373.

Pearson, G., and G. Freeman (1968). Evidence suggesting a relationship between polyoma virus-induced transplantation antigen and normal embryonic antigen. *Cancer Res., 28*:1665–1673.

Sedallion, J. P., and G. Jacob (1967). Cytotoxic and cytolytic effect of mouse anti-embryo sera on Ehrlich ascites cells. *Nature, Lond., 215*:156–157.

Teller, M. N., P. C. Merker, J. E. Palm, and G. W. Woolley (1958). The human tumor in cancer chemotherapy in the conditioned rat. *Ann. N.Y. Acad. Sci., 76*:742–751.

Ting, R. C. (1968). Failure to induce transplantation resistance against polyoma tumor cells with syngeneic embryonic tissues. *Nature, Lond., 217*:858–859.

Toolan, H. W. (1951). Successful subcutaneous growth and transplantation of human tumors in X-irradiated laboratory animals. *Proc. Soc. Exp. Biol. Med., 77*:572–578.

—— (1957). Immunoloical responses elicited in the heterologous host by transplantable human tumors: Use of the conditioned rat as a test medium. *Ann. N.Y. Acad. Sci., 69*:830–841.

—— (1958). The transplantable human tumor. *Ann. N.Y. Acad. Sci., 76*:733–741.

ROLE OF TUMOR ANTIGENS IN THE BIOLOGY OF NEOPLASIA*

RICHMOND T. PREHN

*The Institute for Cancer Research, Philadelphia,
Pennsylvania; Department of Pathology, University
of Pennsylvania, Philadelphia, Pennsylvania*

It is now generally recognized that immunity is an important factor in the biology of neoplasia. Most and perhaps all neoplasms are antigenic in the animal of origin and the evidence supporting the concept of immunologic surveillance as a defense against the development of neoplasia is almost overwhelming. However, the interaction of the immune mechanism with the neoplasm is complex, as is shown most strikingly by the phenomenon of enhancement. Immunization may sometimes lead to enhanced rather than inhibited tumor growth. In view of these and other complexities, the classical view that the homograft reaction is simply a defense mechanism against neoplasia may need reexamination.

Despite the apparent existence of a homograft immunity reaction in the earthworm, it is widely believed that the general capacity for homograft immunity evolved as a peculiarly vertebrate adaptation, or at least not much lower than the vertebrates on the phylogenetic scale. The homograft immunity reaction involves not only the response mechanisms similar to those of delayed hypersensitivity, but an antigenic polymorphism. According to some authorities, neoplasia is also largely a vertebrate condition. Since neoplasms are generally antigenic, they would seem to be analogous to homografts. Because of this apparent conjunction of the two traits, neoplasia and homograft-style immunity, it was natural to propose that the homograft response evolved as a vertebrate defense against neoplastic disease, a type of disease that for unknown reasons is not a serious threat to most lower forms of life (BURNET, 1970; GOOD and FINSTAD, 1969).

This evolutionary view of immunologic surveillance leaves a number of questions unanswered. Why should neoplasia be largely a vertebrate

* Supported by USPHS grants CA-08856, CA-06927, CA-05388, and FR-05539 from the National Cancer Institute, and by an appropriation from the Commonwealth of Pennsylvania.

problem? It is difficult to understand why a seemingly efficient defense in invertebrates should become less efficient and require supplementation by a new mechanism, the immune response, in the vertebrates. Even with the existence of this new mechanism, neoplasia is a leading cause of human death. Furthermore, since some tumors have little or no detectable immunogenicity, why are most of them immunogenic? Strong immunogenicity is apparently not necessary to the neoplastic state, so why is it that most neoplasms invite their own destruction by being immunogenic? These questions suggest to me the probability that the currently prevailing views of the evolutionary impetus for and function of the homograft reaction may need modification.

Let me recapitulate the pertinent facts as they are currently understood:

(1) Most tumors are demonstrably immunogenic.

(2) Little or no immunogenicity is actually required for the neoplastic state.

(3) Invertebrates apparently have little or no neoplasia; they also generally seem to lack a homograft type of reaction.

An unbiased observer, looking at these facts, would probably conclude that the homograft reaction, rather than being a defense against neoplasia, may be a positive contributor to neoplastic development. How better explain the fact that in phylogeny, the absence of the homograft type of immune reaction is associated with an absence of neoplasia? How better explain the fact that tumors are usually immunogenic even though the rare occurrence of nonimmunogenic varieties suggests that there is no intrinsic necessity for this? When one adds the fact that specific immunization seems just as likely to produce enhanced tumor growth as it is inhibition, the concept of the primary role of immunity as a defense against neoplasia seems shakey indeed.

What then is the evidence that supports the surveillance hypothesis— evidence that I have already stated is seemingly overwhelming? This evidence is of several main varieties.

(1) Most tumors are antigenic—an obvious prerequisite for either immunosurveillance or immunostimulation.

(2) Oncogenic agents interfere with immune reactions.

(3) Immunodepressants, both in man and mouse, generally increase the incidence of neoplasia.

(4) Immunostimulants may decrease the incidence of neoplasia.

(5) Neoplasia is most prevalent at the two extremes of the life span when immune capacity is weak.

The data on each of these points are extensive and in aggregate they constitute a quite compelling case (BURNET, 1970).

Thus there is a seeming paradox. On the one hand, the phylogenetic data and the near, but apparently not complete, universality of tumor immunogenicity, suggest that immunity may be an aid to tumor growth. On the other hand, the evidence from immunodepression or stimulation studies suggests a defense or surveillance activity for the immune response. Actually, there is no formal necessity for these seemingly opposite functions to be mutually exclusive. Both could be correct; which of them predominates in a given case at a given time might be a function of a variety of modifying circumstances. Perhaps the most likely hypothesis, and one which has recently been set forth in some detail, is that a little immunity may be tumor-stimulatory, while a greater degree of immune reaction is tumor-inhibitory (PREHN, 1971; PREHN and LAPPÉ, 1971).

If the tumor-immune mechanism has a dual activity, stimulating neoplastic growth at one part of the scale and inhibiting at another, one might anticipate that the surveillance function would not be as efficient as it might otherwise be. Actually, while the evidence for surveillance is, in aggregate, very impressive, it is not conclusive. Each individual piece of evidence consistent with the theory of surveillance could be explained in other ways. Thus the susceptibility of the newborn animal to induced neoplasia may depend upon the general growth potential of the target tissues, rather than on, or perhaps in addition to, the lack of immunologic maturity. Newborn thymectomy, while it appears to increase the susceptibility of the mouse to some types of subsequent oncogenesis, is ineffective in others; in fact, in the mammary tumor system, newborn thymectomy actually lessens tumor incidence. Antilymphocyte serum may also fail to increase tumorigenesis and in at least one system is antioncogenic. It is well established that in humans with natural or induced immunologic deficiency states there is an excessive incidence of reticulum cell sarcomas and related neoplasms. However, these may well be the direct result of damage to the immunologic organ rather than a lack of immunologic surveillance. Some evidence is now accumulating to suggest that kidney transplant patients may additionally suffer from an excess of miscellaneous epithelial tumors, although this evidence is not yet completely convincing (PENN et al., 1971). If true, this would be strong evidence in favor of the normal existence of an immunologic surveillance mechanism, but one cannot help asking whether or not the agents used to produce immunosuppression may not have oncogenic effects via pathways other than, or in addition, to, the immunosuppression they produce. Thus the evidence for surveillance is extensive, but circumstantial.

One of the phenomena that argues most strongly against the existence of a very effective immunologic surveillance is "sneaking through." Conversely, this phenomenon may actually argue in favor of the "low dose" immunostimulation hypothesis. A number of observers have noted that in tumor transplantation an antigenic tumor may fail to grow unless excessively large, or very minute, doses of tumor cells are transplanted. The growth of very small tumor inocula, when somewhat larger are inhibited, has even been demonstrated in previously immunized recipients. If highly antigenic tumor cells can exhibit this "sneaking through" behavior, how can surveillance be very effective against nascent spontaneous tumors that often have very low levels of immunogenicity (PREHN and LAPPÉ, 1971)?

Some observers have argued that the very low levels of immunogenicity usually seen in spontaneous tumors may be the result, by selection, of a very efficient surveillance mechanism. Thus this lack of immunogenicity in most spontaneous tumors is turned into an argument in favor of efficient surveillance rather than an argument against it. According to this reasoning, most spontaneous tumors are really highly immunogenic and are therefore suppressed by the immune surveillance mechanism. The tumors that reach a grossly detectable size are a small minority that, because of low immunogenicity, escaped this defense mechanism. According to this formulation, tumors that arise spontaneously in an immune-deficient environment should therefore be very highly immunogenic.

My colleagues and I have investigated this proposition by examining the immunogenicities of tumors that arise "spontaneously" in cultures of mouse embryo cells or which arise when such cells are grown within the protecting confines of intraperitoneal diffusion chambers. In both situations, almost all of the tumors were found to be virtually nonimmunogenic. We concluded that immunoselection could not be the major reason why most spontaneous tumors possess little demonstrable immunogenicity. Conversely, we found that chemically induced tumors produced in similar immune-free environments were usually highly immunogenic. The high immunogenicity of hydrocarbon-induced tumors is therefore not simply because of a permissive immunodepression by the chemical—although immunodepression undoubtedly helps. Rather, the chemical oncogen apparently imparts a high immunogenic capacity as an intrinsic by-product of the process of neoplastic transformation (PREHN, 1970).

I think that when one looks at all the available data, it is probably safe to conclude that there is a surveillance phenomenon. There are just too many different lines of evidence consistent with this interpretation. The work of TRAININ and LINKER-ISRAELI (1971) is particularly convincing. However, most pieces of evidence concern tumors induced with high dosages of chemicals or with viruses that produce very highly immunogenic tumors. Relatively few such highly immunogenic tumors probably occur in nature, where viral and chemical oncogens are more likely to be present in barely threshold amounts. Under these more natural circumstances immunosurveillance may be a weak and inefficient process.

The possible inefficiency of the surveillance function permits one to seriously consider the hypothesis of "low dose" immunostimulation. As has been pointed out, the two apparently paradoxical functions of immune surveillance and immune stimulation are not mutually exclusive. The arguments in favor of a possible immunostimulation of tumor, some of which have already been mentioned, are (1) the fact that most tumors are immunogenic, although the apparent exceptions suggest there may be no intrinsic necessity for this; (2) the possible correlation in phylogeny between lack of homograft immunity and paucity of tumors; (3) the homograft-stimulating effect of a small degree of immune reactivity; (4) the favorable effects of a little incompatibility between mother and fetus on the course of pregnancy; and (5) the direct evidence from tumor immunity experiments, both *in vivo* and *in vitro*.

All of these points have been discussed in some detail elsewhere but the principal evidence will be briefly reviewed here. This evidence, as is the case with the surveillance hypothesis, is considerable, but at present largely circumstantial. Although alternative explanations can be found for most of the individual items, they constitute, in aggregate, a beguiling case (PREHN, 1971; PREHN and LAPPÉ, 1971).

It is now conceded that most tumors are antigenic in relation to the immunologic mechanism of the animal in which they arise. However, it has also been clear, since the beginnings of modern tumor immunology, that tumor immunogenicity varies greatly from tumor to tumor and is occasionally not detectable. This lack of immunogenicity may not be due to an actual lack of tumor antigens. It has recently become clear that many instances of weak immunogenicity are due to the presence of enhancing antibodies rather than to an antigenic deficiency. However, BALDWIN and EMBLETON (1969) have reported chemically induced rat mammary tumors in which no immune response, humoral or otherwise, was detectable. Whatever the true state of the antigen content, it is apparent that some tumors have little or no detectable immunogenicity. It is also known that tumor cell populations are very variable with regard to immunogenic properties (PREHN, 1970). If this is so, why is it that all tumors do not rapidly evolve, by selection, into the nonimmunogenic variety? Although there is a tendency in this direction, many neoplasms remain highly immunogenic indefinitely. The logical explanation would seem to be that some level of immunogenicity is in some manner beneficial to the tumor cells.

The second argument in favor of immunostimulation is derived from the correlation in phylogeny between the existence of a homograft mechanism and tumor susceptibility. The data supporting this correlation are not entirely convincing. However, the correlation, if such exists, argues in a straightforward manner in favor of the necessity of an immune reaction for tumor formation.

Thirdly, several authors have noted that a mild immune reaction may temporarily stimulate mitosis and function in an allograft (PREHN and LAPPE, 1971). What is true for an allograft should also be true for an immunogenic tumor.

There is also evidence that the natural allograft of pregnancy prospers under a modicum of histoincompatibility and immune reaction. This evidence has been recently summarized and again suggests that the immune reaction can be a stimulatory action upon target cells (PREHN and LAPPÉ, 1971).

Finally, there is scattered evidence from tumor immunity experiments that immunostimulation may exist (PREHN and LAPPÉ, 1971). In this connection, I have reference to data suggesting something more than a simple blocking of cellular immunity by enhancing antibody. These data are still very limited and inconclusive, but if the immunostimulation hypothesis is correct, they should be easily and rapidly amplified. The next few months should tell the story.

I can summarize this paper by noting that in the past 15 years we have come from a viewpoint that tumor immunity is impossible (*horror autotoxicus*) to the belief that it may be the chief defense against neoplasia. It is, in my estimation, now time to be done with both of these extremes and to get on with the business of examining the details of what is obviously a tremendously complicated process. Until these details are thoroughly elucidated, immunotherapy in man will be a frustrating and sometimes even a harmful approach to the problem of cancer.

References

Baldwin, R. W., and M. J. Embleton (1969). Immunology of 2-acetylamino-fluorene-induced rat mammary adenocarcinomas. *Int. J. Cancer, 4*:47–53.

Burnet, F. M. (1970). The concept of immunological surveillance. *Progr. Exp. Tumor Res., 13*:1–27.

Good, R. A., and J. Finstad (1969). Essential relationship between the lymphoid system immunity, and malignancy. *Nat. Cancer Inst. Monogr., 31*:41–58.

Penn, I., C. G. Halgrimson, and T. E. Starzl (1971). *De novo* malignant tumors in organ transplant recipients. *Transplant. Proc., 3*:773–778.

Prehn, R. T. (1970). In: *Immune Surveillance,* R. T. Smith and M. Landy (eds.), New York: Academic Press.

—— (1971). Perspectives on oncogenesis: Does immunity stimulate or inhibit neoplasia? *J. Reticuloendoth. Soc., 10*:1–16.

Prehn, R. T., and M. A. Lappé (1971). An immunostimulation theory of tumor development. *Transplant. Rev., 7*:26–54.

Trainin, N., and M. Linker-Israeli (1971). Increased incidence of spontaneous lung adenomas in mice following neonatal thymectomy. *Israel J. Med. Sci., 7*:36–41.

SOME COMMENTS ON THE ROLE OF BLOCKING SERUM FACTORS IN TOLERANCE TO ALLOGRAFTS*

INGEGERD HELLSTRÖM and KARL ERIK HELLSTRÖM

Departments of Microbiology and Pathology, University of Washington Medical School, Seattle, Washington

As discussed in another paper at this symposium (HELLSTRÖM and HELLSTRÖM, 1972), as well as in a recent review article (HELLSTRÖM and HELLSTRÖM, 1970), there is plenty of evidence to suggest that individuals with progressively growing tumors have a cell-mediated immunity against the tumors' specific antigens and blocking serum factors, which can abrogate tumor destruction by the immune lymphocytes *in vitro*. One may ask whether the blocking phenomenon may serve any useful purpose under more physiological situations. There are several reports indicating that it may be so. "Blocking" effects have been detected in sera from mice bearing the conceptus of an allogeneic mating (HELLSTRÖM *et al.*, 1969), from chimeric dogs (HELLSTRÖM *et al.*, 1970), from human patients into whom foreign kidneys have been successfully transplanted (QUADRACCI *et al.*, 1971), and from rats which have been treated with enhancing antisera and have then undergone allogeneic kidney transplants which have become permanently accepted (STUART *et al.*, 1971). Furthermore, mice rendered tolerant to allografts by neonatal injection of spleen cells have been found to possess a cell-mediated immunity to the alloantigens to which they are "tolerant" *in vivo* and also to blocking serum factors, which can abrogate target cell destruction by the immune lymphocytes (HELLSTRÖM *et al.*, 1971).

Even more interesting data in this respect comes from a recent study, performed in collaboration with Dr. T. G. Wegmann, to test whether there is a coexistence of cell-mediated immunity and blocking serum activity in tetraparental (= allophenic) mice (WEGMANN *et al.*, 1971). Such animals

* The authors' work, on which this summary was based, has been supported by Grants CA-10188 and CA-10189 from the National Institutes of Health, by Contract NIH 69-2061 from the National Institutes of Health and by Grant T-453 from the American Cancer Society.

314

provide a unique experimental model because they are formed from the fusion of two eight-cell-stage embryos, derived from different, usually histoincompatible inbred lines (TARKOWSKI, 1961; MINTZ, 1962). The mice grow to adulthood and have apparently normal longevity, while in general retaining both cell lines in all tissues tested. MINTZ and SILVERS (1967) showed that such mice can permanently accept skin grafts from both parental strains while remaining normally competent to reject third part grafts.

It was studied as to whether tetraparental mice obtained by fusing C3H and C57BL embryos have any cell-mediated immunity directed against C3H and/or C57BL, and whether they have any blocking serum factors capable of abrogating (C3H and C57BL) target cell destruction by immune lymphocytes.

A total of seven technically successful experiments were carried out on a total of four tetraparental mice. Skin or lung fibroblasts from C3H and C57BL mice were used as target cells and plated onto Falcon Microtest plates, on which they were exposed to serum from the tetraparental (or control) mice, followed by lymph node cells from the same animals. The results indicated that lymph node cells from the tetraparental mice can destroy parental strain fibroblasts *in vitro,* as compared to F_1 hybrid and both types of parental strain control lymph node cells. The destruction was indistinguishable from that mediated by previously immunized allogeneic lymph node cells. It could be prevented by serum obtained from the tetraparental mice, but not by F_1 hybrid or parental strain sera. It was concluded that a concomitant immunity and serum blocking, rather than a central failure of the immune response, may mediate some aspects of normal tolerance.

Taken together, recent work indicates that a coexistence of cell-mediated immunity and blocking serum factors may exist under certain conditions not involving neoplasia. Understanding the nature as well as the mechanism of action of the blocking serum factors may have important implications for tissue transplantation. Could the formation of blocking serum factors be directed at will, a specific way of immunosuppression would be available. It is also conceivable that the mechanisms involved in allograft tolerance may explain tolerance to tissue-specific antigens and act as a guard against autoimmunity. This deserves to be further studied.

References

Hellström, K. E., and I. Hellström (1970). Immunological enhancement as studied by cell culture techniques. *Ann. Rev. Microbiol.,* 24:373.

Hellström, K. E., and I. Hellström (1972). Cellular immunity to tumor antigens—possible clinical usefulness of the findings so far obtained. This volume, p. 281.

Hellström, I., K. E. Hellström, and A. C. Allison (1971). Neonatally induced allograft tolerance may be mediated by serum-borne factors. *Nature, Lond., 230*:49.

Hellström, K. E., I. Hellström, and J. Brawn (1969). Abrogation of cellular immunity to antigenically foreign mouse embryonic cells by a serum factor. *Nature, Lond., 224*:914.

Hellström, I., K. E. Hellström, R. Storb, and E. D. Thomas (1970). Colony inhibition of fibroblasts from chimeric dogs mediated by the dogs' own lymphocytes and specifically abrogated by their serum. *Proc. Nat. Acad. U.S.A., 66*:65.

Mintz, B. (1962). Incorporation of nucleic acid and protein precursors by developing mouse eggs. *Am. Zool., 2*:432.

Mintz, B., and W. K. Silvers (1967). "Intrinsic" immunological tolerance in allophenic mice, *Science, 158*:1484.

Quadracci, L. J., I. Hellström, G. E. Striker, T. L. Marchioro, and K. E. Hellström (1971). Immune mechanisms in human recipients of renal allografts. *Cell. Immunol., 1*:561.

Stuart, F. P., F. W. Fitch, D. A. Rowley, J. L. Biesecker, K. E. Hellström, and I. Hellström (1971). Manuscript in preparation.

Tarkowski, A. K. (1961). Mouse chimaeras developed from fused eggs. *Nature, Lond., 190*:857.

Wegmann, T. G., I. Hellström, and K. E. Hellström (1971). Manuscript in preparation.

A CYTOKINETIC FACTOR IN THE SEROMUCOID FRACTION OF SERUM OR ASCITES FLUID IN TUMOR-BEARING HAMSTERS

DEAN F. STEVENS

University of Vermont, Burlington, Vermont

Although a more or less accurate conception of the physical changes occurring during cell division was attained by 1870–1880, the mechanisms that initiate, control and assure successful completion of cell division are still largely unknown. BRACHET (1957) has suggested that study of cell division abnormalities may offer a means for clarifying this problem. Certain characteristic division abnormalities frequently occur in malignant cells. One of these, namely, multinucleation, is readily recognized. It also should be pointed out that multinucleation occurs in several types of normal cells, the most numerous probably being binucleated liver cells. In 1959 a new hamster ascites tumor was developed in which as many as 25% of tumor cells were multinucleated (STEVENS and SCHWENK, 1959). Generally, multinucleation in tumor cells is limited to 1 to 3%.

Studies of hamster ascites multinucleated cells indicated that estrogens caused a decrease in their number either *in vivo* or *in vitro* (STEVENS, 1961a). However, if tumor cells were washed and incubated in artificial culture medium instead of ascites fluid, estrogens did not cause a decrease in the number of multinucleated cells. Subsequently it was found that addition of ascites fluid or serum from cancer-carrying hamsters to the artificial medium restored activity and resulted in a reduction in the number of multinucleated cells (STEVENS, 1961b). Serum from hamsters either undergoing gross wound healing, pregnant, partially hepatectomized, or normal was not effective in decreasing multinucleation when added to the artificial medium.

Multinucleated cells treated with cancer serum and estrogen were observed by phase microscopy. These observations indicated that the reduction in the number of multinucleated cells resulted from cleavage occurring around each nucleus in a multinucleated cell, resulting in mononucleated cells. This occurred without condensation of chromosomes or formation of the usual mitotic apparatus.

These observations suggested to us that the hamster multinucleated cell system might be a useful tool in studying the mechanisms of cell division, especially cytokinesis. Accordingly, we have been studying two questions: (1) What is the mechanism of action of cancer serum and estrogen on multinucleated cells? (2) Can the factor in cancer serum be isolated and identified? This paper deals with the second question.

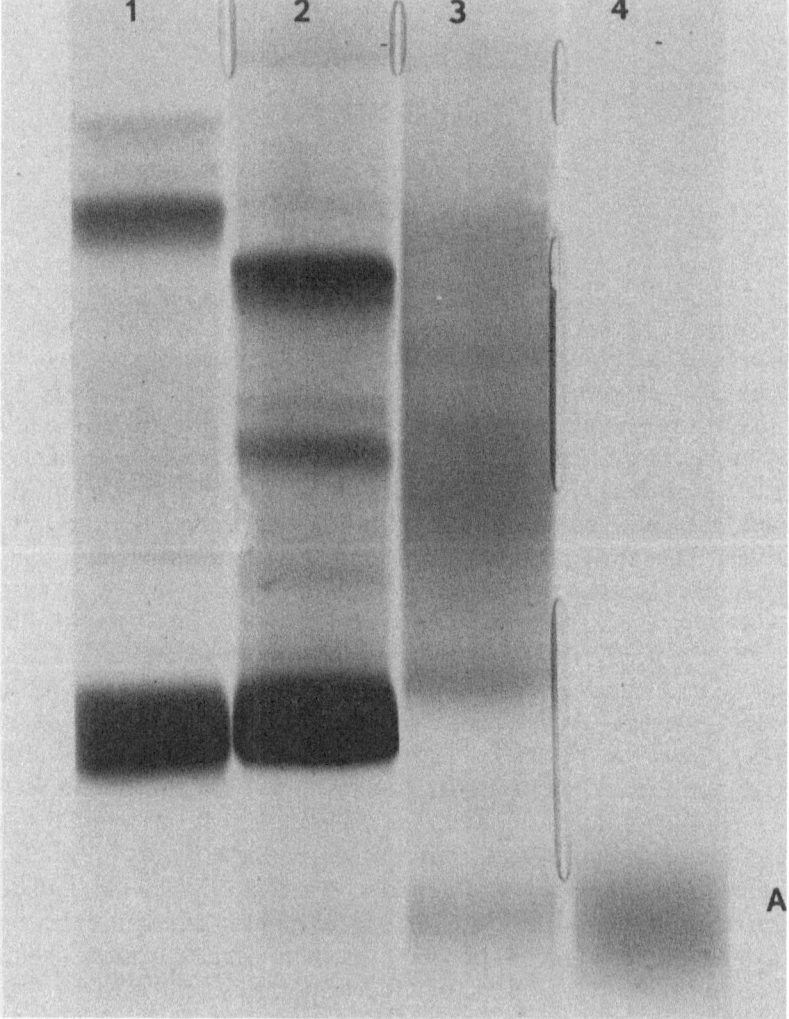

Fig. 1. Disc electrophoresis of (1) total ascites fluid, (2) 65% $(NH_4)_2SO_4$ insoluble, (3) boiled ascites fluid, and (4) pre-albumin band sliced from (3) and rerun. Cytokinetic activity was found only in the bands at point A. Migration is from top to bottom.

Standard protein separation techniques were employed in an attempt to isolate the active factor in cancer serum or ascites fluid.

As shown in Figure 1, the active material always occurred in the pre-albumin band in acrylamide gel electrophoresis and was inactivated by trypsin, carboxypeptidase, as well as by leucine aminopeptidase. Recently a tentative pI of 5.6 and a molecular weight of approximately 40,000 have been established for the factor. A pI of 5.6 ruled out nucleic acid and histone, but did not exclude either a glycoprotein or mucopolysaccharide.

Some literature has implicated serum glycoproteins, especially seromucoids, as being altered in clinical as well as experimental cancer. A number of other pathological conditions also caused increases in seromucoid levels (WINZLER, 1955). Some seromucoids, especially orosomucoid, travel electrophoretically in front of albumin.

On the chance that we might be dealing with a seromucoid, we employed Winzler's technique to isolate seromucoids from hamster cancer serum and ascites fluid. High activity was again obtained in the prealbumin material.

These results suggest that we may be dealing with an orosomucoid or a factor adsorbed to orosomucoid which is specific for malignancy in the golden hamster.

References

Brachet, J. (1957). *Biochemical Cytology*. New York: Academic Press.

Stevens, D. F. (1961a). The *in vivo* and *in vitro* effect of estrone and testosterone on a nuclear abnormality in a hamster ascites tumor. *Exp. Cell Res., 25*:59–64.

—— (1961b). Estrone and a factor in cancer-carrying hamsters. A study on hamster ascites polynucleated tumor cells. *Exp. Cell Res., 25*:654–659.

Stevens, D. F., and E. Schwenk (1959). Amitosis in a new ascites tumor. *Experientia, 15*:470–474.

Winzler, R. J. (1955). Determination of serum glycoproteins. *Methods Biochem. Analysis, 2*:279–311.

SUMMARY OF THE TUMOR
ANTIGENS SESSION

Richmond T. Prehn

*The Institute for Cancer Research and Department of
Pathology, University of Pennsylvania,
Philadelphia, Pennsylvania*

This symposium covered most of the principal features of present-day experimental tumor immunology. Most of the questions and discussion from the floor are adequately dealt with in the published forms of the presentations and need not be reiterated here. Rather, I will use this opportunity to reemphasize what I consider the major trends developed by the speakers and point out a few other items of interest.

The origin of the tumor antigens, *i.e.,* derepression, unmasking, or mutation, was discussed by Dr. Koldovsky. He distinguished carefully between those tumor-associated antigens that are found on the cell surface and which can serve as "tumor-regression antigens" and those found in other portions of the tumor cell. While all changes that characterize the neoplastic cell are of interest to the oncologist, it is natural that the greatest interest should center upon the surface antigens with their potential for influencing tumor behavior. Of particular interest to me is the question of whether or not the diverse array of regression antigens found on the surfaces of hydrocarbon-induced mouse sarcomas represent derepressions of various normal cellular genes. Recently there has been fairly convincing evidence that these tumor regression antigens are indeed present in the normal embryo (BRAWN, 1970). These data, combined with the work discussed by Dr. Koldovsky on SV-40-induced antigens and the work of Phil Gold and his collaborators on the CEA of gut tumors, make tenable the hypothesis that genetic information for most and perhaps all tumor antigens may be coded in the normal cellular genome. Seemingly, such embryonic antigens serve a function, but presumably not as antigens, during normal ontogeny. Normal ontogeny is obviously dependent upon complex interactions among cells. The surface configurations, which are antigenic when found on a tumor cell in an adult animal, very likely are instruments of normal cellular communication and modifiers of cellular behavior in the embryo. Of course, self-tolerance to these antigens would not be expected either to develop or, if developed, to persist if the antigens in question are repressed or masked subsequent to early ontogeny.

One of the most fascinating observations reported at this symposium is that of the Hellströms concerning the immunologic status of allophenic mice. Although these studies are more in the realm of basic immunology than tumor immunology *per se,* the implications for oncology are very great indeed. If the allophenic mouse possesses lymphocytes capable of specifically destroying normal parental strain cells and if this reaction is blocked by allophenic serum, the entire classical concept of immune tolerance is called into question. The usual failure of the animal to exhibit antiself lymphoid reactivity is seen as due to blocking factors (immune complexes and/or enhancing antibody) rather than to a lack of cells capable of responding to autoantigens.

If antiself reactivity is the normal condition, blocked only by serum factors, then one may have to consider carefully the suggestion of Fialkow that an immune reaction can produce neoplastic transformation (FIALKOW, 1967). Perhaps some transient failure in the blocking mechanism may be sufficient to permit the immunologic induction of neoplasia. If my own suggestion that a minimal level of immune reaction may stimulate the growth of tumor cells is correct, then the origin of many cancers might be due to the following sequence of events:

(1) A transient breakdown of the "blocking" mechanism, perhaps as a consequence of viral infection or a diverse multitude of unknown environmental variables.

(2) Chromosomal abnormalities induced by autoimmune reactivity. Associated with this chromosomal damage is neoplastic transformation.

(3) Neoplastic transformation is associated with a plethora of genic derepressions. Some of these result in carcino-embryonic antigens of the "tumor regression" variety.

(4) A low level of immune attack, either against the new regression antigens or a continued reaction against normal antigens because of the postulated defect in blocking mechanisms, could stimulate the transformed cells to overcome the homeostatic inhibitions provided by the normal cell environment (contact inhibition of mitoses, etc.).

(5) With increase in size of the neoplastic clone, antigenic stimulation would increase. The specific immune reactivity of the lymphoid cells against the tumor regression antigens might become sufficient to inhibit further growth or eliminate the clone (immune surveillance).

(6) There will be a tendency of the tumor cell population, presumably by selection, to adjust its antigenic level to just that amount required to keep the immune reaction within the stimulatory range. Surveillance will therefore not be very effective.

(7) The "strength" of the immune reaction will be determined in large measure by the balance between cellular immunity and the serum-blocking factors.

I have recently submitted for publication an article outlining my experimental work to date challenging the hypothesis of tumor cell stimulation by a low level of specific immune reaction. Various numbers of spleen cells from specifically immunized mice were mixed with constant numbers of target tumor cells and inoculated subcutaneously into thymectomized, X-irradiated recipients. It was found that small numbers of admixed immune spleen cells produced a statistically significant and reproducible acceleration of tumor growth in the inoculum as compared with controls in which either nonimmune spleen cells or spleen cells from animals immune to a different noncross-reacting tumor were used. Larger numbers of specifically immune spleen cells produced tumor inhibition. These data imply that the normal immune reaction may have a dual function in relation to neoplasia: (1) stimulation of tumor growth early in the course of the disease or whenever the immune reaction is minimal; and (2) inhibition of tumor growth at other times (PREHN, 1971).

Although the present enthusiasm for immunologic approaches to the cancer problem is entirely justified, I think it is very necessary to keep in mind that the immune response is only one of a complex but little-understood system of homeostatic devices important in oncology. Indeed, it may not be the most important. One interesting possible example is to be found in Burkitt's lymphoma. It is a common observation that a Burkitt lesion may regress, sometimes to reappear in another location. Reoccurrence apparently is seldom, if ever, seen in the location of the original lesion. The location of the primary tumor has apparently become resistant, but this curious behavior is difficult to explain on the basis of classical tumor immunity.

The evidence presented by the Kleins concerning the relationship of a herpes type of virus to Burkitt lymphoma and other tumors has recently been given increased importance by work of another kind. RAPP (1971) has been able to demonstrate the oncologic potential of a herpes type II virus in tissue culture. Thus for the first time, a herpes virus has been shown to actually be a tumor virus. The actual etiologic significance of the EB virus in the Burkitt tumor therefore becomes highly probable.

It will close this discussion by noting that the progress of tumor immunology in the past decade, as exemplified by this symposium, has indeed been extraordinary. Each discovery seems to lead exponentially to new insights and new lines of investigation. The excitement of those of us fortunate enough to be working in this field can hardly be contained. There can be almost no doubt that tumor immunology is developing into a distinct discipline, one which will have a profound impact on biological thought as well as on practical regimens of tumor prevention and therapy. The next decade will be even more exciting than has been the past.

References

Brawn, R. J. (1970). Possible association of embryonal antigen(s) with several primary 3-methylcholanthrene-induced murine sarcomas. *Int. J. Cancer,* 6:245–249.

Duff, R., and F. Rapp (1971). Oncogenic transformation of hamster cells after exposure to herpes simplex type 2. *Nature—New Biol., 233*:48–50.

Fialkow, P. J. (1967). Hypothesis: Immunologic oncogenesis. *Blood, 30*:388.

Prehn, R. T. (1972). The immune reaction as a stimulator of tumor growth. *Science,* in press.

Subject Index

Author Index